WITHDRAWN

Advances in
Physical Organic Chemistry

Advances in Physical Organic Chemistry

Volume 21

Edited by

V. GOLD

Department of Chemistry
King's College London
Strand, London WC2R 2LS

and

D. BETHELL

The Robert Robinson Laboratories
University of Liverpool
P.O. Box 147, Liverpool L69 3BX

ACADEMIC PRESS 1985
(Harcourt Brace Jovanovich, Publishers)

London   Orlando   San Diego   New York
Toronto   Montreal   Sydney   Tokyo

COPYRIGHT © 1985, BY ACADEMIC PRESS INC. (LONDON) LTD.
ALL RIGHTS RESERVED.
NO PART OF THIS PUBLICATION MAY BE REPRODUCED OR
TRANSMITTED IN ANY FORM OR BY ANY MEANS, ELECTRONIC
OR MECHANICAL, INCLUDING PHOTOCOPY, RECORDING, OR
ANY INFORMATION STORAGE AND RETRIEVAL SYSTEM, WITHOUT
PERMISSION IN WRITING FROM THE PUBLISHER.

ACADEMIC PRESS INC. (LONDON) LTD.
24-28 Oval Road
LONDON NW1 7DX

United States Edition published by
ACADEMIC PRESS, INC.
Orlando, Florida 32887

ISSN: 0065 3160

ISBN: 0-12-033521-2

PRINTED IN THE UNITED STATES OF AMERICA

85 86 87 88   9 8 7 6 5 4 3 2 1

Contents

Contributors to Volume 21   vii

Preface   ix

The Discovery of the Mechanisms of Enzyme Action, 1947–1963   1

F. H. WESTHEIMER

1 Introduction 2
2 Coenzymes 4
3 Enzymes 11
4 Rates and specificity 23
5 Conclusion 30

The Spectroscopic Detection of Tetrahedral Intermediates Derived from Carboxylic Acids and the Investigation of Their Properties   37

BRIAN CAPON, MIRANDA I. DOSUNMU, and MARIA de NAZARÉ de MATOS SANCHEZ

1 Introduction 38
2 Specially stabilized tetrahedral intermediates 39
3 The detection of simple tetrahedral intermediates 48
4 Kinetic studies on the breakdown of hemiorthoesters 60
5 Nitrogen-containing tetrahedral intermediates 89
6 Future investigations 93

A General Approach to Organic Reactivity: The Configuration Mixing Model 99
ADDY PROSS

1 Introduction 99
2 Theoretical background 102
3 Applications 139
4 General consequences 177
5 Conclusion 190

Gas-phase Nucleophilic Displacement Reactions 197
JOSÉ MANUEL RIVEROS, SONIA MARIA JOSÉ, and KEIKO TAKASHIMA

1 Introduction 198
2 Experimental techniques 200
3 General features of gas-phase ion-molecule reactions 204
4 Gas-phase S_N2 reactions involving negative ions 206
5 Some examples of gas-phase S_N2 reactions involving positive ions 220
6 Nucleophilic displacement reactions by negative ions in carbonyl systems 222
7 Gas-phase nucleophilic reactions of carbonyl systems involving positive ions 229
8 Nucleophilic displacement reactions in aromatic systems 234
9 Conclusions 237

Author Index 241

Cumulative Index of Authors 251

Cumulative Index of Titles 253

Contributors to Volume 21

Brian Capon Department of Chemistry, University of Hong Kong, Pokfulam Road, Hong Kong

Miranda I. Dosunmu Department of Chemistry, University of Calabar, Calabar, Nigeria

Sonia Maria José Institute of Chemistry, University of São Paulo, São Paulo, Brazil 05508

Maria de Nazaré de Matos Sanchez Departamento de Química, Universidade Federal de Santa Catarina, Florianópolis, Brazil

Addy Pross Department of Chemistry, Ben-Gurion University of the Negev, Beer Sheva, Israel 84105

José Manuel Riveros Institute of Chemistry, University of São Paulo, São Paulo, Brazil 05508

Keiko Takashima State University of Londrina, Londrina, Pr., Brazil

F. H. Westheimer Department of Chemistry, Harvard University, Cambridge, Massachusetts 02138, U.S.A.

Preface

The publication of the twenty-first volume of *Advances in Physical Organic Chemistry* provides an appropriate occasion on which to take stock of physical organic chemistry and, in particular, of the series.

The first volume appeared in 1963. At that time the pioneering work of such chemists as R. P. Bell, Louis Hammett (who, two decades earlier, had first coined the expression physical organic chemistry), and Sir Christopher Ingold (who wrote the Foreword to that volume) had already achieved wide acceptance and incorporation into the student syllabus. However, the subject was then still regarded as being of relatively peripheral interest. In the ten years that followed, the central position of physical organic chemistry within the discipline of chemistry was reappraised. IUPAC and individual academic institutions accorded the subject its deserved place, and *Advances in Physical Organic Chemistry* played its part, providing a focus by publishing authoritative and considered reviews in which the potential for future progress was also discussed.

In recent years the series has continued to adjust to this progress and to maintain a flexible view of the scope of physical organic chemistry. The development of new techniques for studying molecules in solid, liquid, and gaseous phases has led organic chemists to take an increasingly sophisticated interest in the underlying physical processes, while physical chemists have found their new techniques most appropriately applied to organic compounds. Our choice of contributors has aimed to reflect this situation.

Broadening the scope of physical organic chemistry to embrace any or all quantitative studies of organic compounds and their behaviour has its drawbacks. Practitioners now often prefer to define their interests in more restricted ways, describing themselves as organic photochemists, mass spectroscopists, fast-reaction kineticists, and so on, and specialized languages, journals, and review publications have been established to cater to these divisions. Sadly, the development of these specializations has led to the fragmentation of interests among physical organic chemists and it is endangering communication between scientists. The aim of this series of volumes will continue to be to bring together physical, theoretical, and inorganic chemistry insofar as they relate to organic compounds, their properties, and chemical reactions in a unifying manner that, we hope, will be intelligible and useful to all chemists.

With such a diversity of potential, where then lies the immediate future of physical organic chemistry? There still remains a need for physical understanding of areas of organic chemistry that have fallen within the traditional confines of the subject. The details of the origin of the activation barriers to organic reactions are still not well understood, and the present volume includes a theoretical approach to the problem which reveals new facets of some simple reactions and suggests new experiments. We intend to develop this theme further by attracting contributions from the burgeoning field of computational chemistry. The rapid growth in organometallic chemistry has yet to make its full impact on the series, but this will surely come. Bioorganic research is a field in which physical organic methodology achieved notable successes and promises more. To mark the twenty-first issue of *Advances in Physical Organic Chemistry* we include the transcript of a lecture given by Professor F. H. Westheimer about the way that the physical organic approach contributed to the development of molecular enzymology in the years leading up to 1963 when this series was founded. Since then, contributions relevant to bioorganic chemistry have appeared in our pages from time to time, amongst them the reviews by Scheraga (Vol. 6), Fife (Vol. 11), Fendler and Rosenthal (Vol. 13), Kirby, and Kunitake and Shinkai (Vol. 17). This currently fashionable and expanding area will merit further coverage in future volumes.

We hope that the series will continue to flourish. Its success to date has been largely due to the enthusiastic support of the scientific community. This is reflected in the quality of the contributors who have undertaken to present their special interests for a wider readership with an interest in understanding the molecules and reactions of organic chemistry on a quantitative basis. To all of them we offer our sincere thanks. We continue to welcome critical and/or constructive comments from our readers, as well as suggestions for and offers of manuscripts.

January 1985

V. GOLD
D. BETHELL

The Discovery of the Mechanisms of Enzyme Action, 1947–1963*

F. H. WESTHEIMER

Department of Chemistry, Harvard University, Cambridge, Massachusetts, U.S.A.

1 Introduction 2
2 Coenzymes 4
    Pyridoxal phosphate 4
    Nicotinamide adenine dinucleotide 6
    Thiamin pyrophosphate 9
    Biotin 11
3 Enzymes 11
    Sucrose phosphorylase 11
    Serine esterases 13
    Enzymes utilizing ketimine intermediates 18
    Ribonuclease 21
    Formyl GAR amidotransferase 23
4 Rates and specificity 23
    Kinetics 24
    Catalysis 25
    Entropy 26
    Approximation 27
    Strain 28
    Specificity 28
    Induced fit 29
5 Conclusion 30
Acknowledgement 30
References 30

*Based, in part, on a lecture at the Louis Hammett Symposium on the History of Physical Organic Chemistry, 186th A.C.S. National Meeting, Washington, D.C., 31 August 1983.

1 Introduction

The application of the concepts of physical organic chemistry to the understanding of mechanisms of enzyme action constitutes one of the major triumphs of our science. The principles pioneered by Lapworth, Ingold, Hammett and others have led, in a relatively brief period, to a revolutionary change in the field of enzymology. In 1947, nobody knew the mode of action of a single enzyme or coenzyme; by 1963, the catalytic mechanisms were known for four coenzymes and for several enzymes that function without cofactors. Although the chemical community is still unable to synthesize catalysts with the specificity and power of enzymes, by 1963 we were able to rationalize some of the factors that cause their great rate accelerations (Jencks, 1963), and make a start on experimental approaches to the specific binding of substrate to protein. (Earlier work is summarized in Dixon and Webb, 1964.)

In this review, I shall to the best of my ability outline the significant events in this development. I have restricted myself to work completed prior to 1963. This distance should provide at least a little perspective. However, I must start with a disclaimer. Surprisingly, history is much more difficult than chemistry. It is hard to find the first mentions of specific ideas, or the first examples of principles. Furthermore, the first example may not be important to history anyway; what is important is the chemistry that excited others, and so stimulated further work, and unfortunately one cannot go back to check the thinking of the time. What I present here is offered with simultaneous conviction and diffidence.

Many important discoveries in two diverse fields laid the foundation for the understanding of the mechanisms of enzyme action. First, the concepts of reaction mechanisms, and the tools to determine reaction mechanisms had to be worked out. The application of reaction kinetics, of the spectroscopic identification of intermediates, the trapping of intermediates, the understanding of acid and base catalysis, of metal-ion catalysis, of Brønsted and Hammett relationships (Hammett, 1940), the use of isotopes as tracers and for the determination of kinetic isotope effects (Westheimer and Nicolaides, 1949; Melander, 1950, 1960), and much more, provided the intellectual armament needed to solve problems in enzymology. Additionally, however, the purification of enzymes and the determination of their structures were needed to allow chemistry to advance with confidence into enzymology. The purification and crystallization of proteins had been achieved much earlier (Sumner, 1926; Northrop, 1930; Northrop and Kunitz, 1932a,b; Kunitz and Northrop, 1935; Sumner and Dounce, 1937; Negelein and Wulff, 1937) but the first step in establishing the structures of proteins – indeed, the essential step in showing that proteins *had* structures as conventionally defined by organic chemists – was only taken in 1951, when Frederick Sanger determined the amino-acid sequence for insulin (Sanger and Tuppy, 1951a,b; Sanger and Thompson, 1953; Ryle et al., 1955). The determination of the three-dimensional structures of proteins by X-ray analysis can fairly be said to date from the publications of John Kendrew and of Max Perutz in *Nature* (Perutz et al., 1960; Kendrew et al., 1960), although chemists had been convinced (Perutz, 1954), even before digital computers revolutionized the field of X-ray crystallography, that the solution would prove possible. Thus the early progress in enzymic mechanisms occurred simultaneously with the establishment of structure.

Regrettably, most organic chemists in the 1950s regarded enzymes as an unrewarding if not an improper field of investigation, and possibly some still do so today. For many years, synthetic organic chemists were captivated by the determination of structure and by the synthesis of steroids and alkaloids, and physical organic chemists by solvolysis. The intellectual advances were stunning, but the result was to leave the field of mechanistic biochemistry to the biochemists. Yet it certainly is a legitimate part of chemistry. In the late nineteenth century, Emil Fischer – a card-carrying organic chemist – used maltase and emulsin to establish the stereochemistry of the anomeric derivatives of sugars (Fischer, 1898) and clearly regarded enzymology as a part of his territory. In his Nobel Lecture, he wrote in part, "I ... foresee the day when physiological chemistry will not only make extensive use of the natural enzymes as agents, but when it will prepare synthetic ferments for its own purposes" (Fischer, 1902). The application of the principles of physical organic chemistry to enzymology enlarges the knowledge base on which to build Fischer's "synthetic ferments." He probably did not expect it to take us a century.

The first examples of mechanism must be divided into two principal classes: the chemistry of enzymes that require coenzymes, and that of enzymes without cofactors. The first class includes the enzymes of amino-acid metabolism that use pyridoxal phosphate, the oxidation–reduction enzymes that require nicotinamide adenine dinucleotides for activity, and enzymes that require thiamin or biotin. The second class includes the serine esterases and peptidases, some enzymes of sugar metabolism, enzymes that function by way of enamines as intermediates, and ribonuclease. An understanding of the mechanisms for all of these was well underway, although not completed, before 1963.

2 Coenzymes

PYRIDOXAL PHOSPHATE

Pyridoxal phosphate is the coenzyme for the enzymic processes of transamination, racemization and decarboxylation of amino-acids, and for several other processes, such as the dehydration of serine and the synthesis of tryptophan that involve amino-acids (Braunstein, 1960). Pyridoxal itself is one of the three active forms of vitamin B_6 (Rosenberg, 1945), and its biochemistry was established by 1939, in considerable part by the work of A. E. Braunstein and coworkers in Moscow (Braunstein and Kritzmann, 1947a,b,c; Konikova et al., 1947). Further, the requirement for the coenzyme by many of the enzymes of amino-acid metabolism had been confirmed by 1945. In addition, at that time, E. E. Snell demonstrated a model reaction (1) for transamination between pyridoxal [1] and glutamic acid, work which certainly carried with it the implication of mechanism (Snell, 1945).

$$\underset{[1]}{\text{HO}\overset{\text{CHO}}{\underset{\text{H}_3\text{C}\text{N}}{\bigcirc}}\text{CH}_2\text{OH}} + {}^{-}\text{O}_2\text{C}-\text{CH}_2-\text{CH}_2-\underset{\underset{\text{NH}_3^+}{|}}{\text{CH}}-\text{CO}_2^- \;\rightleftharpoons$$

$$\text{HO}\overset{\text{CH}_2\text{NH}_2}{\underset{\text{H}_3\text{C}\text{N}}{\bigcirc}}\text{CH}_2\text{OH} + \text{HO}_2\text{C}-\text{CH}_2-\text{CH}_2-\underset{\underset{\text{O}}{\|}}{\text{C}}-\text{CO}_2^- \tag{1}$$

Nevertheless, the full-blown mechanism that showed the role of the coenzyme was only written out in detail by Braunstein and M. M. Shemyakin in 1953 (Braunstein and Shemyakin, 1952, 1953). Their formulae (2), complete with the curved arrow notation of physical organic chemistry, clearly pointed out the role of the coenzyme as an electron sink in a ketimine mechanism. They showed how the coenzyme can function in transamination, racemization and, with some help from Hanke and his collaborators (Mandeles et al., 1954), in decarboxylation. The mechanisms they advanced were exactly what we would postulate today, and constituted an early and successful application of theory to mechanistic enzymology. But it must be admitted that the theory appealed because it was reasonable; the authors had little or no evidence, in terms of physical organic chemistry, to support their formulation, which is shown in part below.

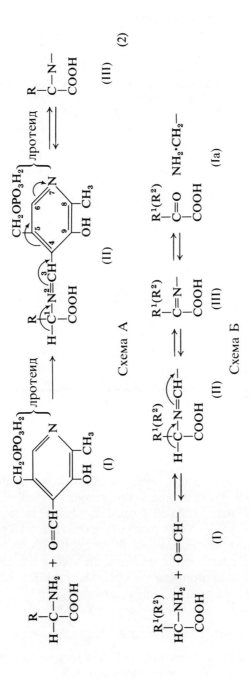

Схема А

Схема Б

At the same time, Snell and coworkers used model systems to achieve most of the reactions of the pyridoxal enzymes (Metzler and Snell, 1952a,b; Olivard et al., 1952; Ikawa and Snell, 1954a,b; Metzler et al., 1954a,b; Longnecker and Snell, 1957). They too developed the modern mechanisms for the series of reactions and demonstrated the role of the coenzyme as an electron sink by substituting alternative catalysts for pyridoxal phosphate. In particular, they showed that 2-hydroxy-4-nitrobenzaldehyde (Ikawa and Snell, 1954) functioned in their model systems just as did the vitamin; its electronic structure is really quite similar (3).

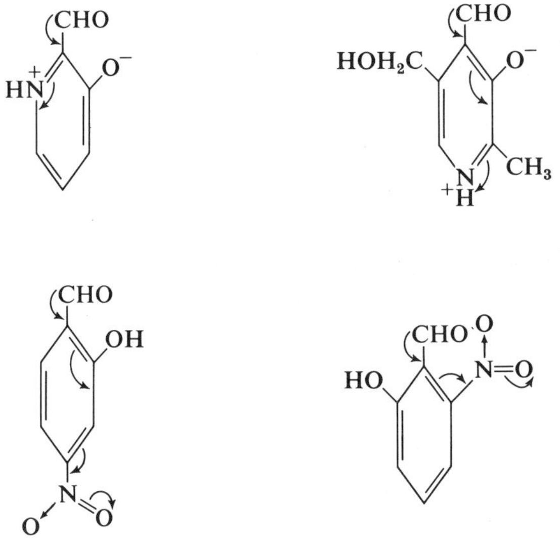

(3)

Some of the details of the actual enzymic processes were also discovered during this early period. Spectroscopic investigations of glutamic-aspartic transaminase suggested that pyridoxal is covalently bound to that enzyme (and presumably to others) by a Schiff-base linkage (Jenkins and Sizer, 1957). Although this early work left unanswered all the questions concerning the role of the protein in accelerating the reactions, the basic mechanistic pathway has been developed according to the electronic principles of physical organic chemistry, and confirmed by studies on model systems.

NICOTINAMIDE ADENINE DINUCLEOTIDE

At more or less the same time, the chemistry of an important class of enzymic oxidation–reduction reactions was partially elucidated by a group of investigators at the University of Chicago (Westheimer et al., 1951). They were concerned with the mechanism of action of the nicotinamide-adenine dinucleotides (NAD^+ [2] and $NADP^+$), which function as coenzymes in many oxidation–reduction processes, and in particular serve in the oxidation of alcohol to acetaldehyde, and in the reduction of pyruvate to lactate. The

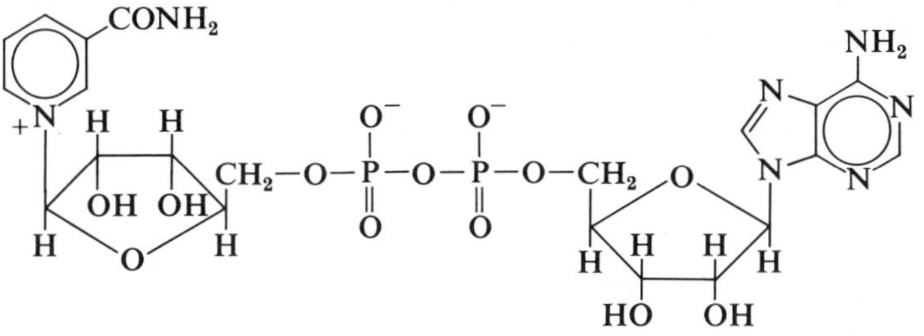

[2]

question they set out to answer was whether the reactions occur with direct transfer of hydrogen from the substrate to the coenzyme (and vice-versa), or whether the reactions involve electron transfer between substrate and coenzyme (perhaps through the protein), with loss of hydrogen ion to the solvent. The first system they studied was that of yeast alcohol dehydrogenase, where the chemical reaction is shown in (4).

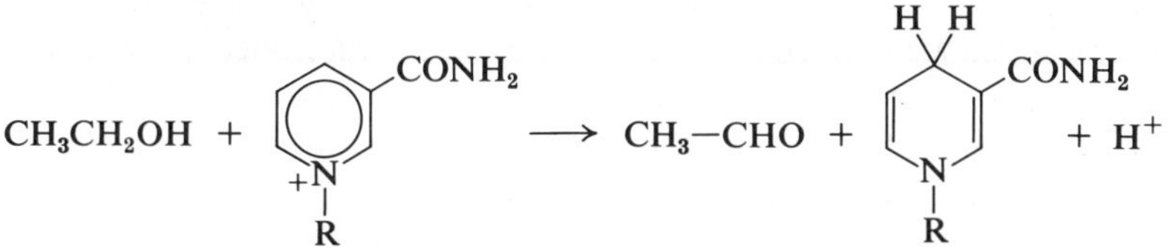

$$CH_3CH_2OH + \underset{R}{\underset{+N}{\text{[nicotinamide]}}}\text{-CONH}_2 \longrightarrow CH_3\text{-CHO} + \underset{R}{\underset{N}{\text{[dihydronicotinamide]}}}\text{-CONH}_2 + H^+ \quad (4)$$

R = ribose-pyrophosphate-ribose-adenine

By using dideuterioethanol, CH_3CD_2OH, as substrate, they demonstrated that one atom of deuterium is transferred from substrate to coenzyme, and further that this atom of deuterium is transferred from the coenzyme, under the influence of lactic dehydrogenase, to pyruvate (Fisher, et al., 1953; Loewus et al., 1953a,b; Vennesland and Westheimer, 1954). The converse experiment was also carried out, that is to say, the reactions were performed with normal substrate and coenzyme, but in D_2O; this resulted in transfer of ordinary hydrogen from substrate to coenzyme. The question they had set out to answer was then settled; the oxidation–reductions proceed by direct transfer of hydrogen between substrate and coenzyme. These results have subsequently been confirmed in numerous investigations with other enzymic reactions that require NAD^+ or $NADP^+$.

But the results provided an exciting bonus. The reactions proved to be stereospecific. NAD⁺ that has been reduced with dideuterioethanol has both a hydrogen atom and a deuterium atom in the 4-position of the pyridine ring, but reductions of pyruvate or of acetaldehyde with NAD(D) prepared enzymically in this way quantitatively transfer deuterium, and leave the hydrogen atom in place. The 1954 diagram showing the stereospecific transfer of hydrogen from the 4-position of NADH is presented in (5).

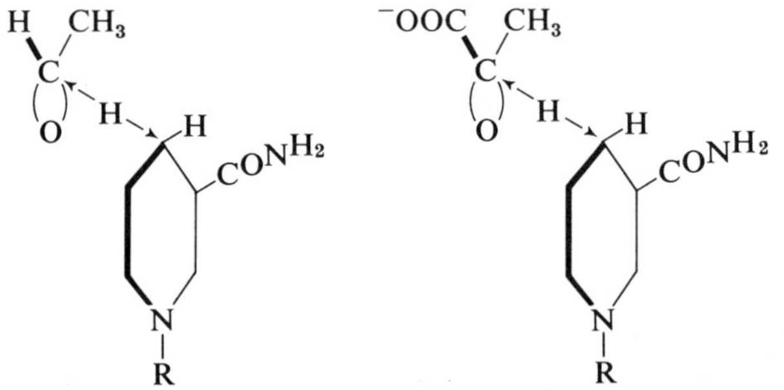

$$\text{(5)}$$

Furthermore, when ordinary acetaldehyde is reduced by deuteriated coenzyme, NAD(D), the resulting monodeuterioethanol, CH_3CHDOH (I), is clearly chiral (Scheme 1). When it is reoxidized by NAD⁺, only deuterium is transferred from the alcohol to the coenzyme. Conversely, when deuterioacetaldehyde is reduced enzymically with NADH, the monodeuterioethanol that is formed is the enantiomer (II); enzymic reoxidation of this monodeuterioalcohol transfers only hydrogen to the coenzyme. The point is brought home when the configuration of the latter chiral monodeuterioalcohol is inverted. The alcohol was converted into its tosylate, and the tosylate group displaced by hydroxide in a standard S_N2 process. The resulting monodeuterioethanol was then enzymically oxidized by NAD⁺ with quantitative transfer of deuterium from the alcohol to the coenzyme (Loewus et al., 1953b).

The two hydrogen atoms of ethanol are now described as enantiotopic, and the stereochemistry of methylene groups with enantiotopic groups is well understood (Mislow and Raban, 1967). It should have been more widely understood in 1951, since A. G. Ogston had, in 1948, discussed the fundamental concept (although without our modern vocabulary) in connection with the enzymic differentiation of the two apparently identical —CH_2—COOH groups of citric acid (Ogston, 1948). In fact, the enzymatic reactions cited above helped to familiarize chemists and biochemists with the idea. The discussion of enantiotopic and diastereotopic atoms in the fourth edition of Morrison and Boyd's popular textbook of organic chemistry

(Morrison and Boyd, 1983) is introduced in terms of the enzymic oxidation of ethanol. Here is an example where an important concept in stereochemistry arose from biochemistry and was subsequently incorporated into physical organic chemistry.

$$CH_3-\underset{D}{\overset{H}{C}}-OH + NAD^+ \cdot \rightleftarrows CH_3CHO + NADD + H^+ \quad \textit{Labeled}$$

I

$$CH_3-\underset{D}{\overset{H}{C}}-OH + NAD^+ \rightleftarrows CH_3CDO + NADH + H^+ \quad \textit{Unlabeled}$$

II

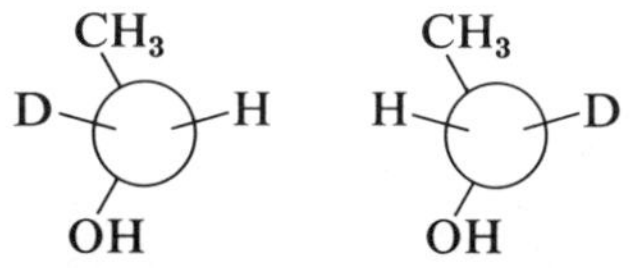

$$\begin{array}{cc} \text{CH}_3 & \text{CH}_3 \\ \text{D}\!\!\!-\!\!\bigcirc\!\!-\!\!\!\text{H} & \text{H}\!\!\!-\!\!\bigcirc\!\!-\!\!\!\text{D} \\ \text{OH} & \text{OH} \\ \text{I} & \text{II} \\ \textit{Loses only D} & \textit{Loses only H} \end{array}$$

Scheme 1

The detailed mechanism of the reactions was not settled at the time, and subsequently a great deal of effort has been spent in trying to decide whether the direct and stereospecific transfer of hydrogen takes place by way of the transfer of a hydride ion, or an electron and hydrogen atom, or an electron, a proton, and another electron. This controversy still rages, although recent work may have established the pathway as hydride ion transfer (MacInnes *et al.*, 1982; Powell and Bruice, 1983). At the time of the work described here, however, this was not known, and is in any event irrelevant to the discovery of *direct* and *stereospecific* transfer of hydrogen in enzymic oxidation–reduction involving NADH.

THIAMIN PYROPHOSPHATE

The next coenzyme for which a mechanism was established was thiamin pyrophosphate [3]. Ronald Breslow used nmr spectroscopy to show that the hydrogen atom at C-2 of a thiazolium salt rapidly exchanges with deuterium in even slightly alkaline solutions (6), so that the coenzyme offers an anionic centre for catalysis (Breslow, 1957). With this established, Breslow could confidently offer the pathway shown in Scheme 2 for the action of the coenzyme in the enzymic synthesis of acetoin; furthermore, similar chemistry is involved in other enzymic reactions (such as the decarboxylation of pyruvate) that require thiamin pyrophosphate (Breslow, 1958).

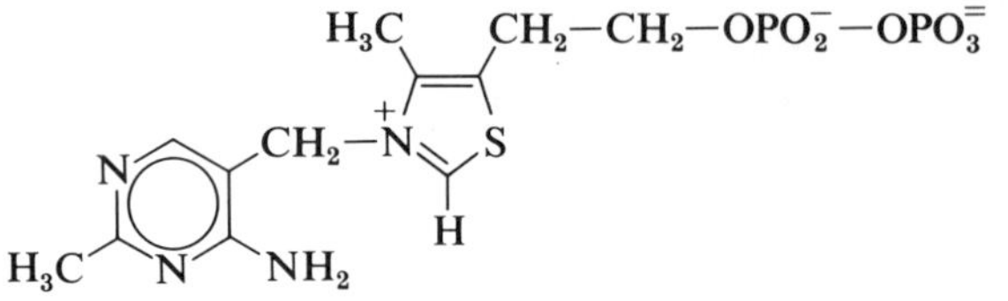

[3]

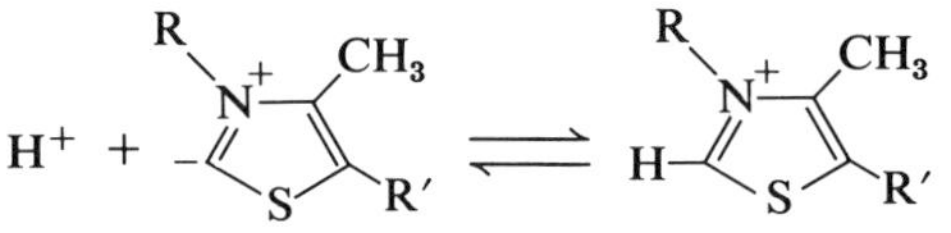

(6)

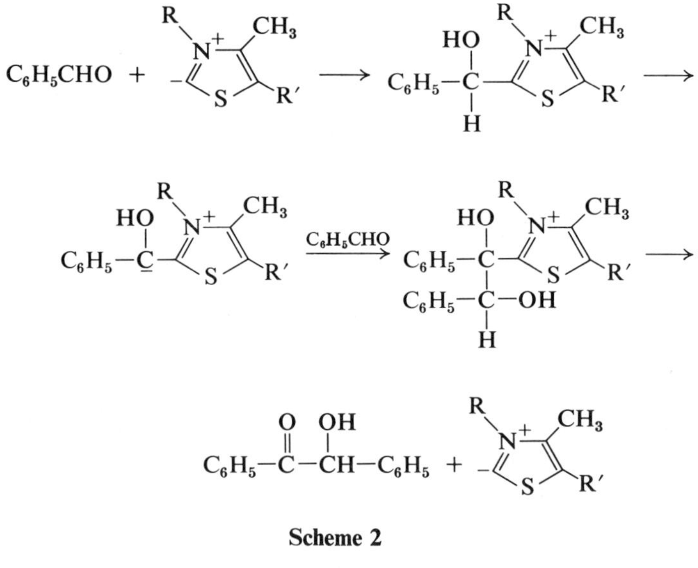

Scheme 2

Thiamin itself (in the absence of enzyme) had previously been shown to catalyse the formation of acetoin from acetaldehyde, albeit in very poor yield (Ukai *et al.*, 1943; Mizuhara *et al.*, 1951; Mizuhara and Handler, 1954). The reaction parallels the formation of benzoin from benzaldehyde, catalysed by cyanide ion. The mechanism of the latter reaction had been suggested in 1903 by Arthur Lapworth, who had shown how an aldehyde, R—CHO, could be converted into the equivalent of the anion R—C=O$^-$ (Lapworth, 1903). It is this idea that Breslow carried over to thiamin pyrophosphate and used to elucidate the way in which the coenzyme functions in co-operation with various enzymes. Much subsequent research has completely verified Breslow's interpretation of his signal findings (Krampitz, 1969). Here is an early example of the application of nmr spectroscopy, one of the major tools of modern physical organic chemistry, to bio-organic chemistry.

BIOTIN

The chemistry of a fourth coenzyme was at least partially elucidated in the period under discussion. F. Lynen and coworkers treated β-methylcrotonyl coenzyme A (CoA) carboxylase with bicarbonate labelled with ^{14}C, and discovered that one atom of radiocarbon was incorporated per molecule of enzyme. They postulated that an intermediate was formed between the enzyme and CO_2, in which the biotin of the enzyme had become carboxylated. The carboxylated enzyme could transfer its radiolabelled carbon dioxide to methylcrotonyl CoA; more interestingly, they found that the enzyme–CO_2 compound would also transfer radiolabelled carbon dioxide to free biotin. The resulting compound, carboxybiotin [4], was quite unstable, but could be stabilized by treatment with diazomethane to yield the methyl ester of N-carboxymethylbiotin (7) (Lynen *et al.*, 1959). The identification of this radiolabelled compound demonstrated that the unstable material is N-carboxybiotin itself, which readily decarboxylates; esterification prevents this reaction, and allows the isolation and identification of the product. Lynen *et al.* then postulated that the structure of the enzyme–CO_2 compound was essentially the same as that of the product they had isolated from the reaction with free biotin, but where the carbon dioxide was inserted into the bound biotin of the enzyme (Lynen *et al.*, 1961). Although these discoveries still leave significant questions to be answered as to the detailed mechanism of the carboxylation reactions in which biotin participates as coenzyme, they provide a start toward elucidating the way in which the coenzyme functions.

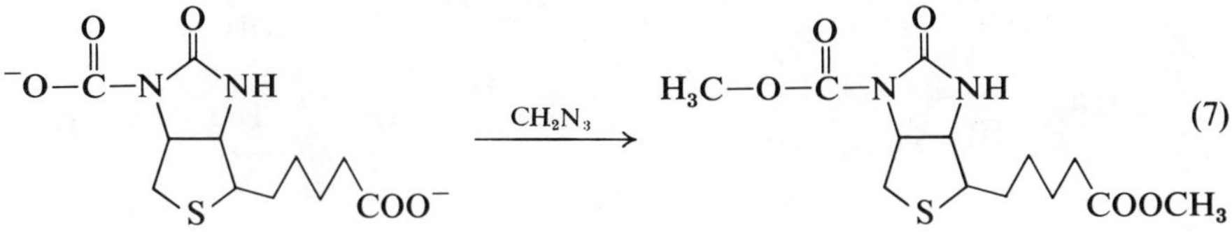

3 Enzymes

SUCROSE PHOSPHORYLASE

At the same time that the mechanisms of action of these four coenzymes was elucidated, the chemistry of four enzymes, or more precisely of four types of enzymes that function without coenzymes, was worked out. The earliest of the investigations concerns the action of sucrose phosphorylase, which catalyzes the reversible reaction (8) of sucrose with inorganic phosphate to yield fructose and glucose-1-phosphate.

$$\text{Sucrose + enzyme} \underset{\pm \text{fructose}}{\rightleftharpoons} \text{glucose–enzyme} \underset{\pm \text{phosphate}}{\rightleftharpoons} \text{glucose-1-phosphate + enzyme} \quad (8)$$

In 1947, Doudoroff, Barker and Hassid (Doudoroff *et al.*, 1947a) postulated that the reaction occurs, as shown above, by way of an intermediate compound of glucose and enzyme. They used ^{32}P to demonstrate that inorganic phosphate exchanges with glucose-1-phosphate in the presence of the enzyme according to (9).

$$\text{GLUCOSE-1-PHOSPHATE} + \text{H}^{32}\text{PO}_4^= \rightleftharpoons \text{GLUCOSE-1-}^{32}\text{PHOSPHATE} + \text{HPO}_4^= \quad (9)$$

Further, Fitting and Doudoroff (1952) showed that fructose labelled with ^{14}C exchanges with sucrose under the influence of the enzyme. Both of these exchanges can be explained best by a mechanism that involves a glucose–enzyme intermediate. The chemistry is reinforced by reactions that involve arsenate. In the presence of arsenate, sucrose phosphorylase catalyzes the hydrolysis of glucose-1-phosphate to glucose and inorganic phosphate, and catalyzes the hydrolysis of sucrose to glucose and fructose. This is readily explained by the lability to hydrolysis of glucose arsenate, together with the postulate of a glucose–enzyme intermediate that can react either with phosphate to yield glucose-1-phosphate or with arsenate to yield glucose-1-arsenate (Doudoroff *et al.*, 1947b; Doudoroff, 1960).

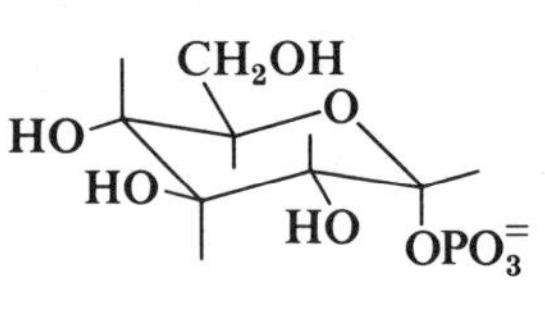

[5] 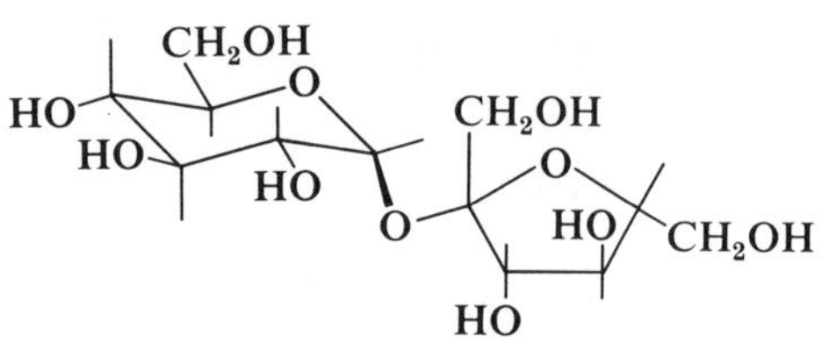   [6]

Another line of evidence that supports this pathway comes from a consideration of stereochemistry. In 1953, D. E. Koshland Jr. pointed out the significance of the fact that the formation of glucose-1-phosphate [5] from sucrose [6] occurs with retention of configuration at C-1 of the glucose moiety. He postulated that a single displacement, in enzymic chemistry as in

ENZYME ACTION

non-enzymic chemistry, should result in inversion, whereas two displacements in series should lead to retention (Koshland, 1953, 1954). The retention of configuration actually observed with sucrose phosphorylase is then consistent with an intermediate in the enzymic pathway. Incidentally, the intermediate has now actually been observed (Voet and Abeles, 1966).

SERINE ESTERASES

Almost simultaneously with the work on sugar phosphorylase, two separate groups developed the two-step mechanism for hydrolyses catalysed by peptidases and esterases. Their mechanism, which is the one that we accept today, postulated an acylated enzyme as intermediate. I. B. Wilson, F. Bergman, and D. Nachmansohn developed this idea for hydrolysis by acetylcholine esterase, while Bryan Hartley did the same for chymotrypsin.

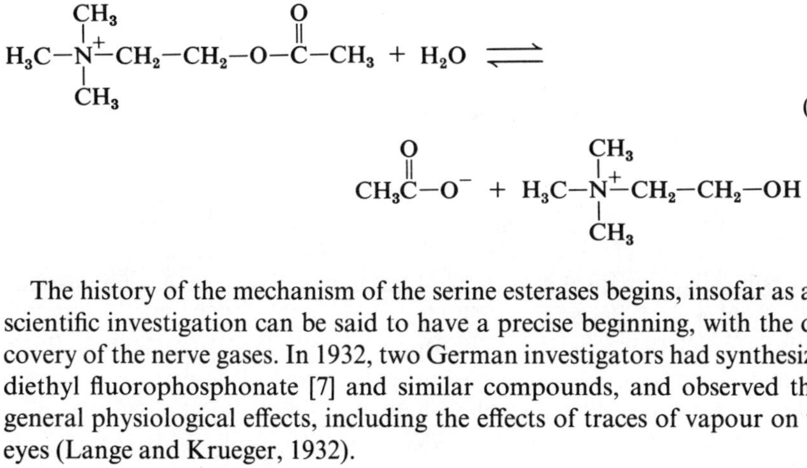

$$H_3C-\overset{CH_3}{\underset{CH_3}{\overset{|}{N^+}}}-CH_2-CH_2-O-\overset{O}{\overset{\|}{C}}-CH_3 + H_2O \rightleftharpoons$$

$$CH_3\overset{O}{\overset{\|}{C}}-O^- + H_3C-\overset{CH_3}{\underset{CH_3}{\overset{|}{N^+}}}-CH_2-CH_2-OH$$

(10)

The history of the mechanism of the serine esterases begins, insofar as any scientific investigation can be said to have a precise beginning, with the discovery of the nerve gases. In 1932, two German investigators had synthesized diethyl fluorophosphonate [7] and similar compounds, and observed their general physiological effects, including the effects of traces of vapour on the eyes (Lange and Krueger, 1932).

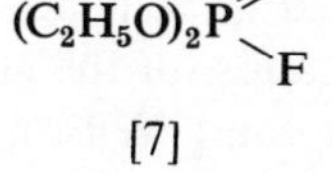

$$(C_2H_5O)_2P\underset{F}{\overset{O}{\diagdown}}$$

[7]

They did not report the minimum lethal dose or any other quantitative measure of toxicity, but nobody reading their paper would have missed the fact that the compounds are violently poisonous. During World War II, research on the fluorophosphonates was carried on for military purposes, and Adrian and his coworkers in Britain noted the similarity between the physiological action of the fluorophosphonates and that of reversible inhibitors of choline esterase (Adrian et al., 1947). This led to a number of scientific investigations of the action of nerve gases on various esterases.

Mazur and Bodansky (1946) found that diisopropyl fluorophosphate (DFP) irreversibly inhibits acetylcholine esterase. In particular, in 1949 Jansen, Balls, and their collaborators demonstrated the stoichiometric reaction of DFP with chymotrypsin (Jansen et al., 1949a,b; Aldridge, 1950).

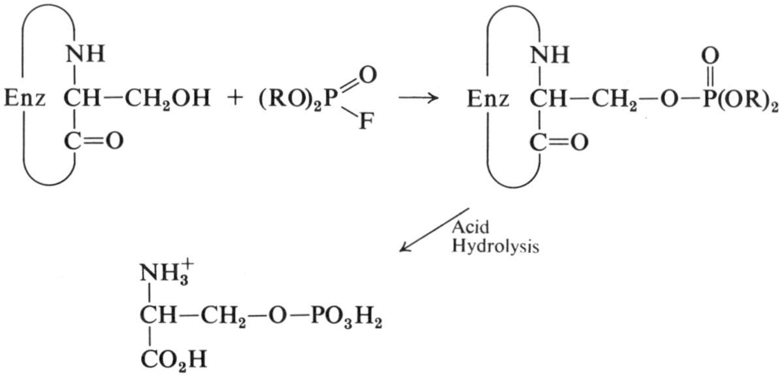

$$\text{Enz}\begin{bmatrix}\text{NH}\\\text{CH}-\text{CH}_2\text{OH}\\\text{C}=\text{O}\end{bmatrix} + (\text{RO})_2\text{P}\overset{\text{O}}{\underset{\text{F}}{\diagup}} \longrightarrow \text{Enz}\begin{bmatrix}\text{NH}\\\text{CH}-\text{CH}_2-\text{O}-\overset{\text{O}}{\overset{\|}{\text{P}}}(\text{OR})_2\\\text{C}=\text{O}\end{bmatrix}$$

$$\swarrow \text{Acid Hydrolysis}$$

$$\begin{array}{l}\text{NH}_3^+\\|\\\text{CH}-\text{CH}_2-\text{O}-\text{PO}_3\text{H}_2\\|\\\text{CO}_2\text{H}\end{array}$$

Scheme 3

In the same year, Nachmansohn and Wilson and their group showed that various active phosphorylating agents, including tetraethyl pyrophosphate (TEPP), react stoichiometrically with acetylcholine esterase to inhibit it irreversibly (Nachmansohn et al., 1948; Augustinsson and Nachmansohn, 1949; Wilson and Bergman, 1950; Bergman et al., 1950; Wilson, 1951). Lastly, around 1954, the site of phosphorylation was tentatively identified (Scheme 3); on hydrolysis, the diisopropylphosphoryl derivative of acetylcholine esterase yielded serine phosphate (Balls and Wood, 1953; Schaffer et al., 1954; Oosterbahn et al., 1955). In 1950, Wilson, Bergman and Nachmansohn published the two-step mechanism for the action of acetylcholine esterase that is presented in Scheme 4 (Wilson et al., 1950). In this formulation, the symbol "G" stands for the group on the enzyme that reacts with the ester (Nachmansohn and Wilson, 1951). Despite the strange double-bond drawn between the inhibitor and the group G of the enzyme, the mechanism has proved to be correct in its essentials.

In 1952, Hartley and Kilby showed that p-nitrophenyl acetate reacts with chymotrypsin, and advanced a two-step mechanism for the process (Hartley and Kilby, 1952). Two years later Hartley showed that a "burst" of p-nitrophenol was produced in the reaction (Hartley and Kilby, 1954). That is to say, a graph of the production of p-nitrophenol from the chymotryptic hydrolysis of p-nitrophenyl acetate does not seem to begin at the origin, but instead a small amount of p-nitrophenol is produced very rapidly. Fur-

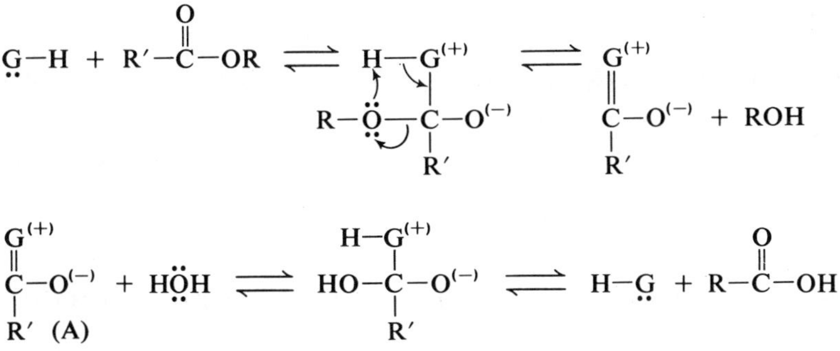

$$\underset{}{\text{G}-\text{H}} + \text{R}'-\overset{\text{O}}{\underset{}{\text{C}}}-\text{OR} \rightleftharpoons \underset{\underset{\text{R}'}{|}}{\underset{\text{R}-\overset{..}{\text{O}}-\text{C}-\text{O}^{(-)}}{\overset{\text{H}-\text{G}^{(+)}}{}}} \rightleftharpoons \underset{\underset{\text{R}'}{|}}{\overset{\text{G}^{(+)}}{\underset{\text{C}-\text{O}^{(-)}}{\|}}} + \text{ROH}$$

$$\underset{\underset{\text{R}'\ (A)}{|}}{\overset{\text{G}^{(+)}}{\underset{\text{C}-\text{O}^{(-)}}{\|}}} + \text{H}\overset{..}{\text{O}}\text{H} \rightleftharpoons \underset{\underset{\text{R}'}{|}}{\overset{\text{H}-\text{G}^{(+)}}{\text{HO}-\text{C}-\text{O}^{(-)}}} \rightleftharpoons \text{H}-\underset{..}{\text{G}} + \text{R}-\overset{\text{O}}{\underset{}{\text{C}}}-\text{OH}$$

Scheme 4

thermore, the amount of *p*-nitrophenol that is formed in the initial fast reaction corresponds to one mole for each mole of chymotrypsin present in solution. The burst is then followed by a much slower, steady catalyzed hydrolysis of the ester. Hartley's original figure is reproduced in Fig. 1.

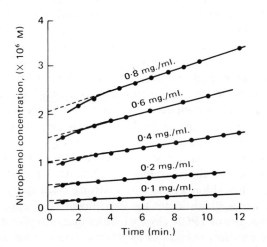

FIG. 1

Stopped flow kinetics (Gutfreund and Sturtevant, 1956) strongly supported the two-step mechanism for the hydrolysis of *p*-nitrophenyl acetate previously advanced from Hartley's laboratory. The reaction proceeds by the acetylation of the enzyme at the active site, followed by slower hydrolysis of the resulting acetylchymotrypsin (Scheme 5). This, of course, regenerates the enzyme for further rapid reaction with the substrate.

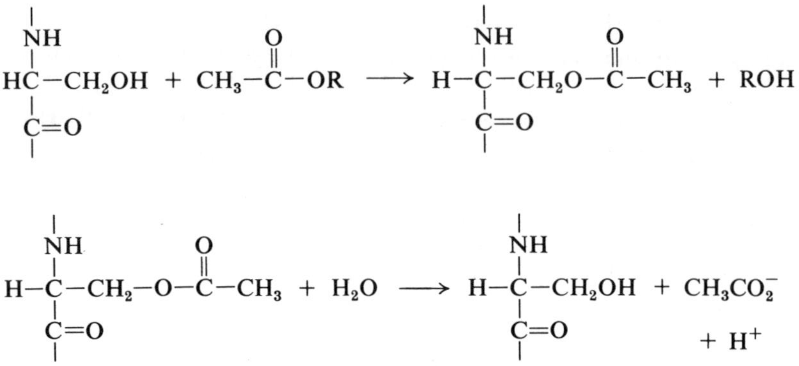

$$\begin{array}{c}|\\NH\\|\\HC-CH_2OH\\|\\C=O\\|\end{array} + CH_3-\overset{O}{\underset{\|}{C}}-OR \longrightarrow \begin{array}{c}|\\NH\\|\\H-C-CH_2O-\overset{O}{\underset{\|}{C}}-CH_3\\|\\C=O\\|\end{array} + ROH$$

$$\begin{array}{c}|\\NH\\|\\H-C-CH_2-O-\overset{O}{\underset{\|}{C}}-CH_3\\|\\C=O\\|\end{array} + H_2O \longrightarrow \begin{array}{c}|\\NH\\|\\H-C-CH_2OH\\|\\C=O\\|\end{array} + CH_3CO_2^- + H^+$$

Scheme 5

The acylation of the serine is presumably assisted by a histidine residue on the enzyme that attacks the proton of the serine hydroxyl group prior to or simultaneously with the acylation reaction. Removal of the proton converts the hydroxyl group into an alkoxide ion, or an incipient alkoxide ion, and so creates a good nucleophile in place of the hydroxyl group. The nucleophilic anion then attacks the carbonyl group of the substrate (or the phosphoryl group of an inhibitor such as diisopropylphosphoryl fluoride). The identification of this essential basic group in chymotrypsin as a histidine residue was made by Whitaker and Jandorf (1956). These investigators treated chymotrypsin with 2,4-dinitrofluorobenzene, and found that the enzyme was then inactivated, with simultaneous destruction of a histidine residue. This finding was confirmed and extended by Schoellman and Shaw, who demonstrated that N-tosylphenylalanyl chloromethyl ketone (TPCK) reacts irreversibly with the enzyme. The reaction inactivates it and stoichiometrically destroys one of the two histidine residues of chymotrypsin. The inhibitor, like other chloroketones, is a strong electrophile and might be expected to react with histidine; the experiment is of special significance because the structure of TPCK closely mimics that of acylated phenylalanine esters that are excellent substrates for chymotrypsin. One would therefore expect the inhibitor to occupy the active site of the enzyme, and undergo reaction in some way parallel to that of a true substrate. In fact, the reaction is enzymic; TPCK does not react readily or specifically with samples of chymotrypsin that have been inactivated with DFP, or denatured with urea. In consideration of the close resemblance of the inhibitor to a natural substrate for the enzyme, and of the enzymic process of inhibition, the reaction of the enzyme with TPCK has been described as affinity labelling. (Schoellman and Shaw, 1962, 1963). Many other examples of affinity labelling have subsequently been discovered.

ENZYME ACTION

The mechanism of the reaction is outlined in Scheme 6. In order to release ammonia or an amino-acid from an amide or peptide, the leaving group must be protonated, perhaps by the histidinium ion that is formed when the proton is removed from the hydroxyl group of the serine residue. This level of detail in the mechanism, however, was not seriously considered in the early days of mechanistic enzymology.

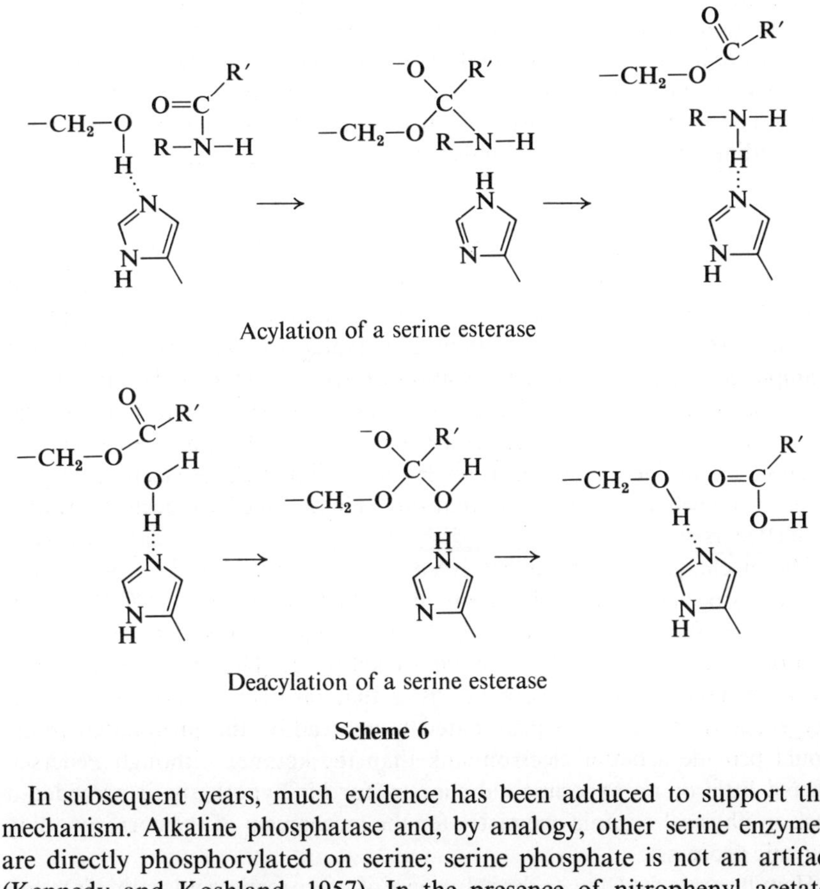

Acylation of a serine esterase

Deacylation of a serine esterase

Scheme 6

In subsequent years, much evidence has been adduced to support this mechanism. Alkaline phosphatase and, by analogy, other serine enzymes, are directly phosphorylated on serine; serine phosphate is not an artifact (Kennedy and Koshland, 1957). In the presence of nitrophenyl acetate, chymotrypsin is acetylated on serine, and the resulting acetylchymotrypsin has been isolated (Balls and Aldrich, 1955; Balls and Wood, 1956). Similarly, the action of *p*-nitrophenyl pivalate gave rise to pivaloyl chymotrypsin, which could be crystallized (Balls *et al.*, 1957). Neurath and coworkers showed that acetylchymotrypsin is hydrolyzed at pH 5.5, but that it is reversibly denatured by 8 M urea; the denatured derivative is inert to hydrolysis and even to hydroxylamine, whereas the renatured protein, obtained by diluting the urea solution, reacts to regenerate active enzyme. Thus the enzyme activity depends on the three-dimensional structure of the enzyme, and not merely on specific groups (Dixon *et al.*, 1956).

p-Nitrophenyl acetate is not a natural substrate for chymotrypsin, but it has now been possible to show that normal substrates react in the same way that it does (Schonbaum *et al.*, 1959). The confusion that surrounded the kinetics of the hydrolyses of esters and amides was resolved, with complete confirmation of the two-step mechanism (Bender *et al.*, 1962; Bender and Zerner, 1962b). Furthermore, modern X-ray structures of the enzymes are in accord with the deductions from chemistry (Matthews *et al.*, 1967; Blow, 1969). The mechanisms postulated in the early 1950s have been expanded and modified, but in their essentials they have been confirmed.

ENZYMES UTILIZING KETIMINE INTERMEDIATES

A third early mechanism for enzymic processes involves the formation of imines between the amino group of a lysine residue on an enzyme and the carbonyl group of a substrate, followed by standard imine chemistry. The first example concerned the decarboxylation of acetoacetic acid (Hamilton and Westheimer, 1959). The mechanism was based on the non-enzymic physical organic chemistry of Kai Pedersen (Pedersen, 1934). He postulated that the catalysis by aniline of the decarboxylation of dimethylacetoacetic acid proceeds by a mechanism parallel to that shown in Scheme 7 for acetoacetic acid itself (Pedersen, 1938).

The mechanism for the uncatalyzed decarboxylation of β-ketoacids had previously been established by Bredt and by Pedersen (Bredt, 1927; Pedersen, 1929; 1936; Westheimer and Jones, 1941). The acid loses CO_2 to form the enol of the product, which subsequently ketonizes. The idea behind Pedersen's mechanism for aniline catalysis is that nitrogen is more basic than oxygen, and so could be protonated more readily; the protonated imine would provide a better electron sink than the ketone. Although Pedersen offered little or no experimental support for his hypothesis, it provided a basis in physical organic chemistry for the mechanism of the corresponding enzymic process.

Hamilton marked the carbonyl group of acetoacetic acid with ^{18}O, and then carried out the enzymic decarboxylation (Hamilton and Westheimer, 1959). The product of the decarboxylation, acetone, contained none of the label. This result is demanded by the ketimine mechanism, whereas the mechanism of uncatalyzed decarboxylation would have required that the label appear intact in the product. Of course, in order to make these statements we had to carry out an elaborate set of control experiments, since ^{18}O is washed out of both acetone and acetoacetic acid by buffers and even more

ENZYME ACTION

rapidly by enzyme. It was possible, however, to show that the decarboxylation occurs much more rapidly than the washing out of the label; during the enzymic process little isotopic label is lost except that lost as a direct result of the enzymic decarboxylation. This work provided a firm basis for an enzymic pathway via a ketimine (Westheimer, 1963).

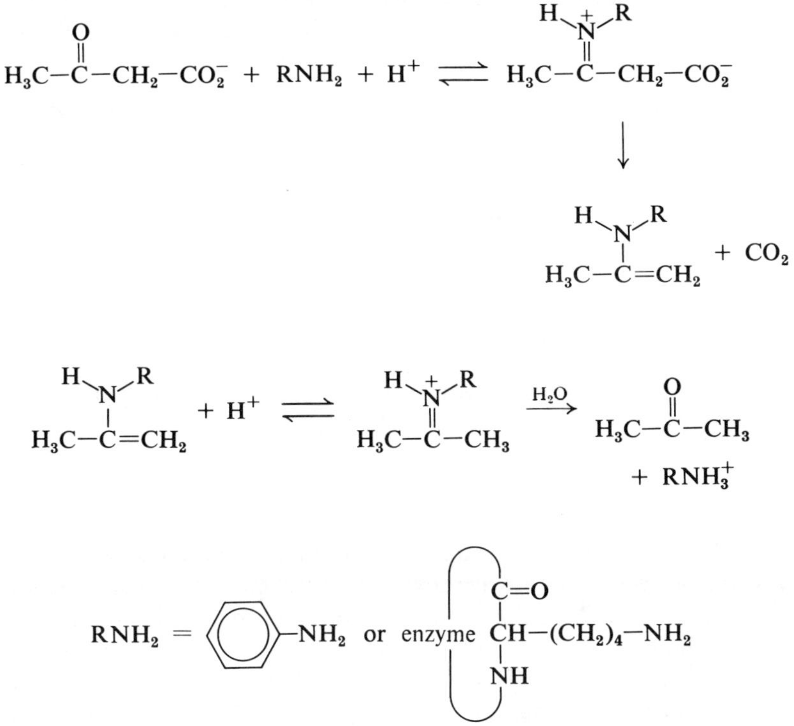

$$H_3C-\overset{O}{\overset{\|}{C}}-CH_2-CO_2^- + RNH_2 + H^+ \rightleftharpoons H_3C-\overset{\overset{H\;\;\;R}{\underset{+}{N}}}{\overset{\|}{C}}-CH_2-CO_2^-$$

$$\downarrow$$

$$H_3C-\overset{\overset{H\;\;\;R}{\underset{|}{N}}}{C}=CH_2 + CO_2$$

$$H_3C-\overset{\overset{H\;\;\;R}{\underset{|}{N}}}{C}=CH_2 + H^+ \rightleftharpoons H_3C-\overset{\overset{H\;\;\;R}{\underset{+}{N}}}{\overset{\|}{C}}-CH_3 \xrightarrow{H_2O} H_3C-\overset{O}{\overset{\|}{C}}-CH_3 + RNH_3^+$$

$$RNH_2 = \langle\!\langle \bigcirc\rangle\!\rangle-NH_2 \text{ or enzyme } \overset{C=O}{\underset{NH}{\overset{|}{CH}-(CH_2)_4-NH_2}}$$

Scheme 7

The general idea of this mechanism was promptly confirmed both for the decarboxylation of acetoacetate, and for the physiologically much more important reaction of transaldolase. In both cases, the development depended upon a new tool introduced in 1958 by Edmond Fischer and E. Krebs (Fischer *et al.*, 1958). They discovered that pyridoxal phosphate is attached to phosphorylase A by a ketimine linkage, and that the C=N bond of this linkage could be irreversibly reduced with sodium borohydride. Pyridoxal phosphate does not participate directly in the enzymic reaction of phosphorylase; the significance of the work rests on the fact that the reduction occurs without inactivating the enzyme. In 1961, Horecker and his coworkers reduced a mixture of glucose-6-phosphate-^{14}C and transaldolase with borohydride. The process inactivated the enzyme; on hydrolysis a product was obtained that contained ^{14}C, and involved the attachment of dihydroxyacetone to the enzyme (Horecker et al., 1961; Pontremoli et al., 1961). A few months later, work from the same laboratory showed that aldolase, like transaldolase, is inactivated by borohydride, and the authors suggested that both inactivations occur by way of the reduction of Schiff bases (Grazi et al., 1962). Presumably the chemistry involved is that presented in Scheme 8.

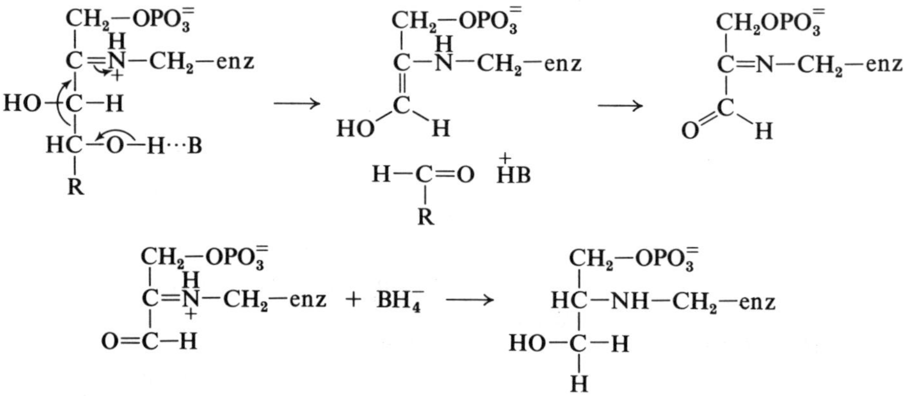

Scheme 8

In 1962 too, Fridovich showed that the addition of sodium borohydride to a mixture of acetoacetate decarboxylase and acetoacetate inactivates the enzyme, whereas the addition of borohydride to a buffered solution of the enzyme alone has no effect on the rate at which it can promote the decarboxylation of acetoacetate (Fridovich and Westheimer, 1962); this work confirmed the ketimine mechanism that had previously been advanced for the decarboxylation. Subsequent work (beyond the scope of this review) showed that the reaction product, on hydrolysis, yielded ε-isopropyllysine [8], formed by the reduction of the ketimine of acetone (11), and control experiments showed that this ketimine was actually an intermediate in the enzymic pathway, as had been postulated (Warren et al., 1966).

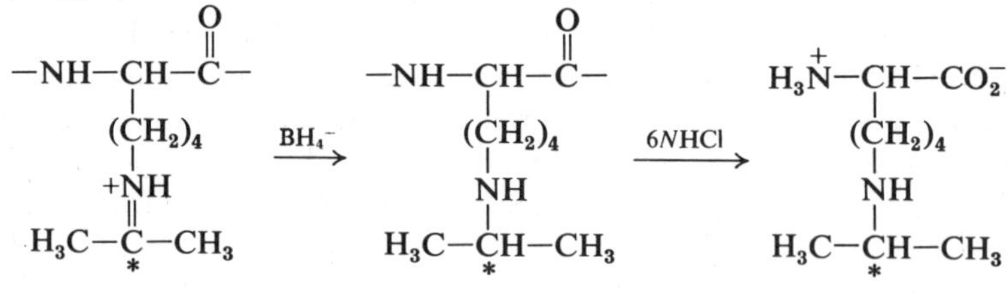

ENZYME ACTION

A model system and mechanism (12) had also been developed for the metal-ion promoted enzymic decarboxylations of dibasic ketoacids, such as oxaloacetic acid (Steinberger and Westheimer, 1949, 1951).

$$^-O_2C-\underset{\underset{M^{++}}{\overset{\|}{O}}}{C}-CH_2-CO_2^- \longrightarrow {}^-O_2C-\underset{\underset{M^{++}}{\overset{|}{O^-}}}{C}=CH_2 + CO_2 \longrightarrow {}^-O_2C-\underset{\underset{M^{++}}{\overset{\|}{O}}}{C}-CH_3 \qquad (12)$$

RIBONUCLEASE

The mechanism of action of pancreatic ribonuclease was sorted out in the early 1960s, and is now reasonably well understood (Richards and Wyckoff, 1971). The enzyme is particularly stable and easy to purify; one of the steps in the preparation involves heating the solution of proteins that contains the enzyme to 95°C at pH 3 (Kunitz and McDonald, 1953). It was the first enzyme – the second protein after insulin – for which the sequence was determined. That feat was accomplished primarily by W. H. Stein, Stanford Moore, and Werner Hirs (Hirs *et al.*, 1960; Spackman *et al.*, 1960; Ryle and Anfinsen, 1957). The investigators at Rockefeller introduced two new features to the sequencing of proteins: the automatic fraction collector and the automatic amino-acid analyser; these innovations considerably enhanced the power of the fundamental method pioneered by Frederick Sanger in 1951. Although some advances in speed and convenience have been added subsequently, the method of Stein and Moore has been standard ever since 1960, and has been supplemented, but not really replaced, by the modern methods of sequencing DNA, where the sequence of a protein is inferred from that of its gene.

An understanding of the mechanism of action of the enzyme came from experiments that showed that the histidine residues at positions 12 and 119 in the amino-acid chain, and lysine-41, are essential to enzymic activity. That histidines may be important was suggested by the observation that photochemical oxidation of ribonuclease destroyed histidines and eliminated enzymic activity (Weil and Seibles, 1955). The two histidines are attacked by iodoacetate, and when either has reacted, the enzyme is nearly, but not entirely inactivated (Gundlach *et al.*, 1959a,b; Barnard and Stein, 1959; Stein and Barnard, 1959; Crestfield *et al.*, 1963a,b). The importance of the two histidines was confirmed by research of Frederick Richards and Vithayathil (1959) who discovered that subtilisin cleaves ribonuclease between amino-acid residues 20 and 21. The two pieces formed by this cleavage are tightly bound to one another and do not spontaneously separate. When they are separated, however, neither is active alone; each contains one of the two essential histidines, both of which are required for enzymic activity. Probably the most interesting part of Richards' research, however, pertains to the strong binding between the two cleavage pieces from ribonuclease, which spontaneously bind (but do not bond) when they are placed together in the same solution. Some lessons concerning specific but non-covalent binding, such as that between enzyme and substrate, are undoubtedly hidden in this research. The essential lysine, amino-acid residue number 41, is attacked by Leuchs anhydrides, and enzymic activity destroyed, but inorganic phosphate inhibits the attack on lysine-41. In fact, in the presence of phosphate, all the lysine residues except lysine-41 are attacked, while the activity of the enzyme remains essentially intact (Cooke et al., 1963). The significance of these facts is set out below.

The pH-rate profile for the action of the enzyme shows a typical pH maximum, with sharply lower rates at either higher or lower pH than the optimum; these facts suggest that both an acidic and a basic group are required for activity (Herries, 1960). The two essential histidine residues could serve as these groups if, in the active site, one were protonated and the other present in its basic form. The simultaneous acid–base catalysis would parallel that of the model system (discussed below) of Swain and J. F. Brown. The essential lysine, which binds phosphate, presumably serves to bind a phosphate residue of the ribonucleic acid. These facts led Mathias and coworkers to propose the mechanism for the action of ribonuclease that is shown in (13) (Findlay et al., 1961).

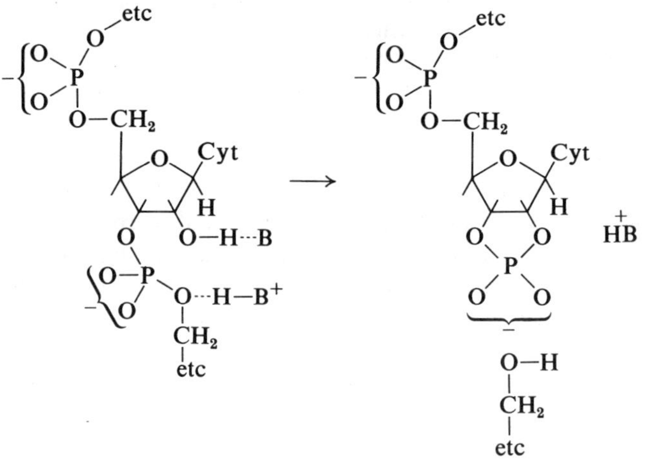

(13)

Subsequent research has confirmed these deductions from chemistry. In particular, X-ray crystallography has shown that the two histidine and the essential lysine residues, which are widely separated one from another along the length of the amino-acid chain, are in fact gathered in a cleft in the three-dimensional structure and do indeed constitute a proper active site (Kartha et al., 1967; Wyckoff et al., 1967, 1970; Richards and Wyckoff, 1971). The protein chain is folded so as to bring them together. The three-dimensional structure is held in place by four disulfide bonds, but when these bonds are opened by reduction, slow re-oxidation in air re-establishes the proper disulfide linkages and restores enzymic activity; information that dictates the proper folding is built into the sequence (White, 1960, 1961; Anfinsen and Haber, 1961). Insofar as mechanism can be equated with pathway, the mechanism of action of ribonuclease is well established.

FORMYL GAR AMIDOTRANSFERASE

A good start was made in 1963 toward the determination of the mechanism of action of formyl GAR amidotransferase (2-formamido-N-ribosyl-acetamide 5-phosphate: L-glutamine amidoligase) by Buchanan and co-workers. They discovered that L-azaserine [9] irreversibly inactivates this enzyme, and that the inhibition is caused by the attack of the diazo-compound on the sulfhydryl group of a cysteine residue. Azaserine mimics glutamate in structure, and can be regarded as an affinity label for the amidotransferase. Buchanan et al. postulated that a nucleophilic sulfhydryl residue at the active site of the enzyme attacks not only the diazo-group of azaserine, but also the carboxyamide group of glutamine [10], and that a glutamyl enzyme is an intermediate in the enzymic pathway (Dawid et al., 1963; French et al., 1963). The full mechanism was not presented until later (Buchanan, 1973), but the essential experiments that opened the way were performed more than 20 years ago.

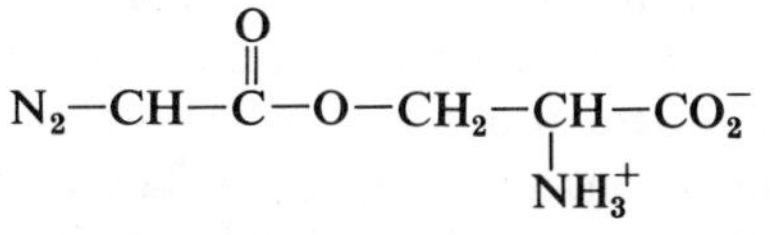

$$N_2-CH-\overset{\overset{\displaystyle O}{\|}}{C}-O-CH_2-\underset{\underset{\displaystyle NH_3^+}{|}}{CH}-CO_2^- \qquad H_2N-\overset{\overset{\displaystyle O}{\|}}{C}-CH_2-CH_2-\underset{\underset{\displaystyle NH_3^+}{|}}{CH}-CO_2^-$$

[9] 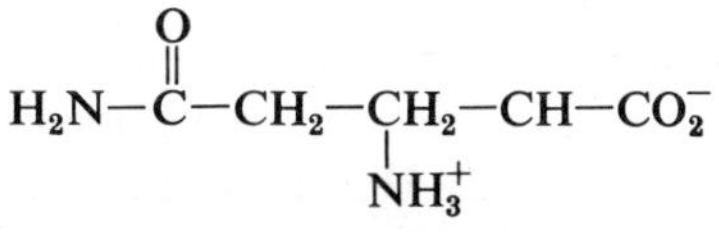   [10]

Obviously, the work reported here constitutes only a part of the great body of research on mechanistic enzymology from the years 1947 to 1963. But they constitute the major part of the history, that is to say, the major discoveries that stimulated the field.

4 Rates and specificity

So far, this discussion has been limited to mechanism, or perhaps to be more specific, to pathway. But what about rates and specificity? These are the essentials of enzymology. How does a protein increase the rate of a reaction by factors of 10^9 to 10^{12}? How do enzymes select their substrates from among the 1000 or so compounds present simultaneously in a cell? These questions – especially that of rates – are in their essence questions in physical organic chemistry. The role of model systems in elucidating the physical organic chemistry underlying enzymic catalysis has previously been elegantly reviewed in this series in more detail than is offered here (Fife, 1975; Kirby, 1980).

KINETICS

In order to put the rates of enzyme reactions on a quantitative footing, and to apply the normal criteria of physical organic chemistry to enzymology, the kinetics of enzyme action had to be explored. The idea of an enzyme–substrate complex had its roots in a number of investigations carried out before or near the turn of the century. When fibrin was treated with papain, and then repeatedly washed, it was nevertheless digested; since the washing did not remove the enzyme, some sort of complex seemed to be required (Wurtz, 1880). Invertase is more stable to heat in the presence than in the absence of sugar; this fact too suggested a complex (O'Sullivan and Tompson, 1890). A. J. Brown studied the production of products from invertase and found that it was linear, rather than logarithmic, with time. Although he did not derive any kinetic equations, he showed qualitatively that this behaviour is consistent with the obligatory formation of an enzyme–substrate complex (A. J. Brown, 1902).

Interestingly, kinetics developed before mechanism (Segal, 1959). In 1902, Victor Henri investigated the rates of several enzymic reactions, including the hydrolysis of sucrose by invertase, and formulated kinetic equations in terms of substrate–enzyme complexes. He discovered the inhibition of the reaction by the products and developed the modern equations for these processes (Henri, 1902). Henri's work, although fundamentally correct, was flawed in that he did not know, in 1902, of the importance of pH, and did not allow time for the complete epimerization of the sugars formed in the hydrolytic process. Fortunately, however, all of his reactions were carried out at a single acidity. In 1913, Leonor Michaelis and Maud Menten repeated Henri's work with control of pH, and again deduced the famous correct equations that now bear their names (Michaelis and Menten, 1913). These investigators, like Henri, assumed that substrate and enzyme are continuously in equilibrium. The reformulation of the reaction process as a steady-state system was later achieved by G. E. Briggs and J. B. S. Haldane (1925). In 1953, S. G. Waley developed the equations for the pH–rate profile in terms of the ionization of catalytic groups on the enzyme and enzyme–substrate complex (Waley, 1953; Alberty, 1956). The pitfalls in this analysis were pointed out shortly thereafter (Bruice and Schmir, 1959), but the method, if applied with caution, is useful and continues to be widely applied, usually without caveat, to this day.

CATALYSIS

The enormous catalytic power of enzymes has come to be explained by several factors (Jencks, 1969): "approximation" (bringing reacting groups together), favourable entropy of activation, simultaneous catalysis by various groups (e.g. acid and base), favourable solvent effects within the cavity or cleft in which the substrate or substrates are contained, "strain" of one sort or another, conformational change in the protein to allow exact fitting of enzyme to transition state, and, of course, just plain "smart" chemistry (Jencks, 1969). "Smart" chemistry includes the use of coenzymes to select a favourable pathway or to introduce new chemistry that was previously unknown, such as the carbon–cobalt chemistry of vitamin B_{12} (Abeles and Lee, 1962; Abeles and Dolphin, 1976). Chemists today, building model systems, have partially imitated the catalytic properties of enzymes. Similarly, a start has now been made on specificity, utilizing synthetic inclusion compounds as well as Schardinger dextrins. In an important paper, Linus Pauling (1948) pointed out that an enzyme must specifically bind the transition state rather than the substrate for an enzymic process, but little was made of this principle at the time; the development of transition state analogues as inhibitors for enzymes, based directly on Pauling's insight, took place after 1963 (Wolfenden, 1972). Although neither an understanding of specificity nor that of rates was well developed in the era that is covered in this review, some beginnings were made.

The idea that simultaneous (i.e., concerted) acid and base catalysis could strongly accelerate reactions was probably introduced by T. M. Lowry in his considerations of general acid–general base catalysis of the mutarotation of glucose (Lowry and Faulkner, 1925). The concept was provided with strong experimental support by C. G. Swain and J. F. Brown (1952a,b), who used 2-pyridone to catalyze the mutarotation of α-tetramethylglucose. They showed that the reaction with 0.001 M pyridone proceeds 700 times as fast as it does in the presence of a mixture of 0.001 M pyridine and 0.001 M phenol, although pyridine is 10,000 times as strong a base, and phenol 100 times as strong an acid, as pyridone. Further, in true enzymic reactions, the concentrations of substrate and enzyme would be much less than 0.001 M, and the advantage of concerted catalysis even greater; this is so because the mutarotation of tetramethylglucose catalyzed by pyridone is second order, whereas that catalyzed by pyridine and phenol is third order. The transition state for the mutarotation is shown in [11].

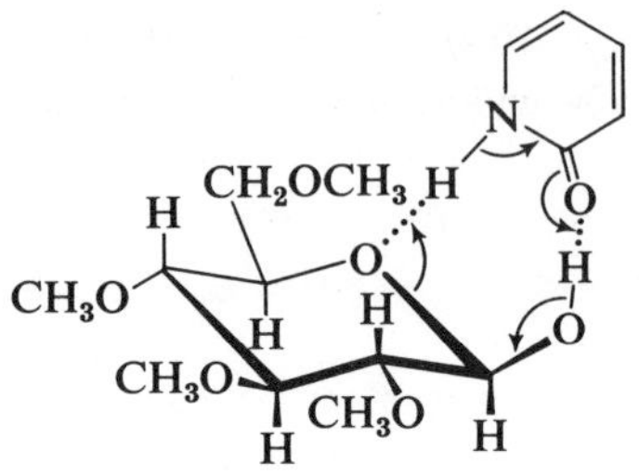

[11]

Swain and Brown pointed out that the complex between pyridone and tetramethylglucose is very much stronger in benzene than in water, and ascribed much of the influence of benzene in increasing the reaction rate over that observed in water to this effect. They also noted, however, that in acid catalysis the substrate is protonated, and acquires a positive charge; similarly, in base catalysis, the substrate acquires a negative charge. Swain and Brown ascribed part of the advantage of simultaneous acid and base catalysis to the avoidance of charged intermediates, a factor that is of especial significance in solvents, or regions, or low dielectric constant. They compared their reaction to enzymic ones. The enormous importance of solvent effects was confirmed in research that was carried out after the period under consideration here (Crosby and Lienhard, 1970). The experiments of Swain and Brown had considerable influence on the thinking of physical organic chemists interested in enzymology.

ENTROPY

An additional factor that favours concerted action in general, including concerted catalysis by acid and base, is that of entropy [considerations of entropy of activation in non-enzymic chemistry were reviewed in the first article of the first volume of this series (Schaleger and Long, 1963)]. When two molecules come together in solution, they lose entropy of translation; at 1 M the loss of entropy is around 33 e.u., while at 10^{-6} M this amounts to about 60 e.u., or 18 kcal mol^{-1} at room temperature (Westheimer and Ingraham, 1956). In order to form an enzyme–substrate complex, the forces of attraction between enzyme and substrate must be sufficient to overcome the loss of entropy. Of course, not all of this entropy disappears; when three degrees of translational freedom are lost, the enzyme–substrate complex gains three degrees of vibration or internal rotation. But the entropy connected with these degrees of freedom can compensate for only a part of the translational entropy. In a third-order reaction, or a reaction where acid and base react successively, the entropy of translation for both acid and base must be surrendered. When, however, the acid and base are contained in the same molecule, the entropy loss is much less. Only three degrees of translational entropy are lost in bringing substrate into contact with the catalyst. Still, after one catalytic group is located correctly with respect to the substrate, the second must still be brought to its proper place. This involves ring formation, again with loss of entropy, but that loss is much less than the entropy of translation that would have to be lost in bringing two molecules together from solution. Concerted catalysis, such as that supplied by pyridone or by an enzyme, saves an enormous amount of entropy, and so reduces the free energy of activation for reaction. In this sense, enzymes were described as "entropy traps" (Westheimer, 1962). The concepts advanced in 1962 have subsequently been enormously refined (Page and Jencks, 1971).

APPROXIMATION

A major source of acceleration in enzymic reactions is "approximation," that is to say, the bringing together of two or more reactants in the active site. Once the reagents are in contact, the subsequent reaction is intra- rather than intermolecular. Comparisons of the rates of intermolecular and intramolecular reactions are, however, difficult because the rate constants for bimolecular reactions have the units of $M^{-1} s^{-1}$, whereas rate constants for unimolecular reactions have the units of s^{-1}. The best one can do in comparing them is to state the molarity at which the reactants would have to be present in the bimolecular reaction to achieve the rate of the unimolecular process; when the "effective molarity" is large – say 1000 M or more – one has some measure of the power of "approximation" to accelerate chemical reaction.

The idea of the importance of approximation goes back into the early experience of organic chemists, who found that intramolecular ring closures occur much more readily than similar intermolecular reactions. Many studies of the quantitative effects of approximation were carried out in the 1950s and early 1960s (Edwards, 1950, 1952; Leach and Lindley, 1953a,b; R. F. Brown and van Gulick, 1956; Bender and Neveu, 1958; Schmir and Bruice, 1958; Bruice and Sturtevant, 1959; Bernard et al., 1959; Bender, 1960; Gaetjens and Morawetz, 1960; Bruice and Benkovic, 1963). Most of the examples in the papers cited here involve nucleophilic catalysis, i.e., the catalyst forms an intermediate with a covalent bond between it and the substrate. The finding that alkyl substituents in a chain of atoms aid ring closure goes back at least to the discovery of the special ease of anhydride formation from tetramethylsuccinic acid (14). (Auwers and Meyer, 1890; Ingold, 1921). Bruice explained this effect as a decrease in the entropy of rotamers (Bruice and Pandit, 1960a,b) for molecules where an alkyl substituent restricts the rotation of the chain that connects the reactive ends.

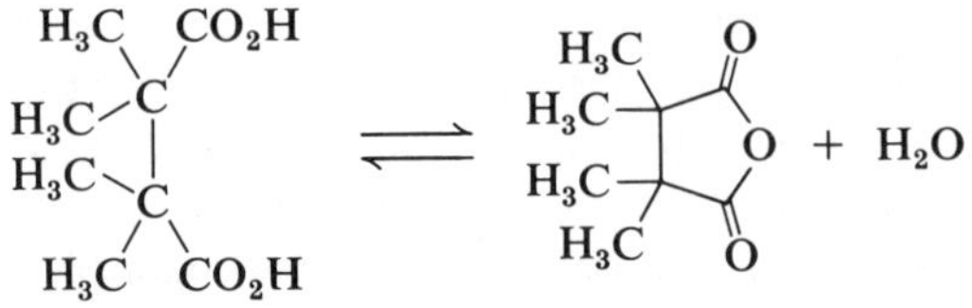

$$\begin{array}{c}\text{H}_3\text{C} \quad \text{CO}_2\text{H} \\ \text{H}_3\text{C}-\text{C} \\ \text{H}_3\text{C}-\text{C} \\ \text{H}_3\text{C} \quad \text{CO}_2\text{H}\end{array} \rightleftharpoons \begin{array}{c}\text{H}_3\text{C} \quad \text{O} \\ \text{H}_3\text{C} \\ \text{H}_3\text{C} \\ \text{H}_3\text{C} \quad \text{O}\end{array}\text{O} + \text{H}_2\text{O} \qquad (14)$$

STRAIN

Among the most spectacular early examples of the effect of alkyl groups on the rate of reactions is that described by Bunnett and Okamoto (1956) who found that a single methyl group increases the rate of the Smiles re-arrangement (15) by a factor of half a million. This effect is far too large to be caused by the loss of entropy due to restriction of rotation, and presumably is caused by steric strain in the starting material that is relieved in the transition state for the rearrangement.

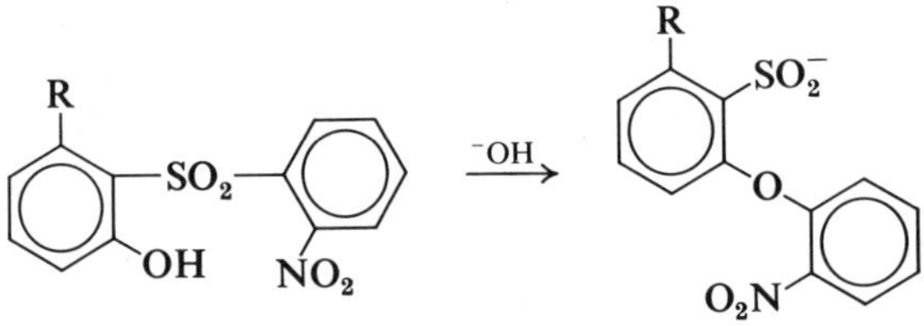

$$(15)$$

[R = H or CH$_3$]

Strain of various sorts (Huisgen and Ott, 1959) has been suggested as the cause of at least part of the acceleration of rate caused by enzymes. The most popular theory described enzymes as a "rack" on which the substrate was stretched. Electrostatic or other forces caused strong interactions between the enzyme and two ends of the substrate and pulled or bent it in such a way as to weaken the fissile bond. The strain was postulated to lower the energy required to reach the transition state (Eyring *et al.*, 1954; Lumry, 1959).

SPECIFICITY

An understanding of enzyme action requires not only a knowledge of pathway and rates, but also an explanation of specificity. The dominant idea in this area was that of Emil Fischer, who described the enzyme–substrate complex in terms of lock and key (Fischer, 1894). In essence, Fischer presented a steric model where a cavity in the enzyme was assumed to be shaped to fit the substrate and to hold it firmly in place. This model served enzymologists well for decades and helped them to visualize the interactions between specific groups (say imidazole residues) on the enzyme and specific groups (say the proton of an hydroxyl group) on the substrate.

Chemists tried to find non-enzymic analogues of the complexation of enzyme with substrate. Tight complexes between various molecules were well known; in particular, Wieland had established the complexation of desoxycholic acids with carboxylic acids (Wieland and Sorge, 1916), and other investigators subsequently found complexes of desoxycholates with many other organic compounds (Reinbolt, 1926; Fieser and Newman, 1935). Many complexes – clathrates – exist only in the solid state, where, for example, urea, thiourea, quinol and several other compounds crystallize with cavities in the lattice that can be filled with other molecules, even including the rare gases (Cramer, 1952, 1956). The most interesting complexing agents that were discovered before 1963 were the Schardinger dextrins (cyclodextrins). These consist of rings of glucose molecules; a series of cyclodextrins with 6, 7 and 8 glucose residues constitute rings with larger and larger cavities in the centre. Furthermore, the size of the molecules that form stable complexes increases with the number of glucose residues in the cyclodextrin, i.e., with the size of the cavity in the ring. The complexes of the cyclodextrins are distinguished from the clathrates in that they are stable not only in the crystal but also in solution (French, 1957; Cramer and Henglein, 1957a,b; 1958). The "guest" molecules really do sit in the cavity of the "host"; the structure of the iodine complex of a cyclodextrin was established in 1954 by X-ray analysis (von Dietrich and Cramer, 1954). These complexes serve as models for enzyme–substrate complexes and illustrate the essential correctness of Fischer's lock and key.

An important feature of the cyclodextrins is that they can also accelerate chemical reactions, and therefore serve as models for the catalytic as well as the binding properties of enzymes. The rapid reaction is not catalysis, since the dextrin enters reaction but is not regenerated; presumably it arises from approximation, where complex formation forces the substrate and the cyclodextrin into intimate contact. In particular, cyclodextrins can increase the rate of cleavage of phenyl pyrophosphate by factors of as much as 100 (Cramer, 1961). More recent work has improved upon this early example.

INDUCED FIT

Although the idea of the steric fit of substrate to enzyme was a powerful one, in detail it presented some problems. One of these, the notion that the enzyme should fit the transition state better than it fits the substrate, has already been mentioned. The most severe of the problems, however, was illustrated with the chemistry of phosphorylase. Why was the enzyme specific for the transfer of glucose from glycogen? Why did it not catalyze the hydrolysis of glucose-1-phosphate in an aqueous solvent? The cavity that would be occupied by glucose in the normal reaction between glucose and glucose-1-phosphate could not remain empty, creating a vacuum; water molecules, smaller than glucose, could surely fit in. Since water presents an hydroxyl group similar to that of glucose, why did the enzyme not use it? Considering this fact, and many others, Koshland in 1958 proposed the idea of an "induced fit" to explain enzymic specificity. He suggested that native enzymes possess cavities that are only approximately correct for the substrate, but that, since proteins are flexible, the forces between substrate and enzyme can pull the enzyme into the exact position for binding and subsequent chemical reaction. The various catalytic groups needed at the active site are only properly aligned for reaction (i.e., properly aligned to bind only the transition state) after the substrate is in place, and has appropriately altered the enzyme structure (Koshland, 1958, 1959a,b).

This idea was immediately appealing. Although most of Koshland's arguments at the time were suggestive rather than compelling, the concept of induced fit has now been verified by X-ray crystallography with three separate enzyme systems (Reeke *et al.*, 1967; Pickover *et al.*, 1979; Alber *et al.*, 1981).

5 Conclusion

Mechanistic enzymology has developed at a rapid pace since 1963, and the artificial catalysts of which Fischer spoke in 1902 appear within the realm of possibility (Breslow, 1982). Enzymologists now have reasonable notions of mechanism and of the factors needed to achieve rapid rates and specificity. Nevertheless, the foundations for mechanism and to some extent for an understanding of the factors relevant to enzymology were laid down in the years 1947–1963. By the end of that time, enzymology had become incorporated into physical organic chemistry.

Acknowledgement

The work in the author's laboratory relevant to the events recorded here was generously supported by the National Institutes of Health, Bethesda, Maryland.

References

Abeles, R. H. and Dolphin, D. (1976). *Accts. Chem. Res.* **2**, 114
Abeles, R. H. and Lee, H. A., Jr. (1962). *Brookhaven Symposium* No. 15, p. 310
Adrian, E. D., Feldberg, W. and Kilby, B. A. (1947). *Brit. J. Pharm.* **2**, 2795

Alber, T., Benner, D. W., Bloomer, A. C., Petsko, G. A., Phillips, D. C., Rivers, P. S. and Wilson, I. A. (1981). *Phil. Trans. Roy. Soc. B* **293**, 159
Alberty, R. A. (1956). *J. Cell and Comp. Physiol.* **47** *(Suppl. 1)*, 245
Aldridge, W. N. (1950). *Biochem. J.* **46**, 451
Anfinsen, C. B. and Haber, E. (1961). *J. Biol. Chem.* **236**, 1361
Auwers, K. and Meyer, V. (1890). *Ber.* **23**, 101
Augustinsson, K. B. and Nachmansohn, D. (1949). *J. Biol. Chem.* **179**, 543
Balls, A. K. and Wood, H. N. (1953). *J. Biol. Chem.* **202**, 67
Balls, A. K. and Aldrich, F. L. (1955). *Proc. Nat. Acad. Sci.* **41**, 190
Balls, A. K. and Wood, H. N. (1956). *J. Biol. Chem.* **210**, 245
Balls, A. K., McDonald, C. E. and Brecher, A. S. (1957). *Proc. Internat. Symposium on Enzyme Chem.*, Tokyo and Kyoto, **2**, 392; *Chem. Abs.* (1957). **54**, 1343b
Barnard, E. A. and Stein, W. D. (1959). *J. Mol. Biol.* **1**, 339
Bender, M. L. (1960). *Chem. Rev.* **60**, 53
Bender, M. L. and Neveu, M. C. (1958). *J. Am. Chem. Soc.* **80**, 5388
Bender, M. L., Schonbaum, G. R. and Zerner, B. (1962). *J. Am. Chem. Soc.* **84**, 2540
Bender, M. L. and Zerner, F. (1962) *J. Am. Chem. Soc.* **84**, 2550
Bergman, F., Wilson, I. B. and Nachmansohn, D. (1950). *J. Biol. Chem.* **186**, 693
Bernard, S. A., Katchalski, E., Sela, M. and Berger, A. (1959). *J. Cell. Comp. Physiol.* **54** *(Suppl. 1)*, 195
Blow, D. M. (1969). *Biochem. J.* **112**, 261
Braunstein, A. E. (1960). *In* "The Enzymes", 2nd edn (Boyer, P. D., Lardy, H. and Myrback, K., eds), Vol. 2, p. 113. Academic Press, New York
Braunstein, A. E. and Kritzmann, M. G. (1947a). *Enzymologia* **2**, 129
Braunstein, A. E. and Kritzmann, M. G. (1947b). *Enzymologia* **2**, 138
Braunstein, A. E. and Kritzmann, M. G. (1947c). *Biokhimiya* **2**, 859
Braunstein, A. E. and Shemyakin, M. M. (1952). *Doklady Akad. Nauk. SSSR* **85**, 1115
Braunstein, A. E. and Shemyakin, M. M. (1953). *Biokhimiya* **18**, 393
Bredt, J. (1927). *Ann. Acad. Sci. Fennica* **29A(2)**, 3. (*Chem. Z.* **98II**, 2298)
Breslow, R. (1957). *J. Am. Chem. Soc.* **79**, 1762
Breslow, R. (1958). *J. Am. Chem. Soc.* **80**, 3719.
Breslow, R. (1982). *Science* **218**, 532.
Briggs, G. E. and Haldane, J. B. S. (1925). *Biochem. J.* **19**, 338
Brown, A. J. (1902). *J. Chem. Soc.* 373
Brown, R. F. and van Gulick, N. (1956). *J. Org. Chem.* **21**, 1046
Bruice, T. C. and Benkovic, S. J. (1963). *J. Am. Chem. Soc.* **85**, 1
Bruice, T. C. and Pandit, U. K. (1960a). *Proc. Nat. Acad. Sci. U.S.A.* **46**, 402
Bruice, T. C. and Pandit, U. K. (1960b). *J. Am. Chem. Soc.* **82**, 5858
Bruice, T. C. and Schmir, G. L. (1959).*J. Am. Chem. Soc.* **81**, 4552
Bruice, T. C. and Sturtevant, J. M. (1959). *J. Am. Chem. Soc.* **81**, 2860
Buchanan, J. M. (1973). *Adv. in Enzym.* **39**, 91
Bunnett, J. F. and Okamoto, T. (1956). *J. Am. Chem. Soc.* **78**, 5363
Cooke, J. P., Anfinsen, C. B. and Sela, M. (1963). *J. Biol. Chem.* **238**, 2034
Cramer, F. (1952). *Angew. Chem.* **64**, 437
Cramer, F. (1956). *Angew. Chem.* **68**, 115
Cramer, F. (1961). *Angew. Chem.* **73**, 49
Cramer, F. and Henglein, F. M. (1957a). *Ber.* **90**, 2561
Cramer, F. and Henglein, F. M. (1957b). *Ber.* **90**, 2572
Cramer, F. and Henglein, F. M. (1958). *Ber.* **91**, 308

Crestfield, A. M., Stein, W. H. and Moore, S. (1963a). *J. Biol. Chem.* **238**, 2413
Crestfield, A. M., Stein, W. H. and Moore, S. (1963b) *J. Biol. Chem.* **238**, 2421
Crosby, J. and Lienhard, G. E. (1970). *J. Am. Chem. Soc.* **92**, 5707
Dawid, I. B., French, T. C. and Buchanan, J. M. (1963). *J. Biol. Chem.* **238**, 2178
Dixon, G. H., Dreyer, W. J. and Neurath, H. (1956). *J. Am. Chem. Soc.* **78**, 4810
Dixon, M. and Webb, E. C. (1964). "Enzymes", Ch. 6. Academic Press, New York
Doudoroff, M., Barker, H. A. and Hassid, W. Z. (1947a). *J. Biol. Chem.* **168**, 725
Doudoroff, M., Barker, H. A. and Hassid, W. Z. (1947b). *J. Biol. Chem.* **170**, 147
Doudoroff, M. (1960). *In* "Enzymes", 2nd edn (Boyer, P. D., Lardy, H. and Myrback, K., eds.) Vol. 5, p. 229. Academic Press, New York
Edwards, L. J. (1950). *Trans. Farad. Soc.* **46**, 723
Edwards, L. J. (1952). *Trans. Farad. Soc.* **48**, 696
Eyring, H., Lumry, R. and Spikes, J. D. (1954). *In* "The Mechanisms of Enzyme Action" (McElroy, W. D. and Glass, B. eds) p. 123, Johns Hopkins Press, Baltimore, Md.
Fieser, L. and Newman, M. S. (1935). *J. Am. Chem. Soc.* **57**, 1602
Fife, T. H. (1975). *Adv. Phys. Org. Chem.* **11**, 1
Findlay, D., Herries, D. G., Mathias, A. P., Rabin, B. R. and Ross, C. A. (1961). *Nature* **190**, 781
Fischer, E. (1894). *Ber.* **27**, 2985
Fischer, E. (1898). *Z. Physiol. Chem.* **26**, 60
Fischer, E. (1902). "Nobel Lectures—Chemistry 1901–1921". Elsevier Publishing Co., Amsterdam (1966) p. 34
Fischer, E. H., Kent, A. B., Snyder, E. R. and Krebs, E. G. (1958). *J. Am. Chem. Soc.* **80**, 2906
Fisher, H. F., Conn, E. E., Vennesland, B. and Westheimer, F. H. (1953). *J. Biol. Chem.* **202**, 687
Fitting, C. and Doudoroff, M. (1952). *J. Biol. Chem.* **199**, 153
French, D. (1957). *Adv. Carbohydrate Chem.* **12**, 247
French, T. C., Dawid, I. B. and Buchanan, J. M. (1963). *J. Biol. Chem.* **238**, 2186
Fridovich, I. and Westheimer, F. H. (1962). *J. Am. Chem. Soc.* **84**, 3208
Gaetjens, E. and Morawetz, H. (1960). *J. Am. Chem. Soc.* **82**, 5328
Grazi, E., Cheng, T. and Horecker, B. L. (1962). *Biochem. Biophys. Res. Commun.* **7**, 250
Gundlach, H. G., Stein, W. H. and Moore, S. (1959a). *J. Biol. Chem.* **234**, 1754
Gundlach, H. G., Stein, W. H. and Moore, S. (1959b). *J. Biol. Chem.* **234**, 1761
Gutfreund, H. and Sturtevant, J. M. (1956). *Proc. Nat. Acad. Sci., USA* **42**, 719
Hamilton, G. and Westheimer, F. H. (1959). *J. Am. Chem. Soc.* **81**, 6332
Hammett, L. P. (1940). "Physical-Organic Chemistry". McGraw-Hill, New York
Hartley, B. S. and Kilby, B. A. (1952). *Biochem. J.* **50**, 672
Hartley, B. S. and Kilby, B. A. (1954). *Biochem. J.* **56**, 288
Henri, V. (1902). *Comptes rendus* **135**, 916
Herries, D. G. (1960). *Biochem. Biophys. Res. Commun.* **3**, 666
Hirs, C. H. W., Moore, S. and Stein, W. H. (1960). *J. Biol. Chem.* **235**, 633
Horecker, B. L., Pontremoli, S., Ricci, C. and Cheng, T. (1961). *Proc. Nat. Acad. Sci., USA* **47**, 1949
Huisgen, R. and Ott, H. (1959). *Tetrahedron* **6**, 253
Ikawa, M. and Snell, E. E. (1954a). *J. Am. Chem. Soc.* **76**, 637
Ikawa, M. and Snell, E. E. (1954b). *J. Am. Chem. Soc.* **76**, 653
Ingold, C. K. (1921). *J. Chem. Soc.* **119**, 305

Jansen, E. F., Nutting, M.-D. F. and Balls, A. K. (1949a). *J. Biol. Chem.* **179**, 201
Jansen, E. F., Nutting, M.-D. F., Jang, R. and Balls, A. K. (1949b). *J. Biol. Chem.* **179**, 189
Jencks, W. P. (1963). *Ann. Rev. Biochem.* **32**, 639
Jencks, W. P. (1969). "Catalysis in Chemistry and Enzymology". McGraw-Hill, New York
Jenkins, W. T. and Sizer, I. W. (1957). *J. Am. Chem. Soc.* **79**, 2655
Kartha, G., Bello, J. and Harker, D. (1967). *Nature* **213**, 862
Kennedy, E. P. and Koshland, D. E., Jr. (1957). *J. Biol. Chem.* **228**, 419
Kendrew, J. C., Dickerson, R. E., Strandberg, B. E., Hoit, R. G., Davies, D. R., Phillips, D. C. and Shore, V. C. (1960). *Nature* **185**, 422
Kirby, A. J. (1980). *Adv. Phys. Org. Chem.* **17**, 183
Konikova, A. S., Dobbert, N. N. and Braunstein, A. E. (1947). *Nature* **159**, 67
Koshland, D. E., Jr. (1953). *Biol. Rev.* **28**, 416
Koshland, D. E., Jr. (1954) "The Mechanisms of Enzyme Action", (McElroy, W. D. and Glass, B., eds) p. 608. Johns Hopkins Press, Baltimore, Md.
Koshland, D. E., Jr. (1958). *Proc. Nat. Acad. Sci. USA* **44**, 98
Koshland, D. E., Jr. (1959a). *In* "The Enzymes", 3rd edn (Boyer, P., ed.), Vol. 1, p. 305. Academic Press, New York
Koshland, D. E., Jr. (1959b). *J. Cell. Comp. Physiol.* **54**, 1st Supplement, 245
Krampitz, L. O. (1969). *Ann. Rev. Biochem.* **38**, 213
Kunitz, M. and Northrop, J. H. (1935). *J. Gen. Physiol.* **18**, 433
Kunitz, M. and McDonald, M. R. (1953). *Biochem. Prep.* **3**, 9
Lange, W. and Krueger, G. (1932). *Ber.* **65**, 1598
Lapworth, A. (1903). *J. Chem. Soc.* **83**, 995
Leach, S. J. and Lindley, H. (1953a). *Trans. Farad. Soc.* **49**, 915
Leach, S. J and Lindley, H. (1953b). *Trans. Farad. Soc.* **49**, 921
Loewus, F. A., Westheimer, F. H. and Vennesland, B. (1953a). *J. Am. Chem. Soc.* **75**, 5018
Loewus, F. A., Offner, P., Fisher, H. F., Westheimer, F. H. and Vennesland, B. (1953b). *J. Biol. Chem.* **202**, 699
Longenecker, J. B. and Snell, E. E. (1957). *J. Am. Chem. Soc.* **79**, 142
Lowry, T. M. and Falkner, I. J. (1925). *J. Chem. Soc.* **127**, 2883
Lumry, R. (1959). *In* "The Enzymes", 2nd edn (Boyer, P., Lardy, H. and Myrback, K., eds) Vol. 1, p. 157. Academic Press, New York
Lynen, F., Knappe, J., Lorch, E., Jutting, G. and Ringelmann, E. (1959). *Angew. Chem.* **71**, 481
Lynen, F., Knappe, J., Lorch, E., Jutting, G., Ringelmann, E. and Lachance, J.-P. (1961). *Biochem. Z.* **335**, 123
MacInnes, I., Nonhebel, D. C., Orszulik, S. T. and Suckling, C. J. (1982). *J. Chem. Soc., Chem. Commun.* 121
Mandeles, S., Koppelman, R. and Hanke, M. E. (1954). *J. Biol. Chem.* **209**, 327
Matthews, B. W., Sigler, P. B., Henderson, R. and Blow, D. M. (1967). *Nature* **214**, 652
Mazur, A. and Bodansky, O. (1946). *J. Biol. Chem.* **163**, 261
Melander, L. (1950). *Arkiv. Kemi* **2**, 211
Melander, L. (1960). "Isotope Effects on Reaction Rates". Ronald Press, New York
Metzler, D. E. and Snell, E. E. (1952a). *J. Am. Chem. Soc.* **74**, 979
Metzler, D. E. and Snell, E. E. (1952b). *J. Biol. Chem.* **198**, 353
Metzler, D. E., Longenecker, J. B. and Snell, E. E. (1954a). *J. Am. Chem. Soc.* **76**, 639

Metzler, D. E., Ikawa, M. and Snell, E. E. (1954b). *J. Am. Chem. Soc.* **76**, 648
Michaelis, L. and Menten, M. L. (1913). *Biochem. Z.* **49**, 333
Mislow, K. and Raban, M. (1967). *In* "Topics in Stereochemistry"(Allinger, N. L. and Eliel, E. L, eds) Vol. 1, p. 1. Interscience, New York
Mizuhara, S., Tamura, R. and Arato, H. (1951). *Proc. Japan Acad.* **27**, 302
Mizuhara, S. and Handler, P. (1954). *J. Am. Chem. Soc.* **76**, 571
Morrison, R. and Boyd, R. (1983). "Organic Chemistry", 4th edn. Allyn and Bacon, Boston
Nachmansohn, D., Rothenberg, M. H. and Feld, E. A. (1948). *J. Biol. Chem.* **174**, 247
Nachmansohn, D. and Wilson, I. B. (1951). *Adv. in Enzym.* **12**, 259
Negelein, E. and Wulff, H.-J. (1937). *Biochem. Z.* **293**, 351
Northrop, J. H. (1930). *J. Gen. Physiol.* **13**, 739
Northrop, J. H. and Kunitz, M. (1932a). *J. Gen. Physiol.* **16**, 257
Northrop, J. H. and Kunitz, M. (1932b). *J. Gen. Physiol.* **16**, 267
Ogston, A. G. (1948). *Nature* **162**, 963
Olivard, J., Metzler, D. E. and Snell, E. E. (1952). *J. Biol. Chem.* **199**, 669
Oosterbaan, R. A., Kunst, P. and Cohen, J. A. (1955). *Biochem. Biophys. Acta* **16**, 299
O'Sullivan, C. and Tompson, F. W. (1890). *J. Chem. Soc.* **57**, 834
Page, M. I. and Jencks, W. P. (1971). *Proc. Nat. Acad. Sci. USA* **68**, 1678
Pauling, L. (1948). *Nature* **161**, 707
Pedersen, K. J. (1929). *J. Am. Chem. Soc.* **51**, 2098
Pedersen, K. J. (1934). *J. Phys. Chem.* **38**, 559
Pedersen, K. J. (1936). *J. Am. Chem. Soc.* **58**, 240
Pedersen, K. J. (1938). *J. Am. Chem. Soc.* **60**, 595
Perutz, M. F. (1954). *Proc. Roy. Soc. A* **225**, 264
Perutz, M. F., Rossman, M. G., Cullis, A. F., Muirhead, H., Will, G. and North, A. C. T. (1960). *Nature* **185**, 416
Pickover, C. A., McKay, D. B., Engelman, D. M. and Steitz, T. A. (1979). *J. Biol. Chem.* **254**, 11323
Pontremoli, S., Prandini, B. D., Bonsignore, A. and Horecker, B. L. (1961). **47**, 1942
Powell, M. F. and Bruice, T. C. (1983). *J. Am. Chem. Soc.* **105**, 1014
Reeke, G. N., Hartsuck, J. A., Ludwig, M. L., Quiocho, F. A., Steitz, T. A. and Lipscomb, W. N. (1967). *Proc. Nat. Acad. Sci. USA* **58**, 2220
Reinbolt, H. (1926). *Ann. Chem.* **451**, 256
Richards, F. M. and Vithayathil, P. J. (1959). *J. Biol. Chem.* **234**, 1459
Richards, F. M. and Wyckoff, H. W. (1971). *In* "The Enzymes", 3rd edn (Boyer, P. D., ed.) Vol. 1, Ch. 24. Academic Press, New York
Rosenberg, H. R. (1945). "Chemistry and Physiology of the Vitamins". Interscience, New York
Ryle, A. P., Sanger, F., Smith, L. F. and Kitai, R. (1955). *Biochem. J.* **60**, 541.
Ryle, A. P. and Anfinsen, C. B. (1957). *Biochem. Biophys. Acta* **24**, 633
Sanger, F. and Tuppy, H. (1951a). *Biochem. J.* **49**, 463
Sanger, F. and Tuppy, H. (1951b). *Biochem. J.* **49**, 481
Sanger, F. and Thompson, E. O. P. (1953). *Biochem. J.* **53**, 366
Schaffer, N. K., May, S. C. and Summerson, W. H. (1954). *J. Biol. Chem.* **206**, 201
Schaleger, L. L. and Long, F. A. (1963). *Adv. Phys. Org. Chem.* **1**, 1.
Schmir, G. L. and Bruice, T. C. (1958). *J. Am. Chem. Soc.* **80**, 1173
Schoellman, G. and Shaw, E. (1962). *Biochem. Biophys. Res. Comm.* **7**, 36
Schoellman, G. and Shaw, E. (1963). *Biochemistry* **2**, 252
Schonbaum, G. R., Nakamura, K. and Bender, M. L. (1959). *J. Am. Chem. Soc.* **81**, 4746

Segal, H. L. (1959). *In* "The Enzymes", 2nd edn (Boyer, P., Lardy, H. and Myrback, K., eds) Vol. 1, p. 1. Academic Press, New York
Snell, E. E. (1945). *J. Am. Chem. Soc.* **67**, 194
Spackman, D. H., Stein, W. H. and Moore, S. (1960). *J. Biol. Chem.* **235**, 648
Stein, W. D. and Barnard, E. A. (1959). *J. Mol. Biol.* **1**, 350
Steinberger, R. and Westheimer, F. H. (1949). *J. Am. Chem. Soc.* **71**, 4158
Steinberger, R. and Westheimer, F. H. (1951). *J. Am. Chem. Soc.* **73**, 429
Sumner, J. B. (1926). *J. Biol. Chem.* **69**, 435
Sumner, J. B. and Dounce, A. L. (1937). *J. Biol. Chem.* **121**, 417
Swain, C. G. and Brown, J. F. (1952a). *J. Am. Chem. Soc.* **74**, 2534
Swain, C. G. and Brown, J. F. (1952b). *J. Am. Chem. Soc.* **74**, 2538
Ukai, T., Tanaka, R. and Dokowa, T. (1943). *J. Pharm. Soc. Japan* **63**, 296
Vennesland, B. and Westheimer, F. H. (1954). "The Mechanisms of Enzyme Action" (McElroy, W. D. and Glass, B., eds) p. 357. Johns Hopkins Press, Baltimore, Md.
Voet, J. and Abeles, R. H. (1966). *J. Biol. Chem.* **241**, 2731
von Dietrich, H. and Cramer, F. (1954). *Ber.* **87**, 806
Waley, S. G. (1953). *Biochem. Biophys. Acta* **10**, 27
Warren, S., Zerner, B. and Westheimer, F. H. (1966). *Biochemistry* **8**, 817
Weil, L. and Seibles, T. S. (1955). *Arch. Biochem. Biophys.* **54**, 368
Westheimer, F. H. and Jones, W. A. (1941). *J. Am. Chem. Soc.* **63**, 3283
Westheimer, F. H. and Nicolaides, N. (1949). *J. Am. Chem. Soc.* **71**, 25
Westheimer, F. H., Fisher, H. F., Conn, E. E. and Vennesland, B. (1951). *J. Am. Chem. Soc.* **73**, 2403
Westheimer, F. H. and Ingraham, L. L. (1956). *J. Phys. Chem.* **60**, 1668
Westheimer, F. H. (1962). *Adv. in Enzymol.* **24**, 441
Westheimer, F. H. (1963). *Proc. Roy. Chem. Soc.* 253
Whitaker, J. R. and Jandorf, B. J. (1956). *J. Biol. Chem.* **223**, 751
White, F. H., Jr. (1960). *J. Biol. Chem.* **235**, 383
White, F. H., Jr. (1961). *J. Biol. Chem.* **236**, 1353
Wieland, H. J. and Sorge, H. (1916). *Z. Physiol. Chem.* **97**, 1
Wilson, I. B. and Bergman, F. (1950). *J. Biol. Chem.* **185**, 479
Wilson, I. B., Bergman, F. and Nachmansohn, D. (1950). *J. Biol. Chem.* **186**, 781
Wilson, I. B. (1951). *J. Biol. Chem.* **190**, 111
Wolfenden, R. (1972). *Acct. Chem. Res.* **5**, 10
Wurtz, A. (1880). *Comptes rendus* **91**, 787
Wyckoff, H. W., Hardman, K .D., Allewell, N. M., Inagami, T., Johnson, L. N. and Richards, F. M. (1967). *J. Biol. Chem.* **242**, 3984
Wyckoff, H. W., Tsernoglou, A. W., Knox, J. R., Lee, B. and Richards, F. M. (1970). *J. Biol. Chem.* **245**, 305

The Spectroscopic Detection of Tetrahedral Intermediates Derived from Carboxylic Acids and the Investigation of Their Properties

BRIAN CAPON

Department of Chemistry, University of Hong Kong, Pokfulam Road, Hong Kong

MIRANDA I. DOSUNMU

Department of Chemistry, University of Calabar, Calabar, Nigeria

MARIA de NAZARÉ de MATOS SANCHEZ

Departamento de Química, Universidade Federal de Santa Catarina, Florianópolis, Brazil

1 Introduction 38
2 Specially stabilized tetrahedral intermediates 39
3 The detection of simple tetrahedral intermediates 48
    Orthoesters and related species as precursors 49
    Amide acetals as precursors 55
    Ketene acetals as precursors 57
    Oxocarbocations as precursors 59
4 Kinetic studies on the breakdown of hemiorthoesters 60
    The acid-catalysed reaction 66
    The hydroxide-ion catalysed reaction 74
    Spontaneous or water-catalysed reaction 80
    Formation of oxocarbocations; equilibrium constants for formation of hemiorthoesters 84
5 Nitrogen-containing tetrahedral intermediates 89
6 Future investigations 93
References 94

1 Introduction

Until a few years ago there was only indirect evidence for the existence of the species postulated as tetrahedral intermediates in reactions of derivatives of carboxylic acids (cf. Kirby, 1972). Now, however, it is possible to generate some of them in solution at sufficiently high concentrations for their uv and nmr spectra to be measured and for their reactions to be studied. It is the purpose of this review to record progress that has been made in this area.

The possibility that the reactions of derivatives of carboxylic acids might pass through tetrahedral intermediates (1) has long been considered (Lowry, 1926). In their important paper in which they classified the mechanisms of hydrolysis of carboxylic esters Day and Ingold (1941) wrote bimolecular

$$RC\overset{O}{\underset{Y}{\diagup}} + HX \rightleftharpoons R-\underset{Y}{\overset{OH}{\underset{|}{C}}}-X \rightleftharpoons RC\overset{O}{\underset{X}{\diagup}} + HY \qquad (1)$$

[1]   [2]

reactions with acyl–oxygen fission as passing through this type of intermediate which they referred to as the "most condensed intermediate state." However they did not attempt to make the distinction as to whether this occurs "at the energy maximum or before or after it." Hammett (1940) took a similar view and suggested that the "monoester of the ortho acid" may be an intermediate in "acid-catalysed esterification and hydrolysis." It took another 10 years before these speculations were reinforced by any experimental evidence, albeit indirect, when Bender (1951) demonstrated that exchange of the carbonyl groups of ethyl, isopropyl, and t-butyl benzoate occurred concurrently with their hydrolyses. This was explained as arising through formation of a tetrahedral intermediate which after proton transfers has two chemically equivalent hydroxyl groups, so that it can revert to starting material either with loss or retention of the oxygen of the original carbonyl group (2). Subsequently many further investigations of this type have been carried out and the ratios of the rate constants for hydrolysis and exchange determined (Bender, 1960; Samuel and Silver, 1965; Shain and Kirsch, 1968; Bender and Heck, 1967; Lane *et al.*, 1968; McClelland, 1975, 1976; Deslongchamps *et al.*, 1977, 1980).

Further indirect evidence for the incursion of tetrahedral intermediates in acyl-transfer reactions has been obtained from experiments which indicate a change in rate-determining step and hence the incursion of an intermediate (Jencks and Gilchrist, 1964, 1968; Jencks, 1969; Johnson, 1967; Schowen *et al.*, 1966; Fedor and Bruice, 1965; Hibbert and Satchell, 1967; Chaturvedi *et al.*, 1967; Taylor, 1968; Blackburn and Jencks, 1968; Barnett and Jencks, 1969; Hallas *et al.*, 1971; Hershfield and Schmir, 1972b; Hershfield *et al.*, 1975; Gensmantel and Page, 1979) and from experiments which demonstrate a change in reaction products with pH which has been interpreted as resulting from a change in the partitioning of a tetrahedral intermediate (Schmir and Cunningham, 1965, 1966; Schmir, 1968; Chaturvedi and Schmir, 1968; Smith and Feldt, 1968, Hershfield and Schmir, 1972a).

This large accumulation of indirect evidence clearly demonstrated the presence of tetrahedral intermediates in acyl-transfer reactions but virtually nothing was known about their chemical and physical properties. In the mid-1970s the view was current that, except in certain special instances, they were very reactive species indeed, and Deslongchamps (1975) proposed for a hemiorthoester that the energy barrier for cleavage is "much lower than that to give other conformers." If this were true for all hemiorthoesters the

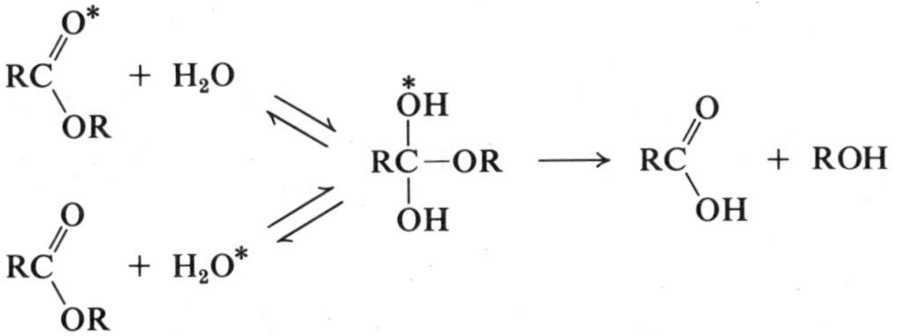

$$RC\begin{matrix}O^*\\\\OR\end{matrix} + H_2O \rightleftarrows \begin{matrix}\overset{*}{O}H\\|\\RC-OR\\|\\OH\end{matrix} \longrightarrow RC\begin{matrix}O\\\\OH\end{matrix} + ROH \qquad (2)$$

$$RC\begin{matrix}O\\\\OR\end{matrix} + H_2O^*$$

chances of ever detecting one by nmr spectroscopy would have been slight. As will be shown later in this review many of the commonly postulated species are much more stable than this, and now quite a lot is known about the properties of hemiorthoesters and some other species.

2 Specially stabilized tetrahedral intermediates

Initially progress in the direct investigation of tetrahedral intermediates derived from carboxylic acids was made by studying species of general structure [2] which possess some special structural feature which makes them particularly stable with respect to the carbonyl derivative to which they are related. Four structural features may be identified which favour the tetrahedral addition structure [2] over the carbonyl form:

(i) a bicyclic or polycyclic structure,

(ii) a strong electron-withdrawing group, e.g., CF_3, attached to the proacyl carbon,

(iii) a group X, for which conjugation with the carbonyl group of the precursor [1] is reduced,

(iv) groups X and Y attached to the central carbon by sulphur atoms.

Many examples of compounds which possess the general structure [2] and have one or more of these structural features are known to exist as stable molecules. Thus, tetrodotoxin [3] has an ionized hemiorthoester group built into a rigid adamantane-like structure (Woodward, 1964; Woodward and Gougoutas, 1964; Goto *et al.*, 1965). Presumably the rigid structure makes the hemiorthoester structure [3] thermodynamically more stable than the alternative structures [4] and [5] that would be formed on its breakdown, since recyclization of the latter would be entropically very favourable. It is also possible that the lower pK_a of the hydroxyl group of the hemiorthoester helps to make its ionized form more stable than the alternative ionized forms (cf. Section 4, p. 79). Another polycyclic structure for which a tetrahedral-intermediate-like structure has been claimed is [6] (Okada *et al.*, 1979).

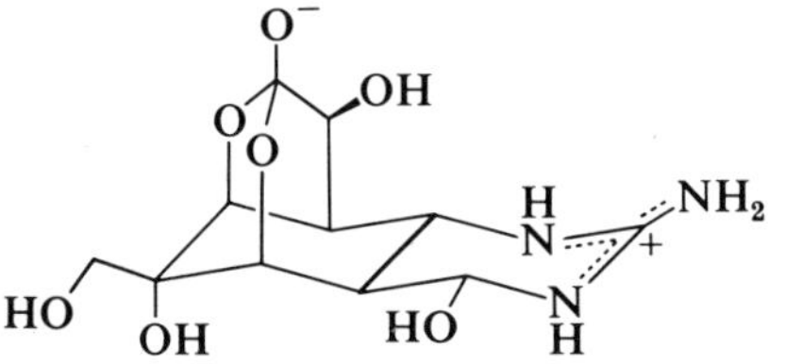

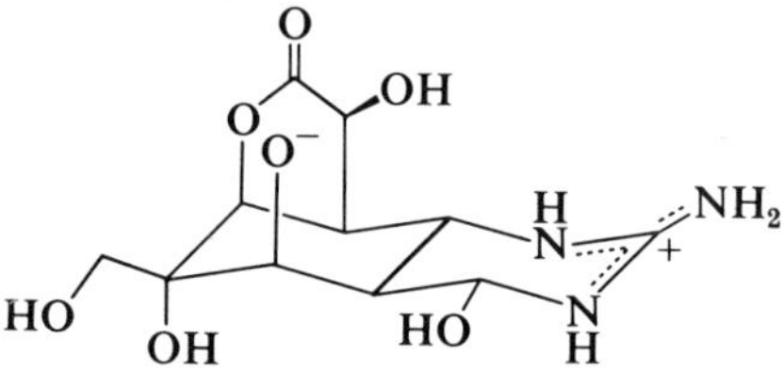

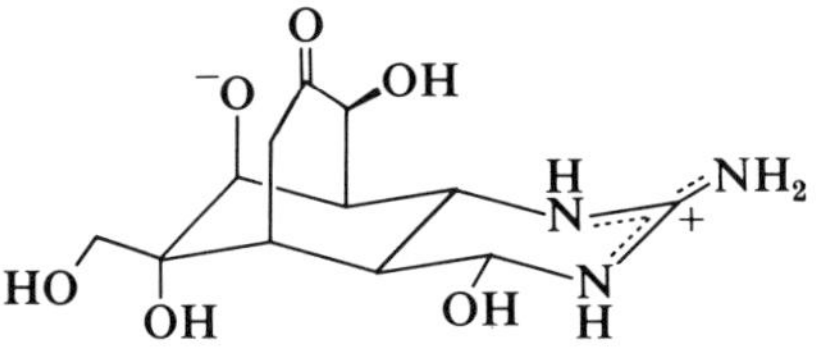

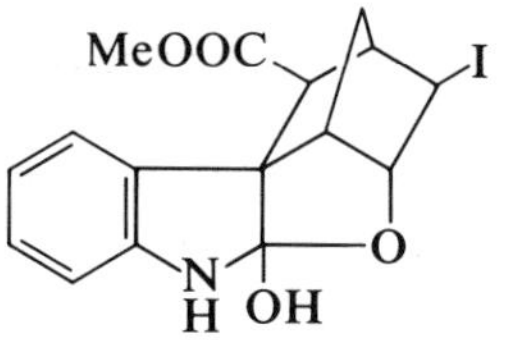

The most widely investigated electron-withdrawing group which has been used to stabilize tetrahedral intermediates is CF_3, and Guthrie (1976) has estimated that the free energy of formation of tetrahedral intermediate [2] from a precursor [1] is 4 to 8 kcal mol^{-1} more favourable when R = CF_3 than when R = H. Some of the compounds which have been detected spectroscopically or even isolated are given in Scheme 1. The ionized forms of the acyclic esters [7] and [8] may be generated in solution by treatment with sodium methoxide or sodium ethoxide in butyl ether (Bender, 1953; Moffat and Hunt, 1959) or sodium methoxide in methanol (Guthrie, 1976). Compound [9] is obtained by treatment of *cis*-3,4-dihydroxytetrahydrofuran with trifluoroacetic anhydride (Bladon and Forrest, 1966). ^{1}H and ^{19}F nmr spectroscopic investigations showed that in CD_3CN solution it is converted partly into its less stable stereoisomer [10] and a small amount (*c*, 7%) of the monotrifluoroacetate of cis-3,4-dihydroxytetrahydrofuran. Similarly, treatment of pinacol with trifluoroacetic anhydride yields [11] (Hine et al., 1973). In solution this is in equilibrium with pinacol monotrifluoroacetate and the equilibrium constant increasingly favours the hemiorthoester on going from less to more polar solvents (CCl_4 < benzene < 1,4-dioxan < MeCN).

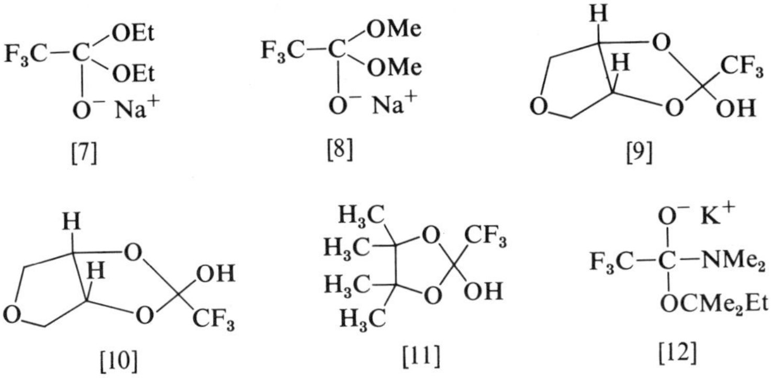

Scheme 1

Similar behaviour is found with the analogous trichloromethyl compound but the hemiorthoester structure is less stable. The monotrifluoroacetates of isobutylene glycol and ethylene glycol exist wholly as esters and the cyclic hemiorthoesters cannot be detected. Compound [12] was reported to be formed when N,N-dimethyltrifluoroacetamide is added to a solution of potassium t-pentoxide in methylcyclohexane or isooctane (Fraenkel and Watson, 1975).

If the conjugation between X and the carbonyl group of [1] is reduced then formation of the tetrahedral adduct [2] becomes easier. Thus, in ergotamine [13] N(4) is conjugated to a carbonyl group and hence the new carbonyl group that would be formed on breaking the C(12)—O(1) bond would be less strongly conjugated with N(4) than in a normal amide. Ring-chain tautomerism of this type (3) has been investigated by Shemyakin and by Frey and their coworkers (Shemyakin et al., 1962, 1965; Antonov et al., 1963, 1965; Griot and Frey, 1963) and the corresponding thio compounds have been studied by Rothe and Steinberger (1968, 1970a,b). Appreciable amounts of the ring tautomer [15] (i.e. the tetrahedral intermediate) are present at equilibrium and the exact proportion depends on R_1, R_2 and n. When, however, the exocyclic carbonyl group of [14] was replaced by a methylene group, the ring form was never detected. In [14] the conjugation of the nitrogen with the carbonyl group in the ring is reduced, and this group becomes more like a ketone carbonyl. Hence, addition to it is easy. When the exocyclic carbonyl is changed to a methylene group, the endocyclic carbonyl becomes a true amide carbonyl, and addition to it becomes difficult. It has been estimated that the "free energy associated with the interruption of conjugation between the carbonyl function ... and the lone pair of the NR_2 group" is 18–19 kcal mol^{-1} (Fastrez, 1977).

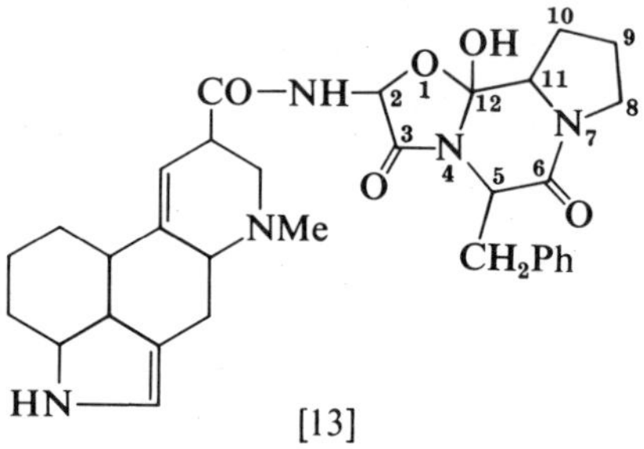

[13]

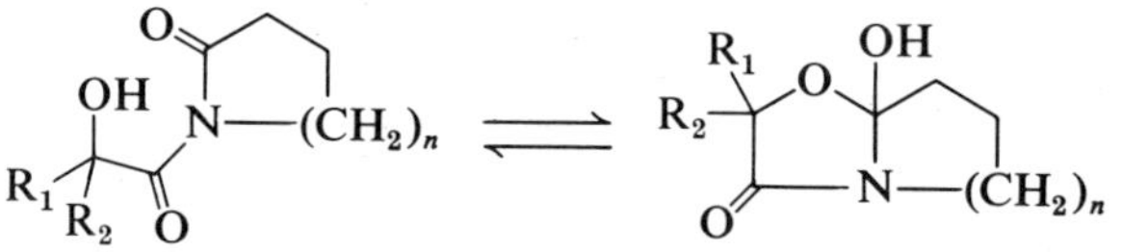

[14] ⇌ [15] (3)

Similar behaviour is found with other peptide derivatives. Thus the peptide lactone [16] on treatment with dry HCl in chloroform yielded the "cyclol" [17] (Sheppard, 1963) and tripeptide oxazoline [18] spontaneously yields "cyclol" [19] when dissolved in ethyl acetate (Jones *et al.*, 1963). The "cyclol" structure was confirmed by an X-ray structure determination for compound [21a] which was obtained by treatment of the peptide–nitrophenyl ester [20a] with a 1:1 carbonate:bicarbonate buffer in aqueous dioxan; [21b] was obtained similarly from [20b] (Lucente and Romeo, 1971; Cerrini *et al.*, 1971).

Other compounds of this type which exist as tetrahedral intermediate-like tautomers include 1-benzyl-5,5-diethyl-3-lactyl barbituric acid (as [22]) (Bobrański and Śladowska, 1972), [23] (Grace *et al.*, 1980), [24] (Patchorik *et al.*, 1957; Grace *et al.*, 1980), [25] (Temple *et al.*, 1965) and [26] (McCapra and Leeson, 1976, 1979; Mager and Addink, 1979).

In addition to these qualitative studies, three compounds of this type have been studied quantitatively and, as discussed in more detail later (p. 89), the kinetics of their breakdown have been investigated. The first of these was [28] which was obtained as a crystalline solid by treating [27] with acetyl chloride and pyridine in chloroform (Rogers and Bruice, 1973, 1974) and the other two, [30] (Gravitz and Jencks, 1974) and [32] (Chahine and Dubois, 1983), were generated in solution from the corresponding cations [29] and [31].

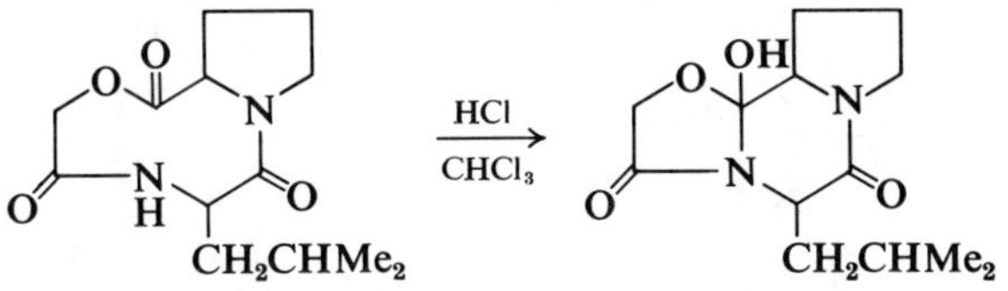

TETRAHEDRAL INTERMEDIATES

[16] → [17] (HCl, CHCl₃)

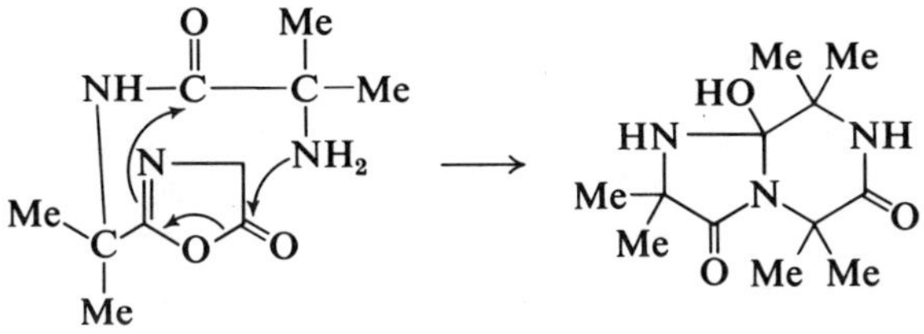

[18] → [19]

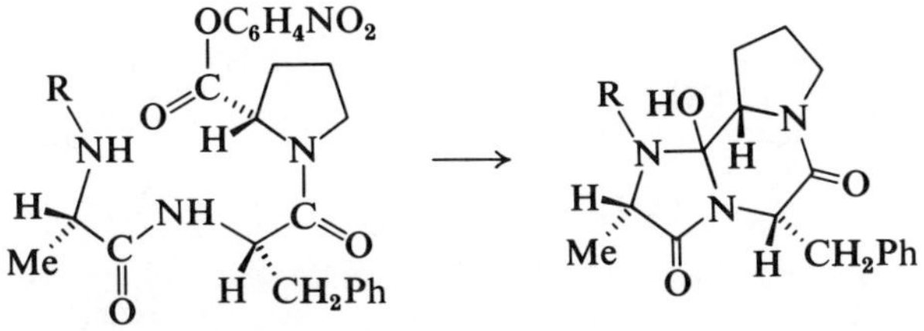

[20] → [21]

(a) R = p-BrC₆H₄CH₂OCO
(b) R = C₆H₅CH₂OCO

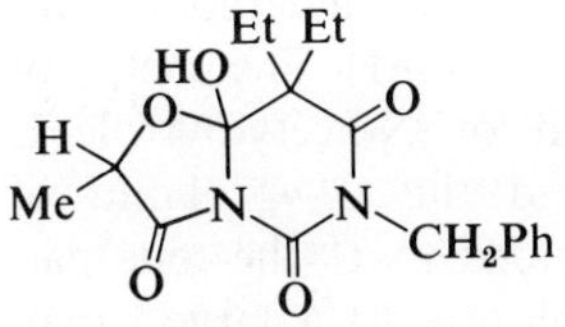

[22] 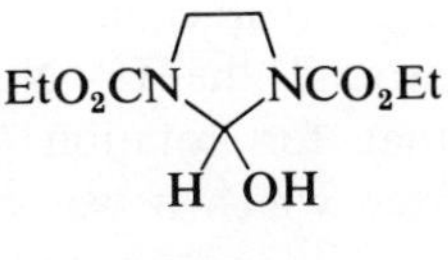   [23] 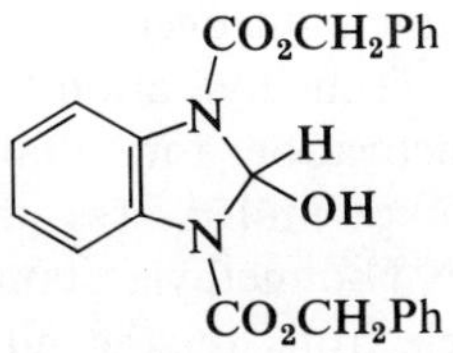   [24]

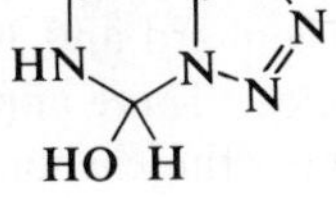

[25] 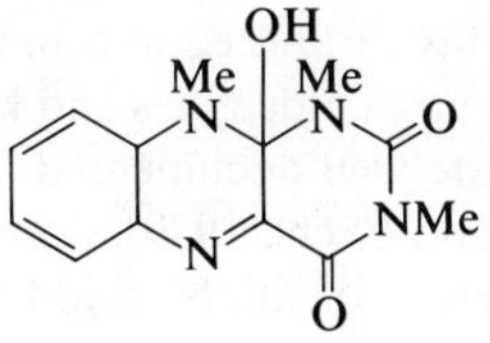   [26]

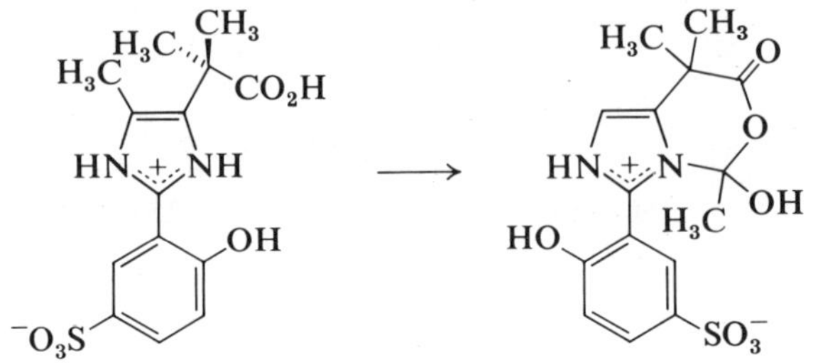

[Structures 27, 28, 29, 30, 31, 32]

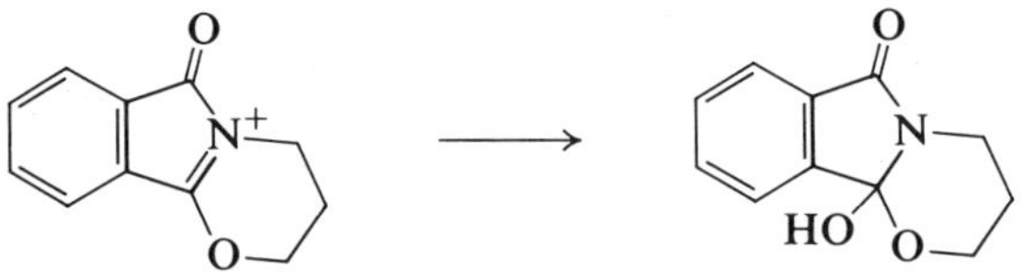

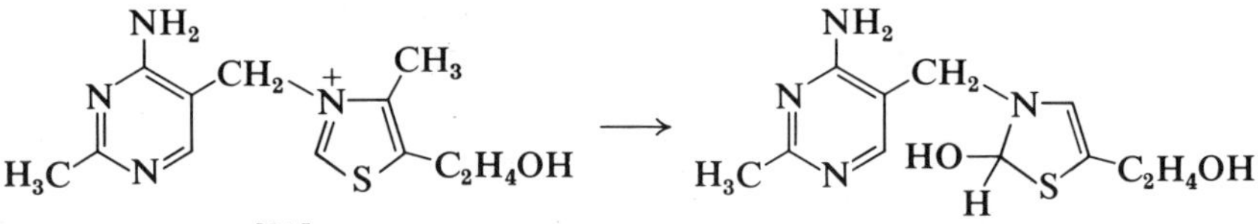

Compounds [34], which have a structure similar to [32], have also been prepared from the corresponding cations [33] and isolated and characterized spectroscopically (Scherowsky and Matloubi, 1978).

N-Trifluoroacetylpyrrole, which has both a trifluoroacetyl group and an amide nitrogen which forms part of a conjugated system, undergoes spontaneous hydration to form [35] (Cipiciani et al., 1977, 1981). The energy of activation for rotation around the C—N bond of N-acetylpyrrole is 7 kcal mol^{-1} less than that for rotation around the C—N bond of N,N-dimethylacetamide from which it was concluded that "the lone pair electrons on the nitrogen atom in N-acetylpyrrole are to a large extent localized in the pyrrole ring" (Dahlqvist and Forsén, 1969). Presumably the same holds for N-trifluoroacetylpyrrole.

The more favourable equilibrium constants for the addition of thiols to the carbonyl group of aldehydes and ketones compared to those for the addition of alcohols are well documented (Jencks, 1964; Lienhard and Jencks, 1966; Kanchuger and Byers, 1979), if not fully understood. There appears to be a similar tendency for thiols to add to the carbonyl (or thiocarbonyl) group of esters (or thioesters) and several tetrahedral intermediates with one or more sulphur atom attached to the central carbon atom are known, e.g. [36] (Takagi *et al.*, 1977), [37] (Olsson *et al.*, 1976), [38] (Levesque *et al.*, 1978), [39] and [40] (Khouri and Kaloustian, 1979).

[33]
X = CH_3OH

[34]

[35]

[36]

[37]

[38]

[39]
R = C_6H_5 or CH_3

[40]
R = C_6H_5 or CH_3

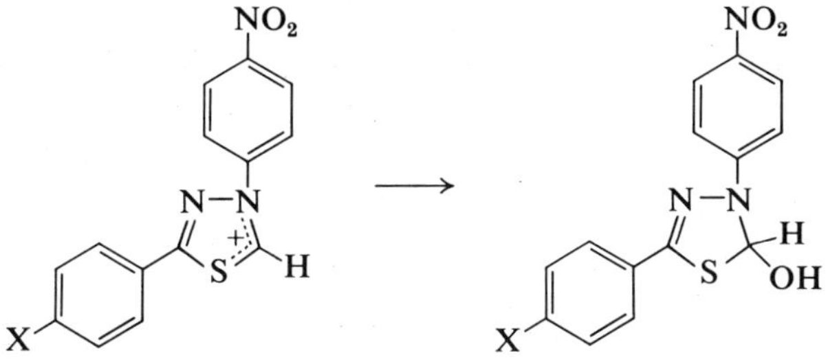

In addition to these well-authenticated structures there are a number of less well-established structures proposed and also structures which have subsequently been shown to be incorrect. Among the latter are [41] (Kuhn and Weiser, 1957), correct structure probably [42] (McClelland *et al.*, 1979b); grisolutein B [43] (Nakamura *et al.*, 1964; see also DeWolfe, 1970), correct structure probably [44] (Challand *et al.*, 1970); saxitoxin [45] (Wong *et al.*, 1971), correct structure [46] (Schantz *et al.*, 1975; Bordner *et al.*, 1975; for a review see Scheuer, 1977); rhetsinine [47] (Chatterjee *et al.*, 1959), correct structure probably [48] (Pachter and Suld, 1960); [49] (Fodor *et al.*, 1970) correct structure probably [50] (Capon *et al.*, 1979) and [51] (Stauffer, 1972; see Guthrie, 1974a).

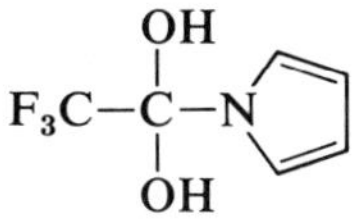

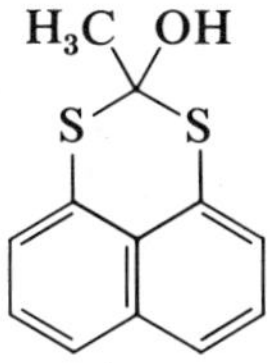

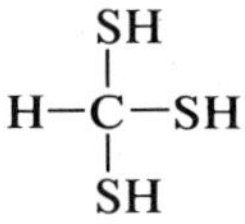

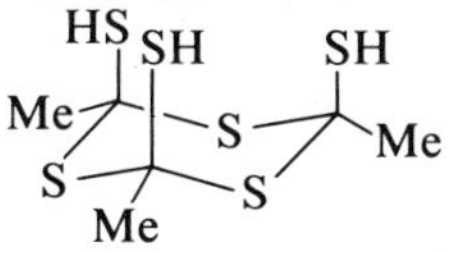

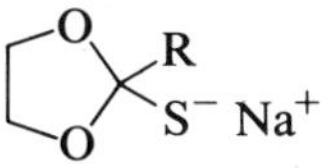

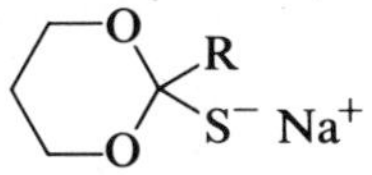

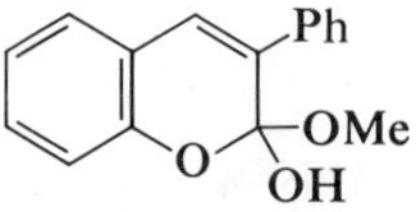

[41] 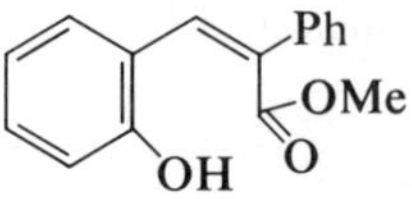 [42] 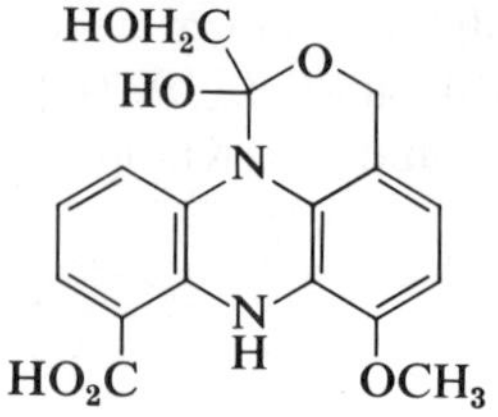 [43]

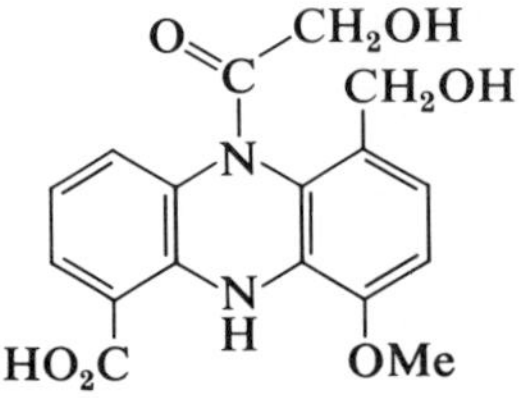

[44] 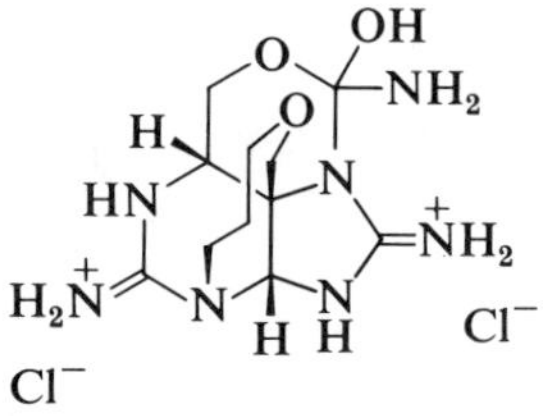 [45] 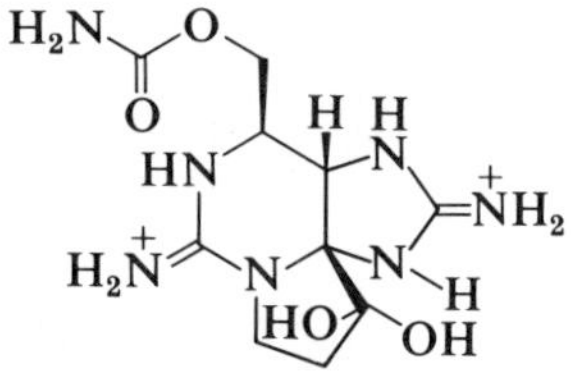 [46]

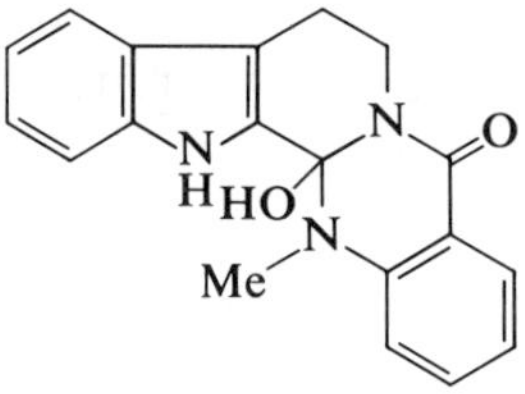

[47] 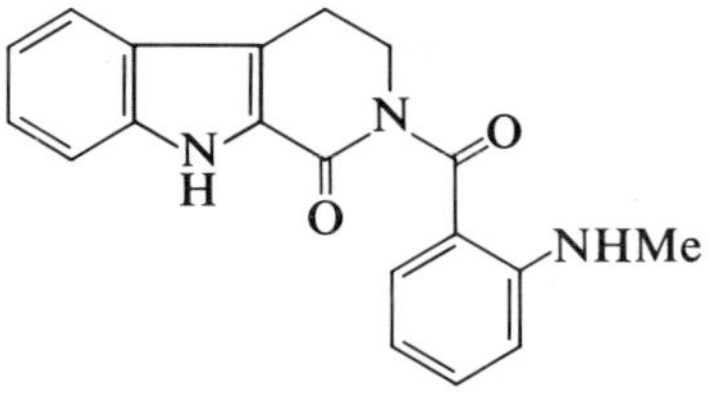 [48]

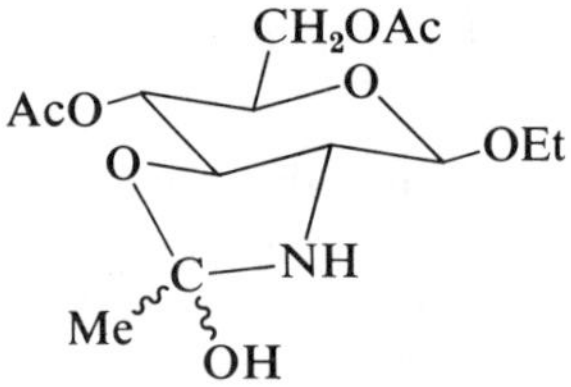

[49] 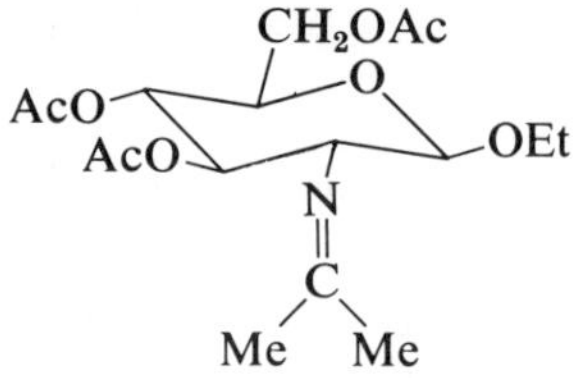 [50]

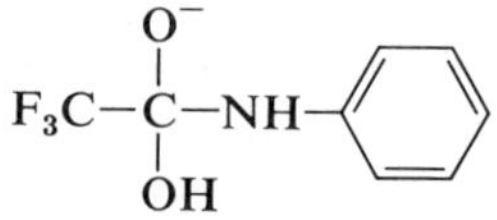

[51] 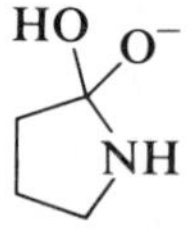 [52]

There are also a number of species for which a tetrahedral intermediate-like structure has been claimed but which require substantiation before the claims can be fully accepted. Thus, on the basis of infrared spectroscopic evidence it was claimed that γ-butyrolactam, on treatment with concentrated (23%) potassium hydroxide solution, is converted into the ionized tetrahedral intermediate [52] characterized by a band at 1555 cm^{-1} (Vinnik and Moiseyev, 1963). However, this interpretation has been criticized (Robinson, 1970), and it is difficult to see why structure [52] should have a band at 1555 cm^{-1} in its ir spectrum with an intensity comparable to that of the carbonyl stretching band at 1665 cm^{-1} of the original amide. Another structure which requires further substantiation is [53] which was claimed to be formed, on the basis of an initial increase in absorbance at 260 nm, when trifluoroacetanilide is added to imidazole buffer (Stauffer, 1974). In view of the failure to detect the analogous intermediate [51] by ^{19}F nmr spectroscopy it would be valuable to have ^{19}F nmr spectroscopic evidence for the existence of [53]. Also the reported rate constant for the breakdown of [53] into acetylimidazole and aniline, 2.22×10^{-3} min^{-1} at 30° (Stauffer, 1974), appears to be rather low for such a species.

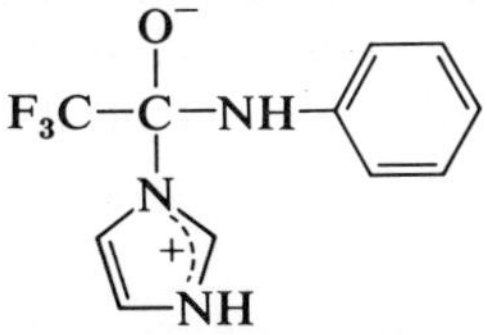

[53]

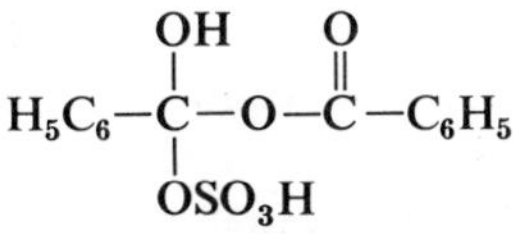

[54]

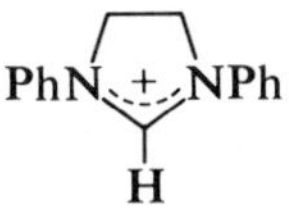

[55]

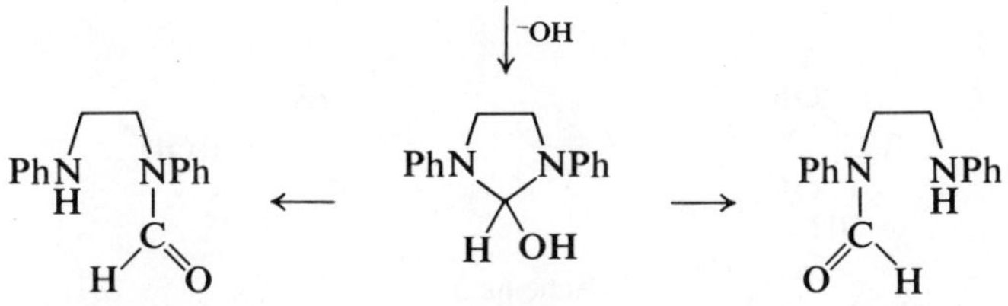

[56]   [57]

Compound [54] was reported to be formed on reaction of benzoic anhydride and sulphuric acid and its structure was assigned on the basis of its ir spectrum and its ability to sulphonate anisole (Collet *et al.*, 1974). It would be useful to have additional evidence, such as an nmr spectrum, to confirm this structure.

3 The detection of simple tetrahedral intermediates

The compounds described in the last section generally have much more complex structures than the tetrahedral intermediates postulated in most acyl transfer reactions. The first convincing direct spectroscopic detection of a simple tetrahedral intermediate (Capon *et al.*, 1981a) without any special stabilizing features was made by Robinson in 1968 (Robinson, 1968, 1970) who showed that the species [57], which would be expected to be the tetrahedral intermediate in the degenerate intramolecular formyl transfer in N-(2-anilinoethyl)formanilide [56], could be generated from the high-energy precursor 1,3-diphenyl-2-imidazolinium chloride [55] and detected by uv spectroscopy. The main evidence that the species which was detected was in fact [57] was that there was a maximum in its uv spectrum ($\lambda = 252 \pm 2$ nm) at a very similar wavelength to that found for N,N-dimethylaniline ($\lambda = 244$ nm) with an extinction coefficient about twice as great. Unfortunately no nmr spectroscopic data for the intermediate were obtained and a subsequent attempt to do this under different conditions was unsuccessful (see Capon *et al.*, 1981a). Robinson's extensive kinetic investigation of the breakdown of [57] is discussed in Section 5 (p. 89).

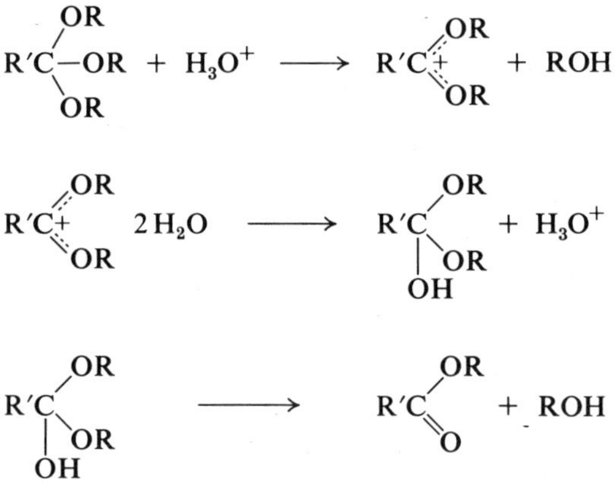

$$R'C(OR)_3 + H_3O^+ \longrightarrow R'C^+(OR)_2 + ROH$$

$$R'C^+(OR)_2 + 2H_2O \longrightarrow R'C(OR)_2(OH) + H_3O^+$$

$$R'C(OR)_2(OH) \longrightarrow R'C(OR)(=O) + ROH$$

Scheme 2

After this important investigation there were no further studies on simple tetrahedral intermediates until 1976 when two groups, one in Glasgow, Scotland and one in Toronto, Canada, independently showed how the tetrahedral intermediates of O,O-acyl transfer reactions could be generated in solution and characterized spectroscopically. These species had long been postulated, but not detected, in two other classes of reactions, the hydrolysis of orthoesters (Scheme 2) and the hydration of ketene acetals (Scheme 3), and it was shown by the above mentioned groups that by a judicious choice of substrate and conditions it was possible to detect the tetrahedral intermediates in these and related reactions.

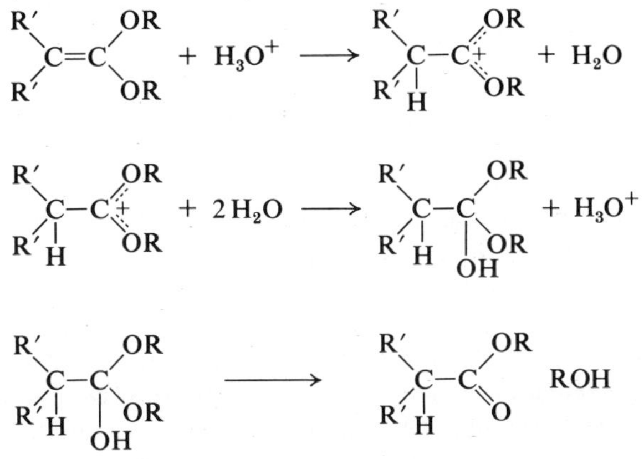

$$\underset{R'}{\overset{R'}{>}}C=C\underset{OR}{\overset{OR}{<}} + H_3O^+ \longrightarrow \underset{R'\ H}{\overset{R'}{>}}C-\overset{+}{C}\underset{OR}{\overset{OR}{<}} + H_2O$$

$$\underset{R'\ H}{\overset{R'}{>}}C-\overset{+}{C}\underset{OR}{\overset{OR}{<}} + 2H_2O \longrightarrow \underset{R'\ H}{\overset{R'}{>}}C-C\underset{\underset{OH}{OR}}{\overset{OR}{<}} + H_3O^+$$

$$\underset{R'\ H}{\overset{R'}{>}}C-C\underset{\underset{OH}{OR}}{\overset{OR}{<}} \longrightarrow \underset{R'\ H}{\overset{R'}{>}}C-C\underset{O}{\overset{OR}{<}} \quad ROH$$

Scheme 3

ORTHOESTERS AND RELATED SPECIES AS PRECURSORS

It had been shown (Capon *et al.*, 1976a) that the hemiacetal could be detected in the hydrolysis of α-acetoxy-α-methoxytoluene [58] and this led to an attempt to carry out a similar experiment with a compound at the carboxylic acid rather than the aldehyde level of oxidation. The compound chosen was

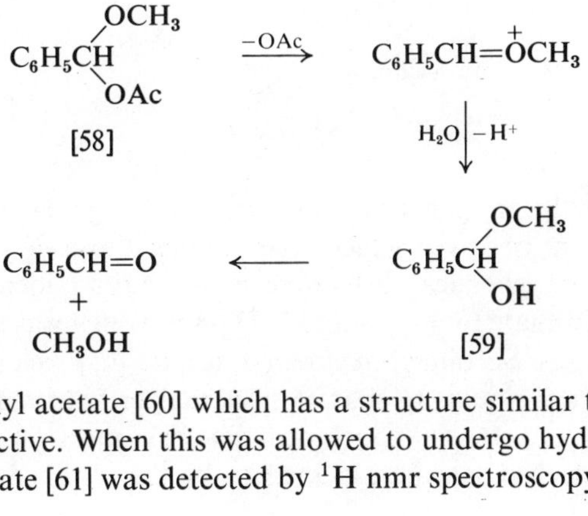

$$C_6H_5CH\underset{OAc}{\overset{OCH_3}{<}} \xrightarrow{-OAc} C_6H_5CH=\overset{+}{O}CH_3$$
[58]

$$\downarrow H_2O\ |\ -H^+$$

$$C_6H_5CH=O + CH_3OH \longleftarrow C_6H_5CH\underset{OH}{\overset{OCH_3}{<}}$$
[59]

dimethoxymethyl acetate [60] which has a structure similar to an orthoester but is more reactive. When this was allowed to undergo hydrolysis dimethyl hemiorthoformate [61] was detected by ^{1}H nmr spectroscopy. An indication

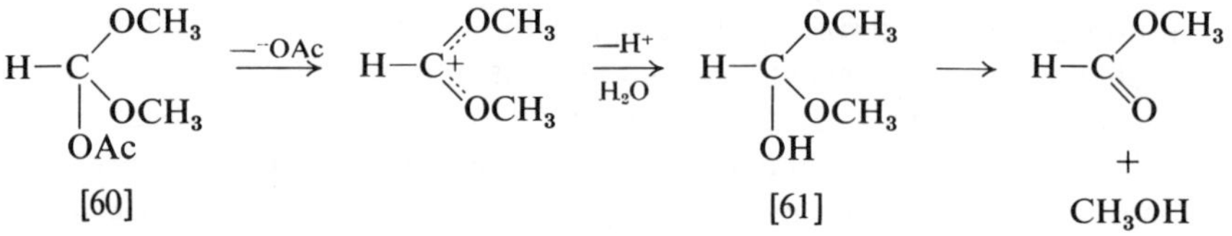

$$\underset{[60]}{\underset{OAc}{H-C\genfrac{}{}{0pt}{}{OCH_3}{OCH_3}}} \xrightarrow{-OAc} H-C^+\genfrac{}{}{0pt}{}{OCH_3}{OCH_3} \xrightarrow[H_2O]{-H^+} \underset{[61]}{\underset{OH}{H-C\genfrac{}{}{0pt}{}{OCH_3}{OCH_3}}} \longrightarrow H-C\genfrac{}{}{0pt}{}{OCH_3}{O} + CH_3OH$$

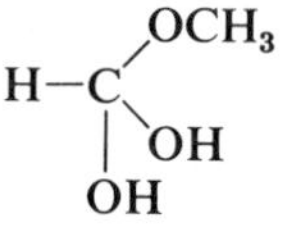

$$\underset{[62]}{\underset{OH}{H-C\genfrac{}{}{0pt}{}{OCH_3}{OH}}}$$

that this should be possible was obtained prior to the experiment from some semi-empirical calculations of Guthrie (1973) on the closely related methyl hemiorthoformate [62]. According to these calculations at pH $c.$ 5, where the minimum of the pH–rate profile occurs, the rate constant for breakdown is $10^{-1}s^{-1}$ at 25°. Hence if [61] reacted at a similar rate it should be possible to detect it if a precursor which reacted fast enough could be found. Dimethoxymethyl acetate was found to be a suitable precursor in a mixture of acetone-d_6 and D_2O (87/13 v/v) at $-35°$. Under these conditions the O-deuterated hemiorthoester could be detected (see Scheme 4) (Capon et al., 1976b; Capon

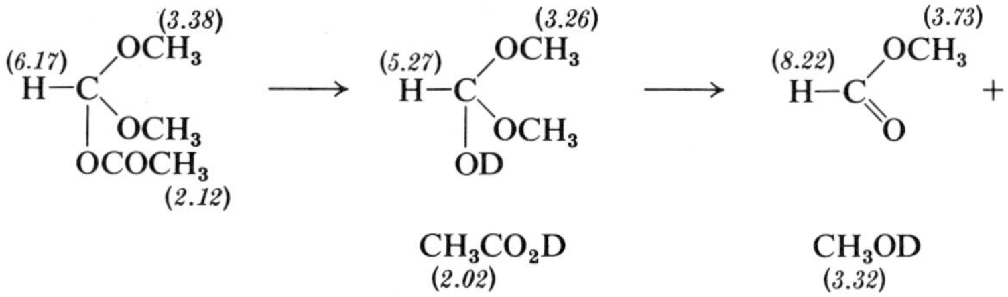

$$\underset{(2.12)}{\underset{OCOCH_3}{\overset{(6.17)}{H}-C\genfrac{}{}{0pt}{}{\overset{(3.38)}{OCH_3}}{OCH_3}}} \longrightarrow \underset{\underset{CH_3CO_2D\;(2.02)}{OD}}{\overset{(5.27)}{H}-C\genfrac{}{}{0pt}{}{\overset{(3.26)}{OCH_3}}{OCH_3}} \longrightarrow \underset{CH_3OD\;(3.32)}{\overset{(8.22)}{H}-C\genfrac{}{}{0pt}{}{\overset{(3.73)}{OCH_3}}{O}} +$$

(The figures in brackets are δ-values)

Scheme 4

and Grieve, 1980). In particular two transient signals at $\delta = 5.27$ and $\delta = 3.26$ ppm were observed which were neither those of starting material nor products. These chemical shifts were similar but not identical to those of trimethyl orthoformate ($\delta = 5.19$ and 3.33 ppm) which was stable under the reaction conditions. So dimethoxymethyl acetate [63] yields a species with a ^1H nmr spectrum similar to that of trimethyl orthoformate but which is much more reactive. It was concluded that this must be dimethyl hemiorthoformate which would be expected to be the tetrahedral intermediate

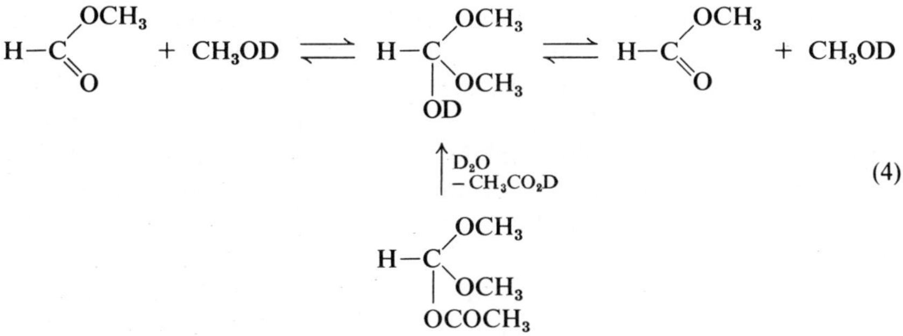

$$H-C\overset{OCH_3}{\underset{O}{\diagdown}} + CH_3OD \rightleftharpoons H-C\overset{OCH_3}{\underset{\underset{OD}{|}}{\diagdown OCH_3}} \rightleftharpoons H-C\overset{OCH_3}{\underset{O}{\diagdown}} + CH_3OD$$

$$\uparrow {D_2O \atop -CH_3CO_2D} \quad (4)$$

$$H-C\overset{OCH_3}{\underset{\underset{OCOCH_3}{|}}{\diagdown OCH_3}}$$

[63]

in the methanolysis of methyl formate with CH_3OD (4). Similar experiments were carried out with 2-acetoxy-1,3-dioxolan and 2-acetoxy-4,4,5,5-tetramethyl-1,3-dioxolan which generated 2-deuteroxy-1,3-dioxolan and 2-deuteroxy-4,4,5,5-tetramethyl-1,3-dioxolan (see Schemes 5 and 6). These species are the expected tetrahedral intermediates for the intramolecular formyl migration of ethylene glycol monoformate and pinacol monoformate.

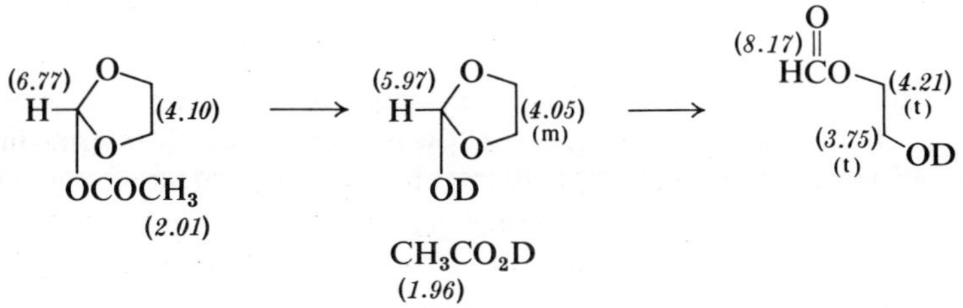

(The figures in brackets are δ-values)

Scheme 5

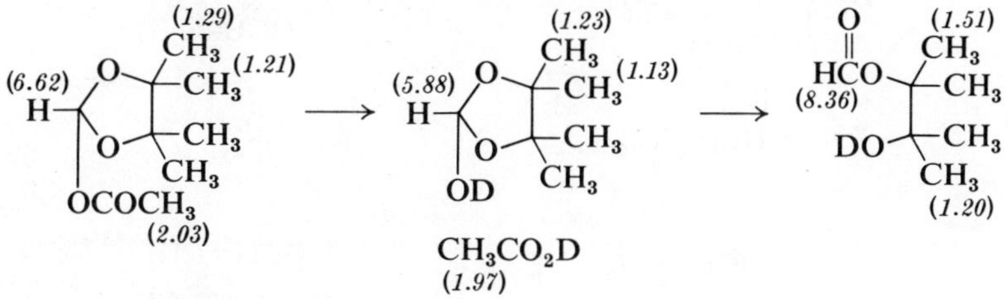

(The figures in brackets are δ-values)

Scheme 6

It was possible to obtain from 2-acetoxy-4,4,5,5-tetramethyl-1-3-dioxolan [64] a solution in which more than 90% of it had been converted into 2-deuteroxy-4,4,5,5-tetramethyl-1,3-dioxolan [65] and it was also possible to follow the reaction by ^{13}C nmr spectroscopy (Scheme 7). The signal of C-2 of [64] occurred at $\delta = 109.6$ ppm and this was replaced by a signal at 109.92 in the intermediate [65] which subsequently decayed with concurrent formation of a signal at $\delta = 162.9$ ppm for the carbon of the carbonyl group of the product pinacol monoformate [66].

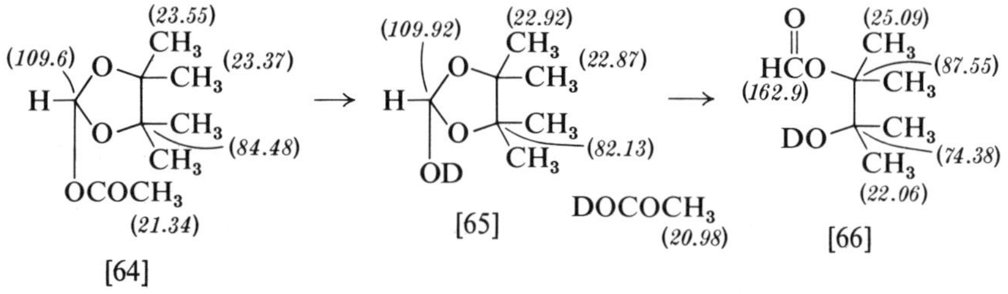

(The figures in brackets are $\delta(^{13}C)$-values)

Scheme 7

In a closely related investigation, designed to generate vinyl alcohol in solution, divinyl hemiorthoformate [68] was detected as an intermediate in the hydrolyses of divinyloxymethyl dichloro- and trichloro-acetate [67] (Capon *et al.*, 1981b). It is interesting that methyl vinyl hemiorthoformate could not be detected in the hydrolysis of methoxyvinyloxymethyl acetate and chloroacetate under similar conditions. So dimethyl and divinyl hemiorthoformate appear to be more stable than methyl vinyl hemiorthoformate.

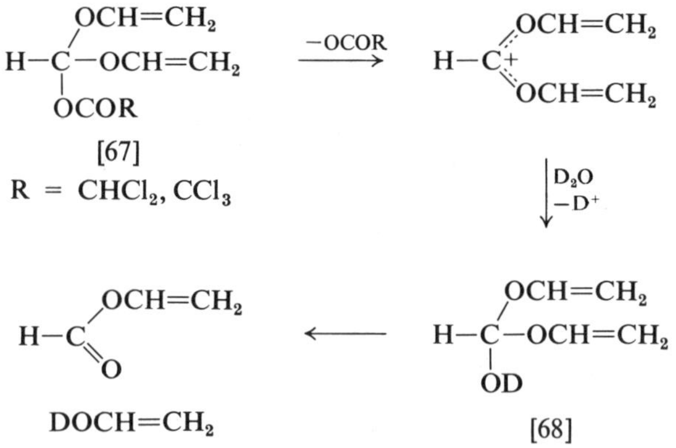

It was speculated that the methoxy group has the stronger "push" for ejecting a leaving group and the vinyloxy group the stronger "pull" to act as a leaving group, so that the intermediate with one good "pushing" group (methoxy) and one good "pulling" group (vinyloxy) was more reactive than the intermediates with two good "pushing" groups or two good "pulling" groups.

At about the same time that the work described above was being carried out at Glasgow, Kresge and McClelland and their coworkers in Toronto observed that whereas the rate constants (k_{H^+}) for the hydronium-ion catalysed formation of 2-hydroxyethyl benzoate from a series of 2-alkoxy-2-phenyl-1,3-dioxolans [69] varied with the alkoxy group at pH 4–7.5, they were independent of the alkoxy-group in 5×10^{-2} M hydrochloric acid.

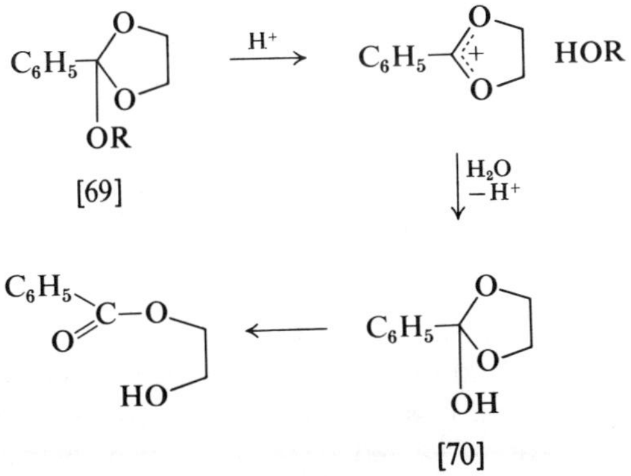

Also k_{H^+} as measured in these acidic solutions was 4 to 30 times less than when it was measured at pH 4.5–7.5 (Ahmad et al., 1977, 1979). Therefore in the acidic solutions the alkoxy group must have been lost before the rate-determining step in the formation of 2-hydroxyethyl benzoate and this step must therefore be breakdown of the hemiorthoester [70]. At higher pH breakdown of the hemiorthoester [70], which is also base-catalysed, becomes very fast, and cleavage of the exocyclic C—O bond of the starting material [69], which depends on R, is rate-determining.

Breakdown of the hemiorthoester intermediate was also found to be rate-limiting in acidic solutions with substituted 2-phenyl-2-methoxy-1,3-dioxolans [71] (substituents p-MeO, p-Me, p-F, p-Cl, p-Br, m-Cl and p-NO$_2$) (Ahmad et al., 1979), 2-phenyl-2-methoxy-4,4,5,5-tetramethyl-1,3-dioxolan [72] (McClelland et al., 1979a), the orthoester [73] derived from exo,exo-2,3-norbornanediol (Santry and McClelland, 1983b), 2-phenyl- and 2-(4-methoxyphenyl)-2-methoxy-1,3-dioxans ([74], [75]), [76] and [77] (McClelland et al., 1981), 2-cyclopropyl-2-methoxy-1,3-dioxolan [78] (Ahmad et al., 1979), and 2-methyl-2-methoxy-1,3-dioxolan [79] (Ahmad et al., 1979, 1982;

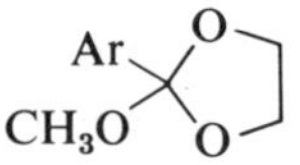

[Structures [71]–[80] shown]

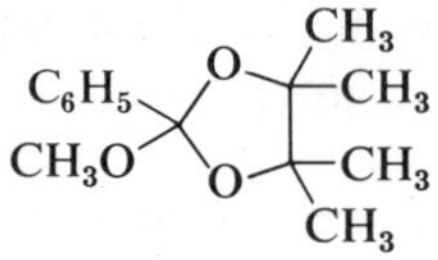

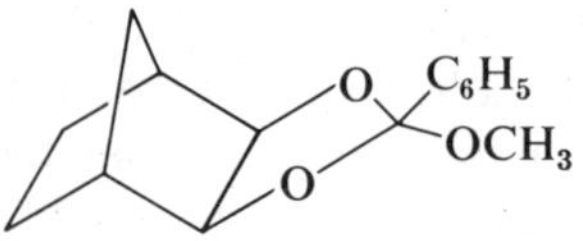

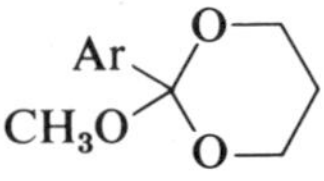

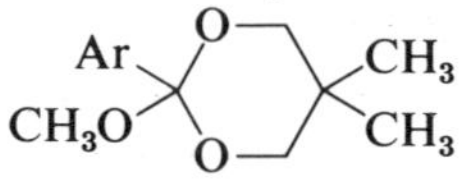

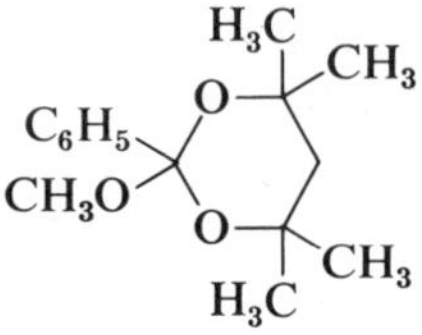

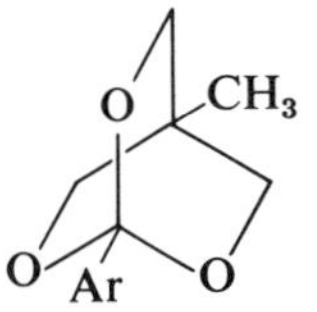

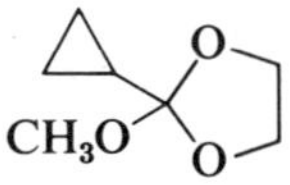

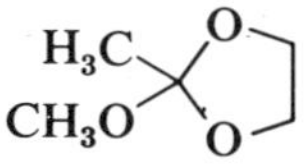

Chiang *et al.*, 1983) but not with 2-methoxy-1,3-dioxolan [80] itself (Ahmad *et al.*, 1979). This method and the method of using acetoxy as leaving group therefore nicely complement one another as 2-hydroxy-1,3-dioxolan which would be the intermediate obtained from [80] can be generated from the 2-acetoxy-compound (see above). On the other hand all attempts to prepare 2-acetoxy-2-phenyl-1,3-dioxolan were unsuccessful (Grieve, 1978), but the intermediate that it would yield can be generated from the 2-alkoxy-compounds [69] (Ahmad *et at.*, 1979). In more acid solutions than were used for the detection of the hemiorthoesters it is sometimes possible to detect the cationic intermediate. Thus when the hydrolysis of 2-(4-methoxy-

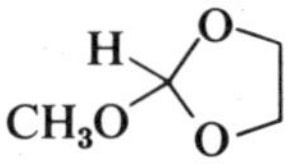

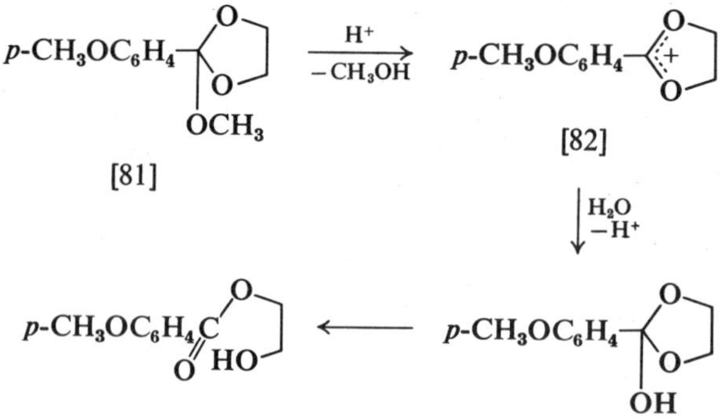

[Reaction scheme showing [81] → [82] → hydroxy intermediate → p-CH$_3$OC$_6$H$_4$ ester, with p-CH$_3$OC$_6$H$_4$ substituent]

phenyl)-2-methoxy-1,3-dioxolan [81] was studied in 0.02 M HClO$_4$ a transient species with a maximum in its uv spectrum near 300 nm was detected. This is almost certainly cation [82] (Ahmad et al., 1979). It was also possible to detect the 2-cyclopropyl-1,3-dioxolenium ion at a similar low acid concentration (<0.1 M) but a higher acid concentration was needed to detect the 2-(p-tolyl)-, 2-phenyl-, and 2-(p-fluorophenyl) ions which are less stable.

Santry and McClelland (1983a) also generated the tetrahedral intermediate of an O,S-acyl transfer reaction from an orthothiolester precursor [83]. At pH < 3 the cation [84] could be detected by its strong absorption at λ = 350 nm and it was concluded from a kinetic analysis that this was in equilibrium with the hemiorthothiolester [85].

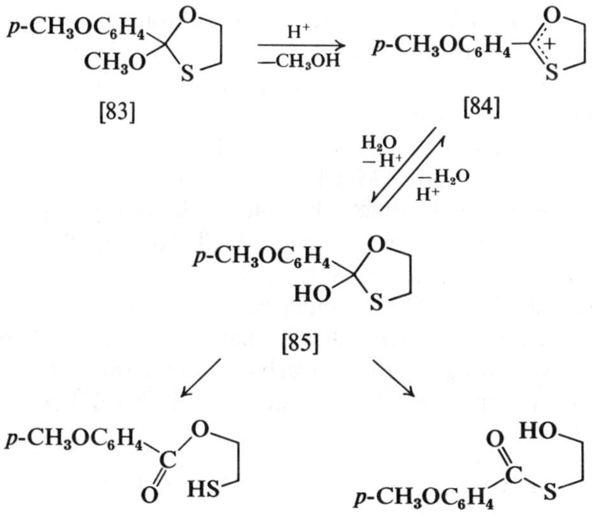

AMIDE ACETALS AS PRECURSORS

The use of orthoesters as precursors for the generation of hemiorthoesters is limited in two ways. Firstly, the hemiorthoester is only generated at high concentrations in acidic solutions, and, secondly, apart from the hemiorthoester [87] generated from the highly reactive trimethyl orthocyclopropanecarboxylate [86] (Burt et al., 1982) only cyclic hemiorthoesters may be generated in this way. McClelland and his coworkers therefore developed new precursors without one or both of these limitations. Firstly they used dimethylamine as leaving group (McClelland and Ahmad, 1979; Ahmad et al., 1979). This is intermediate in leaving group ability between methoxide and acetate and hence 2-dimethylamino-2-phenyl-1,3-dioxolan can be made and it generates

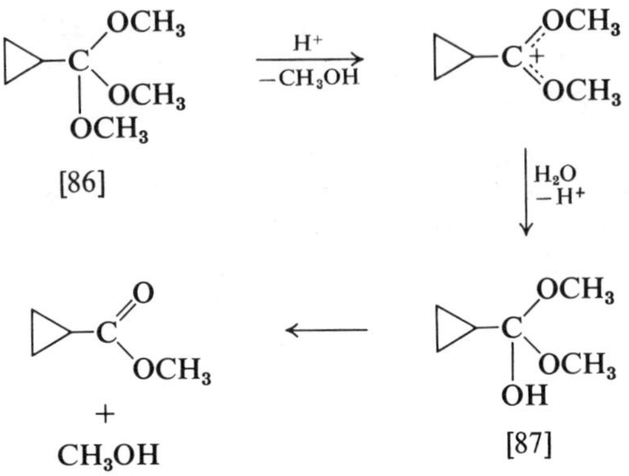

[Scheme showing conversion of [86] to [87] and products]

2-hydroxyl-2-phenyl-1,3-dioxolan at high concentrations at pHs more acidic than c. 3. Acyclic amide acetals, derived from benzoic or substituted benzoic acids, with dimethylamine as leaving group also react with substantial amounts of C—N bond cleavage in acidic solutions (McClelland, 1978) but the rate-limiting step is always fission of the C—N bond and the hemiorthoester never accumulates. McClelland and his coworkers found however that by using substituted N-methylanilines as leaving group acyclic hemiorthobenzoates could be generated (McClelland and Somani, 1979; McClelland and Patel, 1981a). Thus at pH below c. 6 the rate of formation of methyl benzoate from [88] is independent of the substituent in the N-methylaniline portion which indicates that a step after C—N bond breaking must be rate-limiting. As the oxocarbocation intermediate [89] could not be detected it was concluded that this step must be breakdown of the hemiorthoester [90].

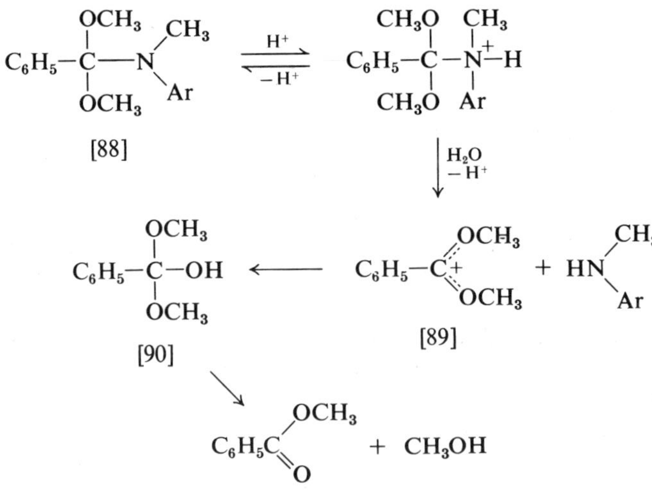

[Scheme showing [88] → [89] → [90] → methyl benzoate + CH_3OH]

TETRAHEDRAL INTERMEDIATES

KETENE ACETALS AS PRECURSORS

Hemiorthoesters may also be detected in the hydration of ketene acetals under favourable circumstances. Although there had been several kinetic investigations of these reactions, hemiorthoesters were not detected as intermediates until 1981 when Capon and Ghosh (1981) showed that the hemiorthoesters [92] and [94] could be detected in the hydration of 2-methylene-1,3-dioxolan [91] and 2-methylene-4,4,5,5-tetramethyl-1,3-dioxolan [93].

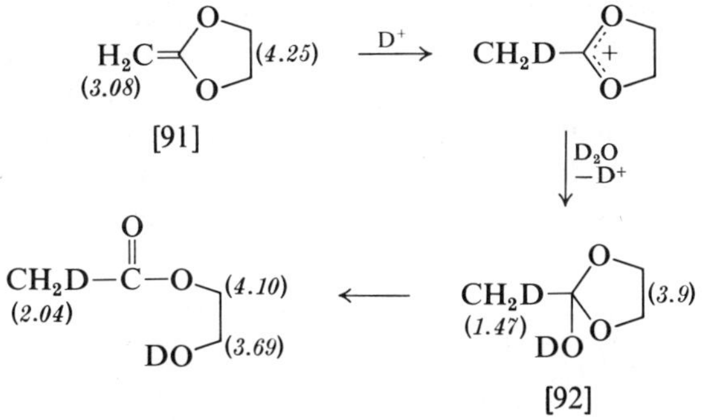

(The figures in brackets are δ-values)

Scheme 8

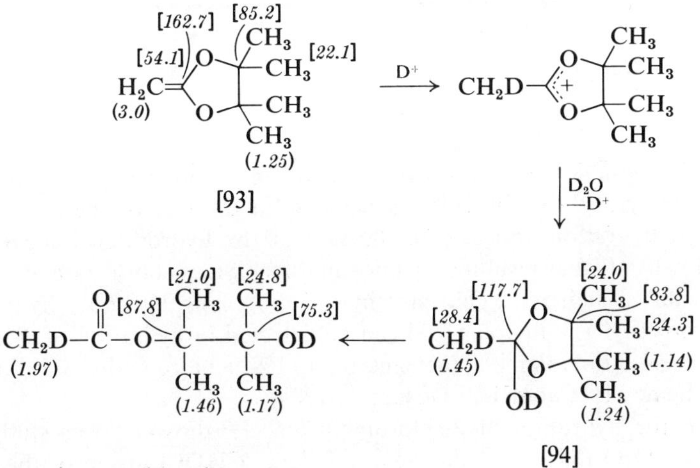

(The figures in curved brackets are $\delta(^1H)$-values and those in square brackets are $\delta(^{13}C)$-values)

Scheme 9

Kankaanperä and Tuominen (1976) had estimated that k_{H^+} for the hydration of 2-methylene-1,3-dioxolan was of the order of 10^6–10^7 $M^{-1}s^{-1}$. Since this was considerably greater than the value of k_{H^+} for the breakdown of some hemiorthoesters already studied it was thought that, under favourable circumstances, it should be possible to detect the hemiorthoesters that were intermediates in the hydration of ketene acetals. This was found to be so with both 2-methylene-1,3-dioxolan (Scheme 8) and 2-methylene-4,4,5,5-tetramethyl-1,3-dioxolan (Scheme 9). Intermediates were detected by ^1H nmr spectroscopy on hydration of both these compounds in mixtures of CD_3CN or CD_3COCD_3 and D_2O (5–10% by volume) which contained acid (Capon and Ghosh, 1981). These had, *inter alia,* signals at δ = 1.47 and 1.45 ppm which correspond to the —CH_2D group of the hemiorthoester. The 2-hydroxy-2,4,4,5,5-pentamethyl-1,3-dioxolan [94] was also detected by ^{13}C nmr spectroscopy (Scheme 9). The signal of C(2) of the dioxolan ring had δ = 117.7 which would be expected for this structure.

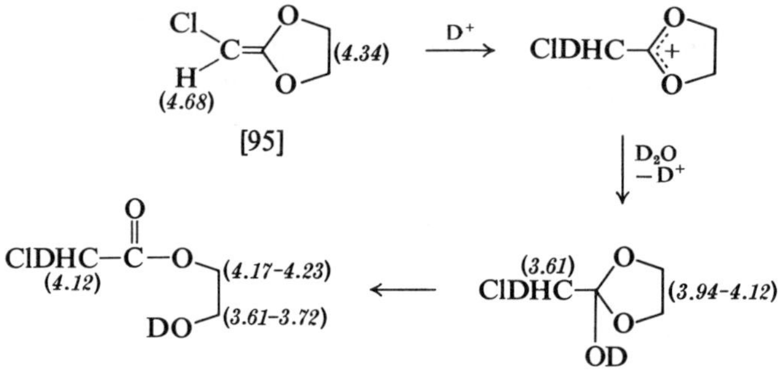

(The figures in brackets are δ-values)

Scheme 10

The introduction of chloro-substituents into the methylene group of 2-methylene-1,3-dioxolan causes a decrease in the rate of the hydronium-catalysed hydration and also in the rate of the hydronium ion-catalysed breakdown of the resulting hemiorthoester. Since both rate decreases are similar, 2-hydroxy-2-chloromethyl-1,3-dioxolan [96] and 2-hydroxy-2-dichloromethyl-1,3-dioxolan [98] can be detected in hydration of 2-chloromethylene-1,3-dioxolan [95] (Scheme 10) and 2-dichloromethyl-1,3-dioxolan [97] (Scheme 11) (Capon and Dosunmu, 1984).

When the hydration of 2-chloromethylene-1,3-dioxolan was studied in CD_3CN—D_2O (85:5 v/v) the signal of the —CHDCl group of the intermediate was a broadened singlet which corresponded to 1 proton at δ = 3.66 pp. When H_2O was used instead of D_2O this signal was sharper and corresponded in intensity to two protons. When a similar set of experiments was

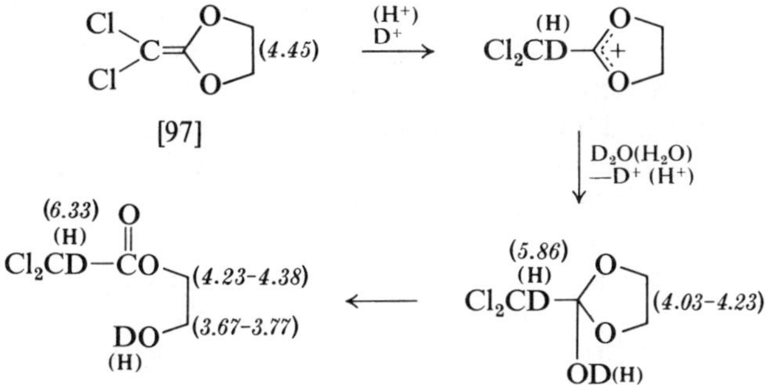

TETRAHEDRAL INTERMEDIATES

Scheme 11

(The figures in brackets are δ-values)

carried out with 2-dichloromethylene-1,3-dioxolan the intermediate could only be detected by the signals of the ring protons when CD_3CN-D_2O (96:4 v/v) was used as solvent; but when H_2O was used instead of D_2O the signal of the $CHCl_2$ group was observed at $\delta = 5.86$.

Similar experiments were also carried out with dichloroketene diethyl and dimethyl acetals but no intermediate could be detected. This is readily explained since the cyclic ketene acetals undergo acid-catalysed hydration about 30 times more rapidly than the corresponding acyclic ones (Straub, 1970; Chiang et al., 1974; Kresge and Straub, 1983) whereas cyclic hemiorthoesters undergo acid-catalysed breakdown 50–60 times more slowly than the corresponding acyclic ones do (see p. 70). Therefore the ratio of rate constants favourable for the detection of the cyclic hemiorthoesters becomes unfavourable with the acyclic hemiorthoesters.

OXOCARBOCATIONS AS PRECURSORS

In the reactions given in the previous sections for the generation of hemiorthoesters oxocarbocations were postulated as intermediates. Since most of these carbocations can be isolated as salts it might be thought that these could also be used as precursors. However, attempts to detect the hemiorthoesters that would be formed on hydration of ions [99]–[102] by ^{1}H nmr spectroscopy were unsuccessful (Grieve, 1978). There were thought to be two reasons for this: *(i)* the heat of reaction of the ion with water causes local heating and hence the hemiorthoester is generated at a higher temperature than that of the nmr probe; *(ii)* the hemiorthoester is generated in the presence of a high concentration of a strong acid (e.g. fluoroboric acid) and this catalyses its breakdown. Attempts to prevent this by using sodium acetate to neutralize the acid were unsuccessful. The mixing used in this investigation was very inefficient and it is possible that with a flow system it would be possible to detect the hemiorthoester by nmr spectroscopy. When uv spectroscopy, which requires much lower concentrations, and a stopped flow mixing device are used it is sometimes possible to detect the hemiorthoesters, and this has been done with those formed by hydration of [101] and [103] (Ahmad et al., 1979; McClelland et al., 1979a).

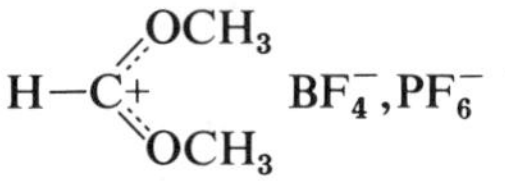

[99] H–C+(OCH$_3$)(OCH$_3$) BF$_4^-$, PF$_6^-$

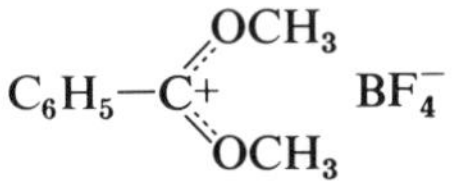

[100] C$_6$H$_5$–C+(OCH$_3$)(OCH$_3$) BF$_4^-$

[101] C$_6$H$_5$–(1,3-dioxolan-2-ylium) BF$^-$

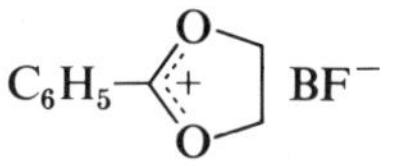

[102] (tetrahydropyran-2-yl)–OCH$_3$ BF$_4^-$

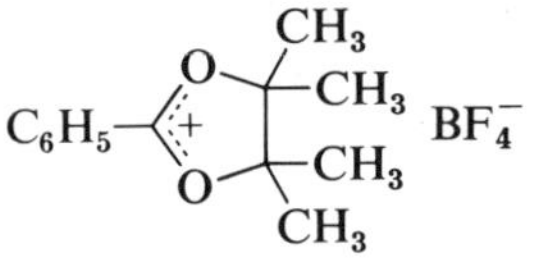

[103] C$_6$H$_5$–(4,4,5,5-tetramethyl-1,3-dioxolan-2-ylium) BF$_4^-$

4 Kinetic studies on the breakdown of hemiorthoesters

The breakdown of hemiorthoesters is catalysed by hydronium ions and hydroxide ions and a spontaneous or water-catalysed reaction can also often be detected. Rate constants for these processes are summarised in Tables 1 to 16. In order to compare rate constants obtained under different conditions, in Tables 7, 8, 11, 15 and 16 the rate constants have been extrapolated to a common set of conditions (water at 25°).

TABLE 1

Rate constants for the hydronium-ion catalysed breakdown of tetrahedral intermediates in CH_3CN—H_2O (c_{H_2O} = 8.33 M) at 15° [a]

Structure	$10^3 k_{H^+}/M^{-1}s^{-1}$
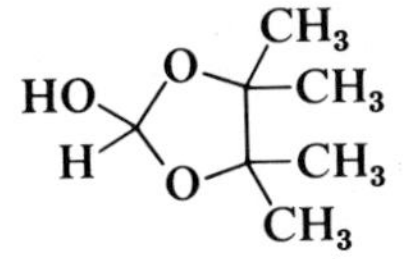 HO, H — C(O-C(CH₃)₃)(O-C(CH₃)₃)	2.03
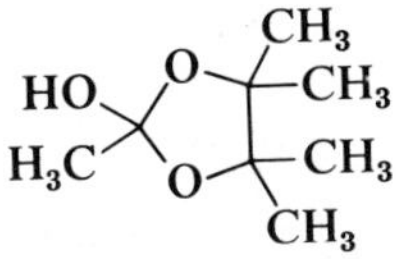 HO, H₃C — C(O-C(CH₃)₃)(O-C(CH₃)₃)	0.310
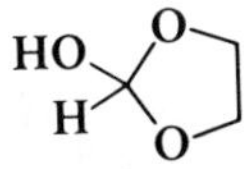 HO, H — C(OCH₂CH₂O) (1,3-dioxolane)	3.94
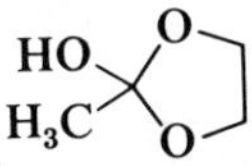 HO, H₃C — C(OCH₂CH₂O)	5.50
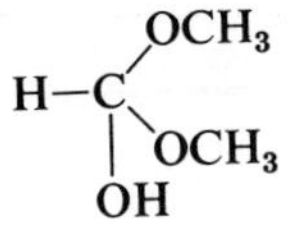 H—C(OCH₃)(OCH₃)(OH)	240
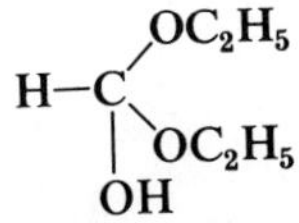 H—C(OC₂H₅)(OC₂H₅)(OH)	186
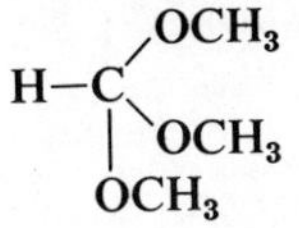 H—C(OCH₃)(OCH₃)(OCH₃)	0.121
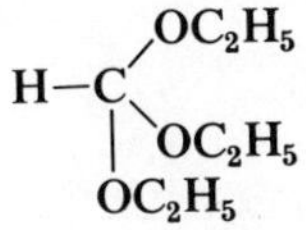 H—C(OC₂H₅)(OC₂H₅)(OC₂H₅)	0.800

[a] Capon and Sanchez, 1983

TABLE 2

Rate constants for the breakdown of 2-substituted-2-hydroxy-1,3-dioxolans (25°; $I = 0.1$ M maintained with NaCl)[a,b]

Substituent	$10^{-2}k_{H^+}/M^{-1}s^{-1}$	k_{H_2O}/s^{-1}
Cyclopropyl	11	1.0
p-Methoxyphenyl	7.5(13)	1.4
p-Tolyl	5.6(10)	1.4
Phenyl	3.0(6.1)	1.5
p-Fluorophenyl	2.4(4.3)	1.6
p-Chlorophenyl	1.7(2.5)	1.7
p-Bromophenyl	1.6	1.9
p-Chlorophenyl	1.1	2.3
p-Nitrophenyl	0.37	2.4

[a] Ahmad et al., 1979
[b] Rate constants in brackets were measured with $I = 1.0$ M maintained with NaClO$_4$

TABLE 3

Rate constants for the breakdown of tetrahedral intermediates in water (25°, $I = 0.1$ M, maintained with NaCl)[a]

Intermediate	$k_{H^+}/M^{-1}s^{-1}$	k_{H_2O}/s^{-1}
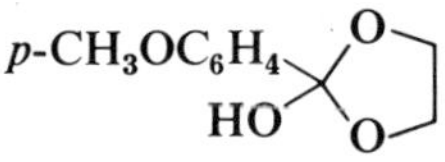 p-CH$_3$OC$_6$H$_4$, HO — 1,3-dioxolane	7.5×10^2	1.4
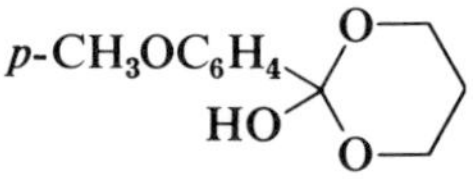 p-CH$_3$OC$_6$H$_4$, HO — 1,3-dioxane	7.6×10^3	1.0
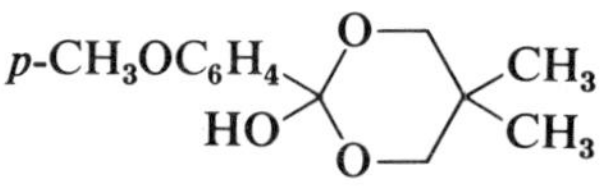 p-CH$_3$OC$_6$H$_4$, HO — 1,3-dioxane with CH$_3$, CH$_3$	2.8×10^3	1.1
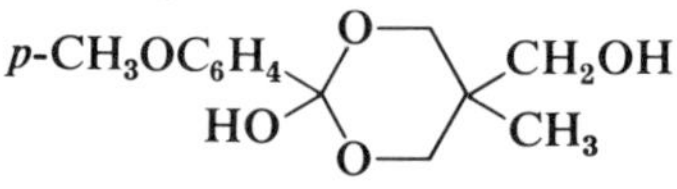 p-CH$_3$OC$_6$H$_4$, HO — 1,3-dioxane with CH$_2$OH, CH$_3$	1.1×10^3	0.7

[a] McClelland et al., 1981

TABLE 4

Rate constants for the breakdown of tetrahedral intermediates in water (25°; $I = 0.1$ M, maintained with NaCl)[a]

Intermediate	$k_{H^+}/M^{-1}s^{-1}$	k_{H_2O}/s^{-1}
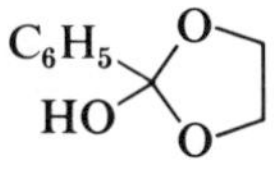 C_6H_5, HO — 1,3-dioxolane	3.0×10^2	1.5
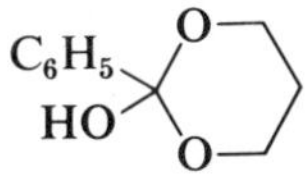 C_6H_5, HO — 1,3-dioxane	3.0×10^3	1.0
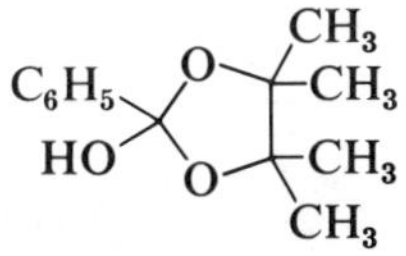 C_6H_5, HO — tetramethyl-1,3-dioxolane	1.3×10	0.074
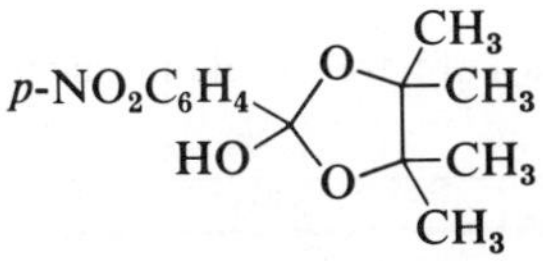 $p\text{-}NO_2C_6H_4$, HO — tetramethyl-1,3-dioxolane	1.7^b	0.15^b
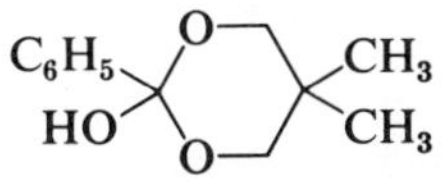 C_6H_5, HO — dimethyl-1,3-dioxane	1.1×10^3	1.3
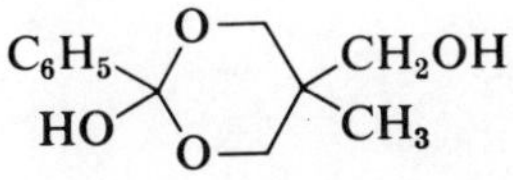 C_6H_5, HO — (CH$_2$OH, CH$_3$)-1,3-dioxane	4.5×10^2	0.7
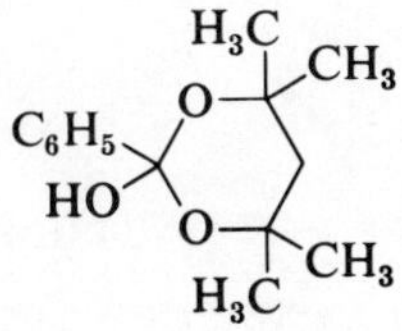 C_6H_5, HO — tetramethyl-1,3-dioxane	3.4×10^3	0.16

[a] McClelland et al., 1981   [b] McClelland et al., 1984

TABLE 5

Rate constants for the hydronium-ion catalysed breakdown of tetrahedral intermediates in water at 15°, $I = 0.100$ M[a]

	$k_{H^+}/M^{-1}s^{-1}$
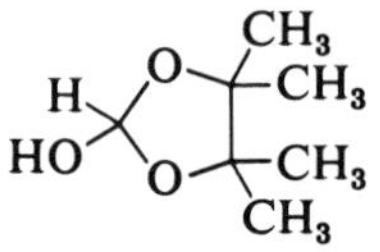 H, HO, O-C(CH₃)₂-C(CH₃)₂-O (pinacol hemiorthoformate)	0.191
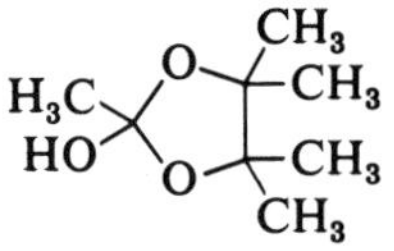 H₃C, HO, O-C(CH₃)₂-C(CH₃)₂-O (pinacol hemiorthoacetate)	0.0180

[a] Capon and Sanchez, 1983

TABLE 6

Rate constants for the hydronium-ion catalysed breakdown of tetrahedral intermediates in CH₃CN—H₂O ($c_{H_2O} = 2.22$ M) at 15°[a]

	$10^{-4}k_{H^+}/M^{-1}s^{-1}$
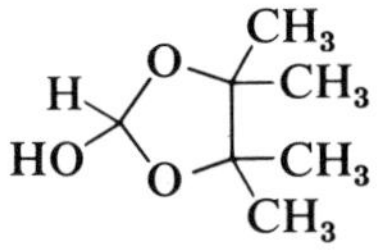 H, HO, O-C(CH₃)₂-C(CH₃)₂-O	5.86
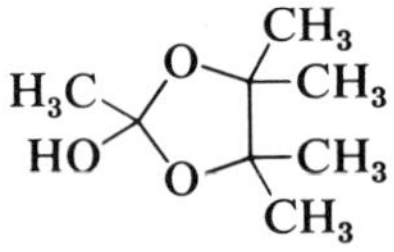 H₃C, HO, O-C(CH₃)₂-C(CH₃)₂-O	1.24
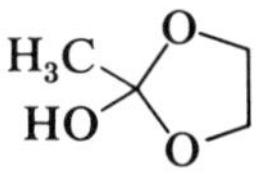 H₃C, HO, O-CH₂-CH₂-O	25.8

[a] Capon and Sanchez, 1983

TABLE 7

Rate constants for the acid-catalysed cleavage of the exocyclic C—O bond of 2-methoxy-1,3-dioxolans and the endocyclic C—O bond of 2-hydroxy-1,3-dioxolans (in H_2O at 25°)

	$k_{H^+}(R = Me)$ for exocyclic C—O bond cleavage	$k_{H^+}(R = H)/M^{-1}s^{-1}$ for endocyclic C—O bond cleavage
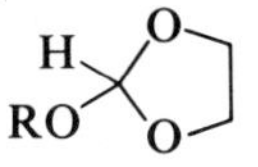 H, RO (dioxolan)	1.75×10^2 [a]	1.15×10^3 [b]
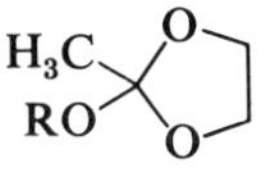 H_3C, RO (dioxolan)	1.5×10^4 [c]	1.4×10^3 [c] (1.6×10^3) [d] (1.24×10^3) [e]
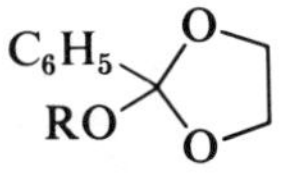 C_6H_5, RO (dioxolan)	7.37×10^3 [a] (5.4×10^3) [e]	3.0×10^2 [c]
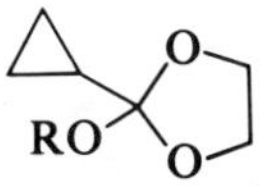 cyclopropyl, RO (dioxolan)	5.0×10^4 [c]	1.1×10^3 [c]

[a] Chiang et al., 1974
[b] Extrapolated from the results of Capon and Sanchez (1983). Ahmad et al., (1979) estimated this rate constant to be greater than $9 \times 10^2 \, M^{-1}s^{-1}$
[c] Ahmad et al., 1979, 1982
[d] Extrapolated from the results of Capon and Sanchez (1983)
[e] Chiang et al., 1983

TABLE 8

Rate constants for the acid-catalysed cleavage of the C—OMe (or C—OEt) bonds of orthoesters and hemiorthoesters (in H_2O at 25°)

	R = Me or Et $k_{H^+}/M^{-1}s^{-1}$	R = H $k_{H^+}/M^{-1}s^{-1}$
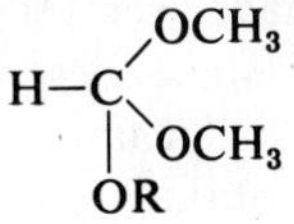 $H-C(OCH_3)(OCH_3)(OR)$	2.63×10^2 [a,c]	7.0×10^4 [d]

continued

TABLE 8 (continued)

	R = Me or Et $k_{H^+}/M^{-1}s^{-1}$	R = H $k_{H^+}/M^{-1}s^{-1}$
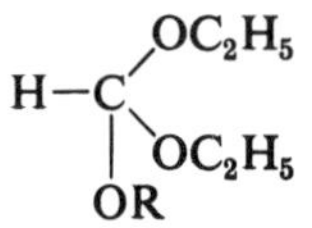 H–C(OC₂H₅)(OC₂H₅)(OR)	$1.73 \times 10^{3\ b,d}$ (1.30×10^3)[e]	$5.45 \times 10^{4\ d}$
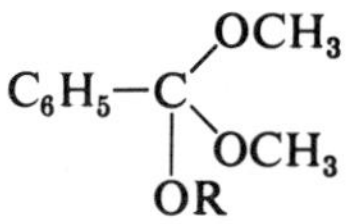 C₆H₅–C(OCH₃)(OCH₃)(OR)	$6.97^{a,c}$	$(1.9 \times 10^4)^f$
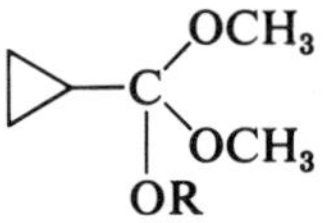 cyclopropyl–C(OCH₃)(OCH₃)(OR)	$8.1 \times 10^{4\ a,g}$	$5.3 \times 10^{3\ g}$

[a] R = Me
[b] R = Et
[c] Chiang et al., 1974
[d] Extrapolated from the results of Capon and Sanchez (1983)
[e] Calculated from the activation parameters of DeWolfe and Jensen (1963)
[f] Solvent 50% aqueous dioxan (McClelland and Patel, 1981b)
[g] Burt et al., 1982

THE ACID-CATALYSED REACTION

General acid catalysis

General acid catalysis has been demonstrated for the breakdown of a series of ring-substituted dimethyl hemiorthobenzoates with H_3PO_4, $NCCH_2CO_2H$, $ClCH_2CO_2H$, and HCO_2H in 50% aqueous dioxan (McClelland and Patel, 1981b). This implies that the reverse reaction, that of methyl benzoate with methanol, is also (kinetically) general acid catalysed and suggests that the chemically similar reaction of water with methyl benzoate should also be (kinetically) general acid catalysed. There have been very few investigations of general acid catalysis of ester hydrolyses since these reactions are normally very slow in the buffer solutions where such catalysis can be studied, and as pointed out by Bender (1960) the proof of the occurrence of general acid catalysis in reactions of carboxylic acid derivatives has had a "stormy history". Hinshelwood and his coworkers demonstrated catalysis by acetic acid molecules in the esterification of alcohols (Rolfe and Hinshelwood; 1934; Hinshelwood and Legard, 1935) and several workers have demonstrated general acid catalysis in lactonization reactions (Weeks, 1965; Milstein and Cohen, 1969, 1970; Tomoto et al., 1972; Hershfield and Schmir, 1973).

However, in these reactions it is likely that breakdown of the tetrahedral intermediate is at least partly rate determining so the general acid catalysis is probably associated with this step rather than with attack of the nucleophile on the carbonyl group. The only example of general acid catalysis of ester hydrolysis where the rate limiting step is probably nucleophilic attack on the carbonyl group appears to be the hydrolysis of tetra-O-methyl-D-glucono-δ-lactone and tri-O-methyl-2-deoxy-D-glucono-δ-lactone which by analogy to the behaviour of D-glucono-δ-lactone were thought not to undergo exchange of the oxygen of the carbonyl group with water (Pocker and Green, 1974). Nevertheless the results quoted above for the breakdown of dimethyl hemiorthobenzoate indicate that (kinetic) general acid catalysis should be detectable in the methanolysis of methyl benzoate (Brønsted $\alpha = 0.49$) and probably in the analogous hydrolysis of methyl benzoate. Therefore any mechanism proposed for these reactions must be able to account for this.

The effect of substituents attached to the pro-acyl carbon atom

The acid-catalysed decomposition of hemiorthoesters is less sensitive to the substituent attached to the pro-acyl carbon atom than the cleavage of the C—O bond of orthoesters. Thus, replacement of the hydrogen at C(2) of 2-methoxy-1,3-dioxolan by methyl, phenyl or cyclopropyl causes an increase in the rate of cleavage of the exocyclic C—O bond (Table 7), although the effect of phenyl is less than that of methyl, possibly as a result of steric inhibition to conjugation, an effect which is much more important in the acyclic series (Chiang et al., 1974). On the other hand similar replacement of hydrogen at C-2 has a much smaller or even rate decreasing effect on the cleavage of the endocyclic bond of 2-hydroxy-1,3-dioxolan (Table 7). This suggests that there is less positive charge development in the transition state than in that for breaking of the exocyclic C—O bond of the 2-methoxy-1,3-dioxolans and is consistent with the breakdown of the hemiorthoesters not passing through a cationic intermediate but being a β-elimination reaction to form the carbon–oxygen double bond directly. Similar results are found with acyclic compounds (Table 8) where both orthoester and hemiorthoester react with cleavage of a non-ring C—O bond. As discussed by Chiang et al., (1974), introduction of a phenyl group into trimethyl orthoformate causes a decrease in the rate of hydrolysis which was ascribed to the electron-withdrawing inductive effect of the phenyl group being more important than the

TABLE 9

The effect of substituents on k_H^+ for the C—O bond cleavage of orthoesters and hemiorthoesters derived from benzoic acid

	Solvent	ρ (s.d.)	No. of points
Orthoesters			
Trimethyl 4-substituted orthobenzoates[a]	H_2O	−1.12 (0.035)	7
Trimethyl 4-substituted orthobenzoates[b]	70% aq.MeOH	−1.96 (0.024)	5
Trimethyl 4-substituted orthobenzoates[a]	98% aq.MeOH	−2.26 (0.088)	5
2-Methoxy-2-(4′-substituted-phenyl)-1,3-dioxolans[c]	H_2O	−1.58 (0.081)	7
Hemiorthoesters			
Dimethyl-4-substituted-hemiorthobenzoates[d]	50% aq.dioxan	−1.58 (0.076)	4
2-Hydroxy-2-(4′-substituted-phenyl)-1,3-dioxolans[c]	H_2O	−1.24 (0.043)	7

[a] Bull et al., 1971
[b] Kwart and Price, 1960
[c] Ahmad et al., 1979
[d] McClelland and Patel, 1981b

mesomeric effect which is sterically inhibited. On the other hand a cyclopropyl group causes approximately the same increase (c. 300-fold) in the rate of cleavage of the C—OMe bond of the orthoester in the acyclic as well as in the cyclic series, but a small rate-decreasing effect on the breakdown of the hemiorthoesters. These results also suggest that the transition state for the breakdown of hemiorthoesters has less cationic character than that for C—O bond cleavage of orthoesters.

The Hammett ρ-value for cleavage of the exocyclic bond of 2-methoxy-2-substituted-phenyl-1,3-dioxolans (-1.58 ± 0.06) is a little larger than that for cleavage of the endocyclic C—O bond of 2-hydroxy-2-substituted-phenyl-1,3-dioxolans (-1.24 ± 0.04) (Table 9) (Chiang et al., 1983). A direct comparison between the ρ-values for C—OMe bond cleavage of trimethyl orthobenzoates and dimethyl hemiorthobenzoates is not possible at present since they have not been measured in the same solvent. However, that based on k_{H^+} for the breakdown of the hemiorthobenzoates (-1.58) is less than that based on the equilibrium constants for their conversion into methyl benzoates and methanol which is -1.9 (derived from the equilibrium constants for formation of the hemiorthobenzoates, McClelland and Patel, 1981b). This implies that the development of positive charge in the transition state is less than in the final product, the ester.

TABLE 10
Experimental rate constants (k_{H^+}) for the breakdown of 2-alkyl-2-hydroxy-1,3-dioxolans and calculated rate constant (k_{H^+}) for the breakdown of methyl hemiorthoacetates

	k_{H^+} (experimental)^a/M^{-1}s^{-1}		k_{H^+} (calc)^b/M^{-1}s^{-1}
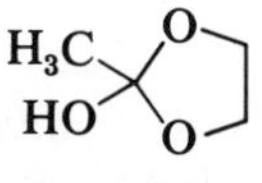 H₃C, HO (dioxolane)	2.58×10^5	CH$_3$C(OH)$_2$OMe	1.78×10^4
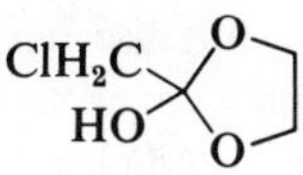 ClH₂C, HO (dioxolane)	36.3	ClCH$_2$C(OH)$_2$OMe	3.47×10^2
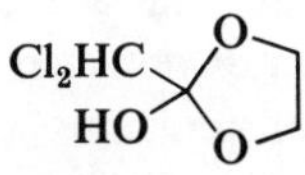 Cl₂HC, HO (dioxolane)	1.87	Cl$_2$CHC(OH)$_2$OMe	4.68, 0.27^c

^{a} Measured at 15° in CH$_3$CN : H$_2$O (c_{H_2O} = 2.22 M) by Capon and Dosunmu (1984)
^{b} In water at 25°. Based on the calculations of Guthrie and Cullimore (1980); see Capon and Dosunmu, 1984
^{c} Based on the two values of log k_{H^+} for the acid-catalysed hydrolysis of the methyl esters quoted by Guthrie and Cullimore (1980, table 4)

Chloro-substituents cause a decrease in k_{H^+} for the breakdown of 2-methyl-2-hydroxy-1,3-dioxolan (Table 10). The effect of introducing a second chlorine is considerably less than that of introducing the first, and the values of k_{H^+} are not correlated well by the σ* constants. This suggests that steric effects may be important as previously proposed (Hine et al., 1973) for the cyclization of pinacol monotrichloroacetate. Alternatively if OH breaking is occurring concertedly with O—C bond breaking the substituent being varied would be acting perpendicularly (Thornton, 1967), not parallel, to the reaction co-ordinate in which case a non-linear free energy relationship between log k_H and σ* would be expected (cf. Capon and Nimmo, 1975). These results may be compared with the semi-empirical calculations of Guthrie and Cullimore (1980) from which it is possible to calculate values of k_{H^+} for the decomposition of methyl hemiorthoacetates (Table 10) which show a trend qualitatively similar to that found for the decomposition of 2-hydroxy-2-methyl-1,3-dioxolans.

The effect of structure

The acyclic hemiorthoesters dimethyl and diethyl hemiorthoformate undergo hydronium ion catalysed breakdown 47 and 61 times faster than the analogous cyclic hemiorthoester with a five-membered ring, 2-hydroxy-1,3-dioxolan (Table 1). This difference is similar to that found in the rate of hydrolysis of acetals. Thus benzaldehyde diethyl acetals are hydrolysed 30 and 35 times faster than the corresponding dioxolan (Fife and Jao, 1965). In the reactions of the dioxolans (hemiorthoester and acetal) the leaving group is a β-hydroxyalcohol rather than methanol or ethanol in the acyclic compounds, and part of the decrease may be due to the unfavourable electron withdrawing effect of the β-alkoxy group on the positively charged oxygen in the transition state. In addition there may be some stereoelectronic constraint on the opening of the five-membered ring (cf. Fife and Jao, 1965). It is difficult to estimate the relative importance of these two (and of any other) effects. An acyclic orthoester with a β-alkoxyethyl group has been studied, viz. tris-(2-methoxyethyl)-orthoacetate (Chiang et al., 1983) but the hemiorthoester could not be detected in its hydrolysis nor in the hydrolysis of trimethyl orthoacetate. The value of k_{H^+} for the initial C—O bond cleavage of the methyl compound was 6.9 to 8.6 times greater than that for the 2-methoxyethyl compound, but as the hemiorthoester was not detected in either reaction it is not possible to estimate the effect of replacing the methoxyl groups of dimethyl hemiorthoacetate by 2-methoxyethoxyl groups on the rate of breakdown.

The ring-opening of cyclic hemiorthoesters with six-membered rings occurs about 10 times more rapidly than analogous compounds with five-membered rings (Tables 3 and 4). This may arise from the leaving group now having an oxygen in the γ- rather than the β-position or from the stereoelectronic constraints being less.

The introduction of methyl groups into both the five- and six-membered rings generally causes a rate decrease. This effect is particularly noticeable when all the carbons of the 1,3-dioxolan ring are fully substituted. Thus 2-hydroxy-2,4,4,5,5-pentamethyl-1,3-dioxolan reacts about 20 times more slowly than 2-hydroxy-1,3-dioxolan.

Hemiorthoesters and orthoesters

The relative rates of the hydronium ion-catalysed cleavage of the C—O bonds of analogous orthoesters and hemiorthoesters, as with analogous acetals and hemiacetals, depends markedly on structure. Thus dimethyl and diethyl hemiorthoformate react respectively 1980 and 233 times faster than trimethyl and triethyl orthoformate (Table 1) and dimethyl hemiorthobenzoate reacts 2700 times faster than trimethyl orthobenzoate (McClelland and Patel, 1981a). This behaviour is similar to that found with formaldehyde dimethyl acetal which undergoes hydronium ion catalysed C—OMe bond cleavage 2600 times more slowly than formaldehyde methyl hemiacetal does (Funderburk et al., 1978). In contrast to these results k_{H^+} for C—OMe bond cleavage of trimethyl orthocyclopropanecarboxylate is 15.3 times greater than k_{H^+} for C—OMe cleavage of dimethyl hemiorthocyclopropanecarboxylate (Burt et al., 1982) and k_{H^+} for C—OEt cleavage of orthoester [104a] ($3.1 \times 10^3 M^{-1} s^{-1}$ at 25°) is very similar to that for C—OEt cleavage of hemiorthoester [104b] ($2-4 \times 10^3 M^{-1} s^{-1}$) (McClelland and Alibhai 1981). This behaviour lies closer to what is found with acetals and hemiacetals of benzaldehyde where the ethyl hemiacetal reacts only 5 times faster than the diethyl acetal (Przystas and Fife, 1981) and the methyl hemiacetal reacts 18.7 times faster than the dimethyl acetal.

[104a] [104b]

This variation in the relative rates of C—O bond cleavage of analogous orthoesters and hemiorthoesters (and of acetals and hemiacetals) is of course a reflection of the different effect of substituents on the rates of these two classes of reaction. As discussed above, substituents capable of stabilizing carbocations have a greater effect on the rates of C—O bond cleavage of orthoesters (and acetals) than of hemiorthoesters (and hemiacetals).

Hemiorthoesters and hemiacetals

Hemiorthoesters are much more reactive than hemiacetals as would be expected from the much greater difficulty in detecting them. Thus k_{H^+} for the breakdown of dimethyl hemiorthoformate and diethyl hemiorthoformate is about 10^5 times greater than k_{H^+} for the breakdown of the corresponding hemiacetals (Table 11), and the cyclic hemiorthoesters [106] breaks down about 200 times faster than cyclic hemiacetal [105]. Presumably the additional conjugation of the extra methoxyl group with the developing carbonyl group stabilizes the transition state for breakdown of the hemiorthoester. It has been estimated that the delocalization energy of an ester which results from conjugation of the alkoxy group with the carbonyl group is approximately 17 kcal mol^{-1} (Fastrez, 1977).

TABLE 11

Rate constants ($k_{H^+}/M^{-1}s^{-1}$) for the hydronium-ion catalysed breakdown of hemiacetals and hemiorthoesters at 25°

Hemiacetal		Hemiorthoester	
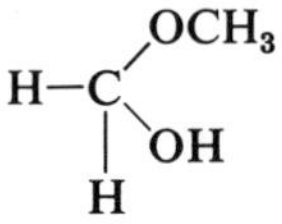 H–C(OCH$_3$)(OH)(H)	0.58[a]	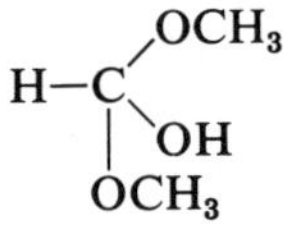 H–C(OCH$_3$)(OH)(OCH$_3$)	7×10^{4} [b]
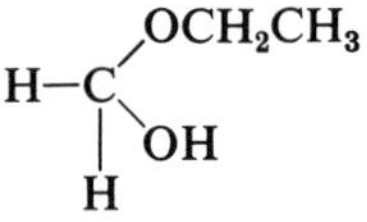 H–C(OCH$_2$CH$_3$)(OH)(H)	0.74[a]	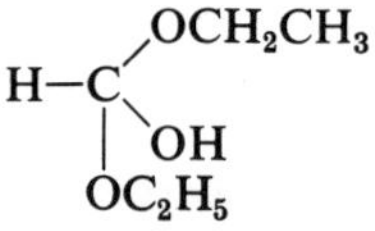 H–C(OCH$_2$CH$_3$)(OH)(OC$_2$H$_5$)	5.5×10^{4} [b]
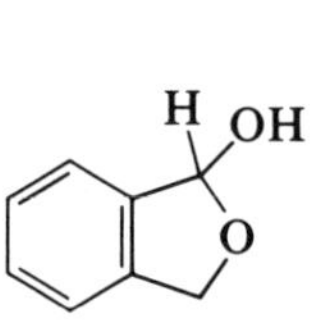 [105] (cyclic hemiacetal, H, OH, O)	43[c]	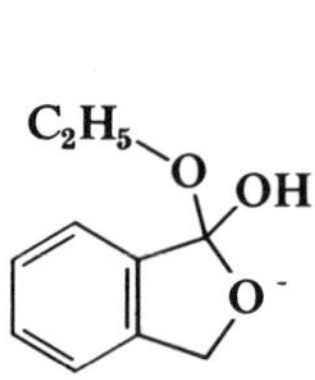 [106] (cyclic hemiorthoester, C$_2$H$_5$O, OH, O$^-$)	$6\text{–}12 \times 10^{3}$ [d]

[a] Funderburk *et al.*, 1978
[b] Extrapolated from the value at 15° (Capon and Sanchez, 1983)
[c] Harron *et al.*, 1981
[d] Endocyclic C—O bond cleavage (McClelland and Alibhai, 1981)

The mechanism of the acid-catalysed breakdown of hemiorthoesters

Electron releasing substituents when attached to the pro-acyl carbon have a smaller effect on k_{H^+} for the breakdown of hemiorthoesters than of orthoesters (see p. 67 above) which suggest that these two classes of compounds react by different mechanisms and that the transition state for breakdown of the hemiorthoesters has less carbocationic character. Jencks and his coworkers (Funderburk *et al.*, 1978) proposed mechanism (5) for the breakdown of hemiacetals and a similar mechanism (6) can be written for the breakdown of hemiorthoesters. This would explain *(i)* the general acid catalysis observed

$$\begin{array}{c}\text{H} \quad \text{OCH}_3 \\ \diagdown \text{C} \diagup \\ \text{H} \quad \text{OH}\end{array} \xrightarrow{\text{HA}} \begin{array}{c}\text{H} \quad ^+\overset{H}{\text{O}}\text{CH}_3 \\ \diagdown \text{C} \diagup \\ \text{H} \quad \text{OH} \quad \text{A}^-\end{array} \rightleftharpoons \begin{array}{c}\text{H} \quad \delta+\overset{H}{\text{O}}\text{CH}_3 \\ \diagdown \text{C} \diagup \\ \text{H} \quad \text{O}\cdots\text{H}\cdots\text{A}^{\delta-}\end{array} \rightleftharpoons \begin{array}{c}\text{H} \quad \quad \text{HOCH}_3 \\ \diagdown \text{C}=\text{O} \\ \text{H} \quad \quad \text{HA}\end{array} \quad (5)$$

$$\begin{array}{c}\text{OR} \\ \text{R}'-\text{C} \diagdown \text{OR} \\ |\phantom{\text{C}}\text{OH}\end{array} \xrightarrow{\text{HA}} \begin{array}{c}^+\overset{H}{\text{O}}\text{R} \\ \text{R}'\text{C} \diagdown \text{OR} \\ |\phantom{\text{C}}\text{OH}\end{array} \rightleftharpoons \begin{array}{c}\delta+\overset{H}{\text{O}}\text{R} \\ \text{R}'\text{C} \diagdown \text{OR} \\ |\phantom{\text{C}}\text{O}\cdots\text{H}\cdots\text{A}^{\delta-}\end{array} \rightleftharpoons \begin{array}{c}\text{OR} \quad \text{HOR} \\ \text{RC} \diagdown \text{O} \\ \text{HA}\end{array} \quad (6)$$

for the breakdown of dimethyl hemiorthobenzoates; *(ii)* the smaller sensitivity of k_{H^+} to electron-releasing substituents in R' than found for the analogous reaction of orthoesters since the rate-limiting step is an E2 elimination not a unimolecular heterolysis to form an oxocarbocation; and *(iii)* the greater rate of breakdown of hemiorthoesters compared to hemiacetals since there would be conjugation between the developing carbonyl group and the additional alkoxyl group in the transition state for breakdown of the hemiorthoester.

The mechanism has been considered previously (McClelland and Patel, 1981b) for the breakdown of dimethyl hemiorthobenzoate but rejected on the basis that, when the catalyst was formic acid, the rate constant for the reaction of the conjugate acid of the substrate (assumed $pK_a = -6$) with formate would have to have a value of $10^{12} M^{-1} s^{-1}$. The possibility was therefore considered of a "one-encounter" mechanism "in which the acid that donates the proton to the departing alkoxy group also acts as a general base to remove the OH proton before diffusional separation can occur." However, while it is necessary to postulate this kind of mechanism for the formic acid catalysed breakdown of dimethyl hemiorthobenzoate, for many hemiorthoesters this is not necessary since on the basis of the mechanism (6) k_{A^-} would not have to be as large as $10^{12} M^{-1} s^{-1}$. Thus, for dimethyl hemiorthoformate k_{A^-} would have to be $5 \times 10^8 M^{-1} s^{-1}$ in water on the assumption that the pK_a of the conjugate acid is -6. Mechanism (6) is therefore a valid one.

On the basis of the mechanism (6) the alcoholysis (and hydrolysis?) of esters should be general-acid catalysed but as explained above (p. 66) there have been so few investigations that it is not known if this is so.

THE HYDROXIDE-ION CATALYSED REACTION

The experimental values for k_{HO^-} for the breakdown of hemiorthoesters are collected in Tables 12–14. As these were determined under several different sets of conditions they have to be extrapolated to 25° with water as solvent and collected together with values of k_{HO^-} for the breakdown of some hemiacetals in Table 15.

TABLE 12

Rate constants for the hydroxide-ion catalysed (k_{HO^-}) and spontaneous (or water-catalysed) (k_{H_2O}) breakdown of hemiorthoesters in CH_3CN-H_2O ($c_{H_2O} = 8.33$ M) at 15°

	$10^{-9}k_{HO^-}/M^{-1}s^{-1}$	$10^2 k_{H_2O}/s^{-1}$
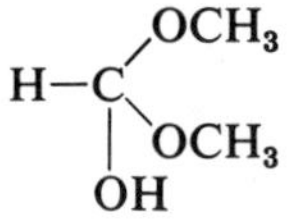 H–C(OCH₃)(OCH₃)(OH)	32.1[a]	2.57[a]
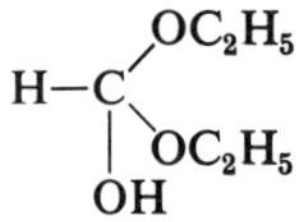 H–C(OC₂H₅)(OC₂H₅)(OH)	6.53[a]	4.36[a]
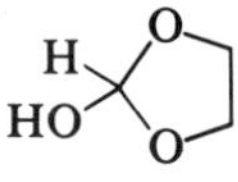 H, HO – dioxolane	600[a]	2.70[a]
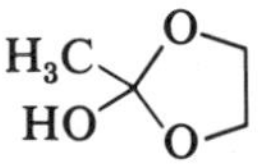 H₃C, HO – dioxolane	19.7[a]	1.53[a]
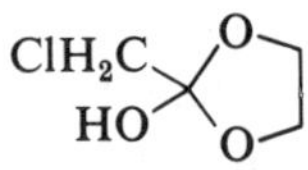 ClH₂C, HO – dioxolane	5360[b]	0.89[b]

TABLE 12 *(continued)*

	$10^{-9}k_{HO^-}/M^{-1}s^{-1}$	$10^2 k_{H_2O}/s^{-1}$
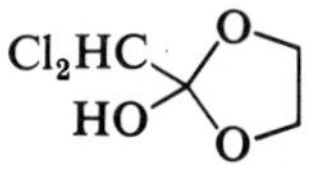 Cl$_2$HC–, HO–, (dioxolane)	24600[b]	1.60[b]
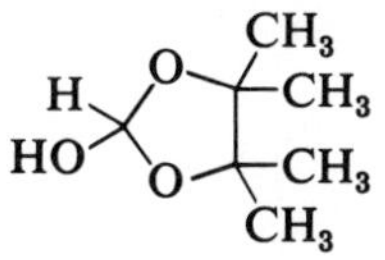 H–, HO–, (dioxolane with 4 CH$_3$)	8.39[a]	0.783[a]
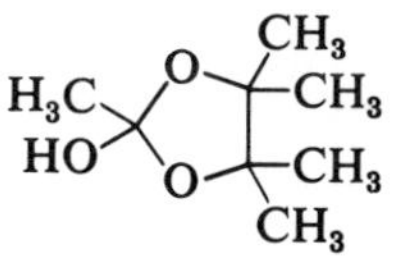 H$_3$C–, HO–, (dioxolane with 4 CH$_3$)	0.00395[a]	0.149[a]

[a] Capon and Sanchez, 1983
[b] Capon and Dosunmu, 1984

TABLE 13

Rate constants for the spontaneous (or water-catalysed), k_{H_2O} and hydroxide-ion catalysed breakdown of hemiorthoesters (in H_2O at 15°)[a]

	$10^{-6}k_{HO^-}/M^{-1}s^{-1}$	k_{H_2O}/s^{-1}
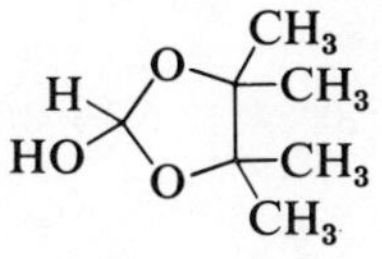 H–, HO–, (dioxolane with 4 CH$_3$)	1020	0.191
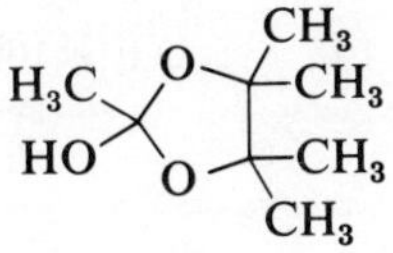 H$_3$C–, HO–, (dioxolane with 4 CH$_3$)	2.87	0.0180

[a] Capon and Sanchez 1983

TABLE 14
Rate constants for the hydroxide-ion catalysed breakdown of hemiorthoesters in CH_3CN—H_2O (c_{H_2O} = 2.22 M) at 15° [a]

	$10^{-5}k_{HO^-}$/$M^{-1}s^{-1}$
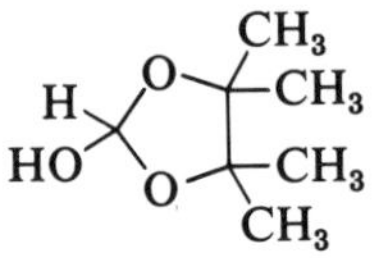 H, HO, O, O, C(CH_3)_3, CH_3 (hemiorthoester with H and HO on central C, OC(CH_3)_3 and OCH_3... actually: H–C(OH)(O–)(O–C(CH_3)_3))	1700
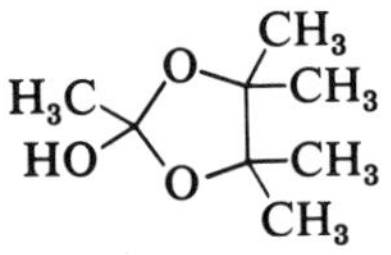 H_3C, HO, O, O, C(CH_3)_3	3.7
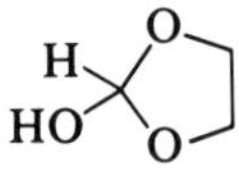 H, HO, dioxolane	899

[a] Capon and Sanchez, 1983

TABLE 15
Extrapolated rate constants for the hydroxide-ion catalysed breakdown of hemiorthoesters and hemiacetals in water at 25° and rate constants estimated on the basis of the mechanism of eqns (7) and (8)

	k_{HO^-} [a,b]	pK_a [c]	k_2 [d,b]	k_{-1} [e,f]
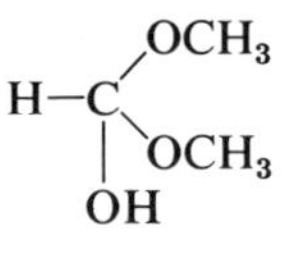 H–C(OCH_3)_2(OH)	3.4×10^{10}	11.79	2.1×10^8	6.2×10^8
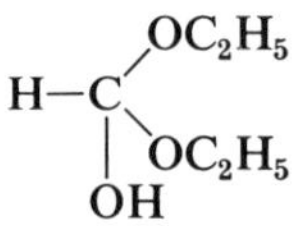 H–C(OC_2H_5)_2(OH)	6.85×10^9	11.48	2.1×10^7	3.0×10^8
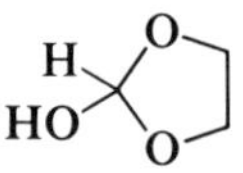 H, HO, dioxolane	6.5×10^{11}	11.48	2.0×10^9	3.0×10^8

TABLE 15 *(continued)*

Structure	k_{HO^-} [a,b]	pK_a [c]	k_2 [d,b]	k_{-1} [e,f]
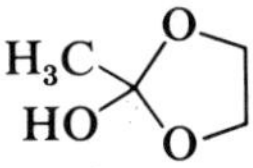 H₃C, HO — (1,3-dioxolane)	2.1×10^{10}	12.12	1.8×10^8	1.3×10^9
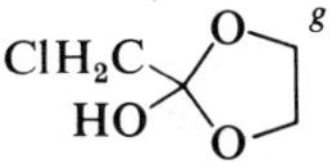 ClH₂C, HO — (1,3-dioxolane) [g]	5.6×10^{12}	10.74	3.1×10^9	5.5×10^7
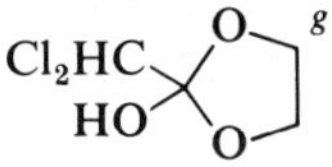 Cl₂HC, HO — (1,3-dioxolane) [g]	2.6×10^{13}	9.57	9.7×10^8	3.7×10^6
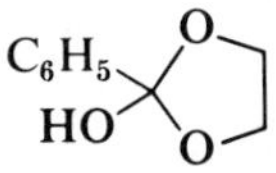 C₆H₅, HO — (1,3-dioxolane)	6×10^{10} [h]	11.33	1.3×10^8	2.2×10^8
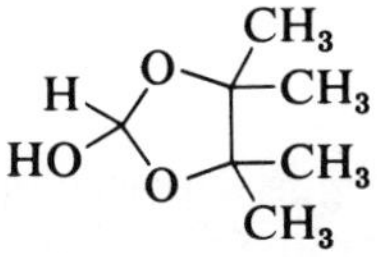 H, HO — (tetramethyl dioxolane)	2.5×10^9	11.37	5.9×10^6	2.4×10^8
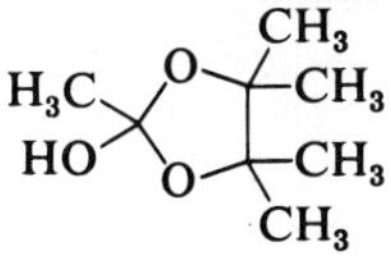 H₃C, HO — (tetramethyl dioxolane)	1.6×10^7	12.02	1.7×10^5	1.1×10^9
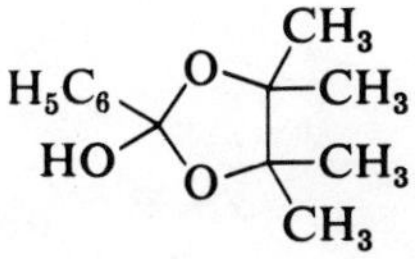 H₅C₆, HO — (tetramethyl dioxolane)	2×10^7 [i]	11.23	3.4×10^4	1.7×10^8
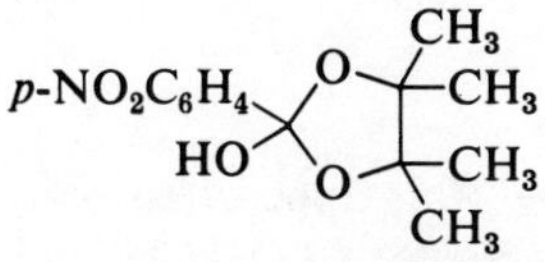 p-NO₂C₆H₄, HO — (tetramethyl dioxolane)	2×10^7 [j]	10.4 [i]	5.1×10^3	2.5×10^7
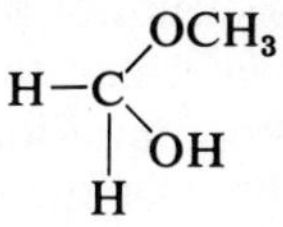 H—C(OCH₃)(OH)H	2.34×10^3	13.20	3.7×10^2	1.6×10^{10}

continued

TABLE 15 *(continued)*

	k_{HO^-} a,b	pK_a c	k_2 d,b	k_{-1} e,f
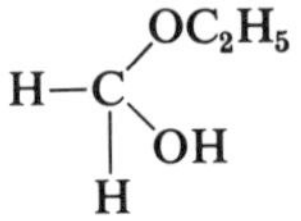 H—C(OC$_2$H$_5$)(OH)(H)	1.3 × 10^3 k	13.04	1.4 × 10^2	1.1 × 10^{10}

^a The extrapolated values at 15° from Capon and Sanchez (1983) multiplied by 2.5
^b Units; M^{-1}s^{-1}
^c Estimated according to Capon and Ghosh (1981); cf. Capon and Sanchez (1983)
^d Estimated on the basis of eqns (7) and (8); $k_2' = k_{\text{HO}^-} K_w / K_a$; p$K_w$ (25°) = 13.996 (Robinson and Stokes, 1959)
^e Estimated from K_a and an assumed value of $k_1 = 10^{11}$ M^{-1}s^{-1} – cf. eqn (7); $k_{-1} = k_1 K_w / K_a$; pK_w (25°) = 13.996 (Robinson and Stokes, 1959)
^f Units: s^{-1}
^g Capon and Dosunmu, 1984
^h Experimental value (Ahmad *et al.*, 1979)
ⁱ Experimental value (McClelland *et al.*, 1979a)
^j Experimental value (McClelland *et al.*, 1984)
^k Experimental value (Funderburk *et al.*, 1978)

Jencks and his coworkers (Funderburk *et al.*, 1978) proposed a mechanism for the breakdown of hemiacetals catalysed by hydroxide-ion which consists of the reversible formation of the anion followed by its unimolecular breakdown. A similar mechanism can be written for the breakdown of hemiorthoesters (7), (8). For at least one hemiorthoester viz. 2-hydroxy-2-

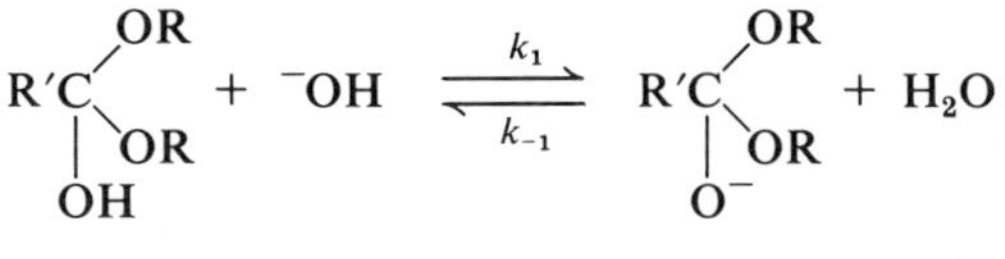

$$\text{R'C(OR)(OR)(OH)} + {}^-\text{OH} \underset{k_{-1}}{\overset{k_1}{\rightleftharpoons}} \text{R'C(OR)(OR)(O}^-) + \text{H}_2\text{O} \quad (7)$$

$$\text{R'C(OR)(OR)(O}^-) \xrightarrow{k_2} \text{R'}-\text{C(OR)}=\text{O} + \text{RO}^- \quad (8)$$

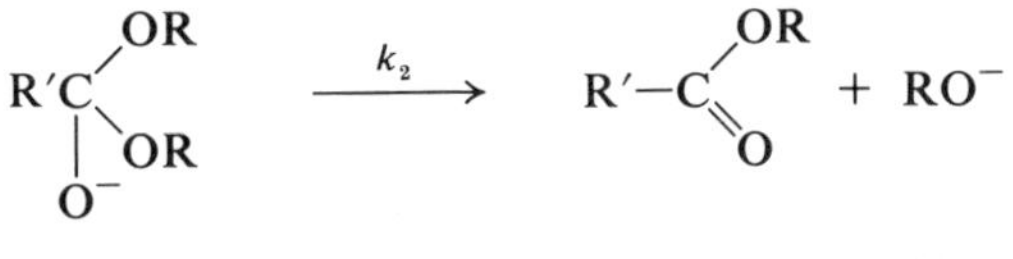

(*p*-nitrophenyl)-4,4,5,5-tetramethyl-1,3-dioxolan [109], this appears to be a valid mechanism since the anionic form [110] has been detected (McClelland *et al.*, 1984). It was shown that in 1 M NaOH the uv spectrum of pinacol *p*-nitrobenzoate [107] was different from what it was in water and that the spectrum changes in a regular manner with hydroxide ion concentration with an apparent K_b of 0.76 (determined at 310 nm) or 0.72 (determined at

TETRAHEDRAL INTERMEDIATES

$$p\text{-NO}_2\text{C}_6\text{H}_4\overset{\text{O}}{\underset{\|}{\text{C}}}-\text{OC}(\text{CH}_3)_2-\text{C}(\text{CH}_3)_2-\text{OH} \underset{+\text{H}^+}{\overset{-\text{H}^+}{\rightleftharpoons}} p\text{-NO}_2\text{C}_6\text{H}_4\overset{\text{O}}{\underset{\|}{\text{C}}}-\text{OC}(\text{CH}_3)_2-\text{C}(\text{CH}_3)_2-\text{O}^-$$

[107] $\text{p}K_a = 17\text{–}19$ [108]

$K_0 = 8 \times 10^{-4}$ $\dfrac{-\text{OH}, -\text{H}_2\text{O}}{-\text{OH}, \text{H}_2\text{O}}$ $K_b = 0.72, 0.76$ $K_- = 10^3\text{–}10^5$

[109] cyclic hemiorthoester with p-NO$_2$C$_6$H$_4$, HO, O, and C(CH$_3$)(CH$_3$)–C(CH$_3$) ring

$\underset{+\text{H}^+}{\overset{-\text{H}^+}{\rightleftharpoons}}$ $\text{p}K_a = 10.4$

[110] cyclic anion with $^-$O in place of HO

260 nm). This was interpreted as arising from formation of the anion [110] of the tetrahedral intermediate and K_b is then the equilibrium constant for formation of this species from the hydroxyester [107]. The equilibrium constant, K_0, for formation of the unionized tetrahedral intermediate from the hydroxyester was estimated to be $8 \pm 1 \times 10^{-4}$. This leads to a value of 10.4 ± 0.2 for the pK_a of the tetrahedral intermediate [109] which agrees quite well with the values estimated for similar species on the basis of linear free energy relationships. The pK_a of the hydroxyester was estimated to be 17–19 and so the equilibrium constant for cyclization of its anion [108], K, is 10^3–10^5. The cyclic form [110] is therefore more stable than the acyclic form [108] since it is the anion of a stronger acid. The rate constant for the hydroxide-ion catalysed breakdown of the neutral tetrahedral intermediate [109] is $2 \times 10^7 \text{M}^{-1}\text{s}^{-1}$ at 25°, which on the basis of a pK_a of 10.4 and the mechanism of eqns (7) and (8) leads to a value for the rate constant for breakdown of the ionized form, k_2, of $6.3 \times 10^3 \text{s}^{-1}$.

Although the mechanism of eqns (7) and (8) may be a reasonable one for the breakdown of [109], for some of the more reactive species, especially those derived from chloroacetic or dichloracetic acid, it may not be, as shown by the following considerations. For this mechanism to be valid the value of k_2 calculated from the relationship $k_{\text{HO}^-} = K_1 k_2$ must be *(i)* less than the time constant for a molecular vibration (10^{12}–10^{14}s^{-1}) *(ii)* less than the value of k_{-1} based on the estimated K_a and the assumption that the ionization equilibrium (7) is diffusion controlled in the thermodynamically favourable direction, i.e. $k_1 = 10^{11} \text{M}^{-1}\text{s}^{-1}$. These two conditions are easily fulfilled with the hemiacetals and the less reactive hemiorthoesters derived from pinacol monoesters (see Table 15), but with the more reactive hemiorthoesters the calculated values of k_2 lie close to or are greater than the calculated value of k_{-1}. Therefore this mechanism may no longer be a valid one for the hydroxide-ion catalysed breakdown of the more reactive hemiorthoesters and the experimental results may be explained by a mechanism in which there is a rate-limiting ionization or one in which there is concerted breaking of O—H and C—O bonds.

The much greater value of k_{HO^-} for the breakdown of the hemiorthoester compared to the hemiacetals may be attributed partly to the more favourable pK_a of the former, but must be mainly due to conjugation of the additional alkoxy group with the developing carbonyl group in the transition state.

The value of k_{HO^-} for the breakdown of 2-hydroxy-1,3-dioxolan is more than 10 times greater than for the breakdown of dimethyl and diethyl hemiorthoformate (Table 12). This contrasts with what is found with the hydronium-ion catalysed breakdown when acyclic hemiorthoesters react faster than cyclic ones, and may be due either to a greater ease of ionization of the hydroxyl group when attached to the dioxolan ring or to the leaving group in the breakdown of the 2-hydroxy-1,3-dioxolan being β-hydroxyalkoxide rather than an alkoxide as in the breakdown of dimethyl and diethyl hemiorthoformate.

The presence of four methyl groups at positions 4 and 5 of the dioxolan ring causes a larger decrease in k_{HO^-} (Tables 12, 13) than in k_{H^+}. It is possible that in the hydroxide-ion catalysed reaction there is, superimposed on the steric effect, an electronic effect of the methyl groups which causes a decrease in the leaving group ability of the alkoxide ion due to their electron-releasing inductive effect. A similar effect is observed in the hydroxide-ion catalysed breakdown of benzaldehyde t-butyl and methyl hemiacetals for which $k_{HO^-}(Me)/k_{HO^-}(Bu^t) = 37$.

SPONTANEOUS OR WATER-CATALYSED REACTION

Experimental values of k_{H_2O} are collected in Tables 2, 3, 4, 12 and 13. These have been extrapolated to a common set of conditions and collected together with the values of k_{H_2O} for the breakdown of hemiacetals in Table 16.

There are at least three possible mechanisms for the spontaneous breakdown of hemiorthoesters, hemiacetals, and related species. Firstly, there may be a rapid and reversible ionization equilibrium followed by hydronium-ion catalysed breakdown of the anion (9) (Gravitz and Jencks, 1974). A necessary condition for this mechanism to be valid is that $k_2{}^*$ calculated from k_{H_2O} and K_a should fall below the diffusion controlled limit of c. $10^{10} M^{-1} s^{-1}$. The second mechanism (10) is similar to this but involves formation of the anion and hydronium ion in an encounter pair which react to give products faster than the diffuse apart (Capon and Ghosh, 1981). With this mechanism therefore the ionization equilibrium is not established and the rate constant for

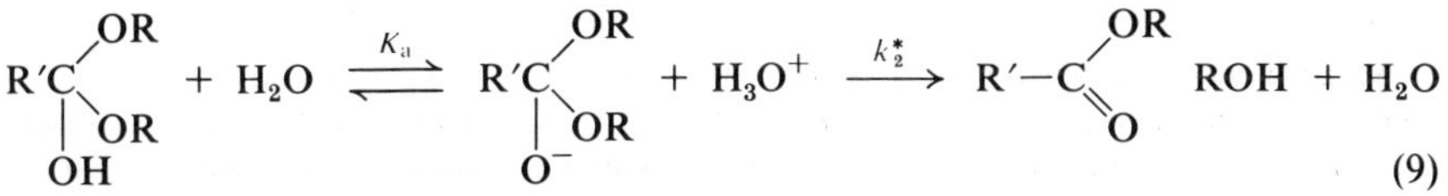

$$R'C\begin{matrix}OR\\OR\\OH\end{matrix} + H_2O \xrightleftharpoons{K_a} R'C\begin{matrix}OR\\OR\\O^-\end{matrix} + H_3O^+ \xrightarrow{k_2^*} R'-C\begin{matrix}OR\\O\end{matrix} \;\; ROH + H_2O \quad (9)$$

$$R'C\begin{matrix}OR\\OR\\OH\end{matrix} + H_2O \xrightarrow{k_1^*} \left[R'C\begin{matrix}OR\\OR\\O^-\end{matrix} \; H_3O^+ \right] \longrightarrow R'C\begin{matrix}OR\\O\end{matrix} + ROH + H_2O \quad (10)$$

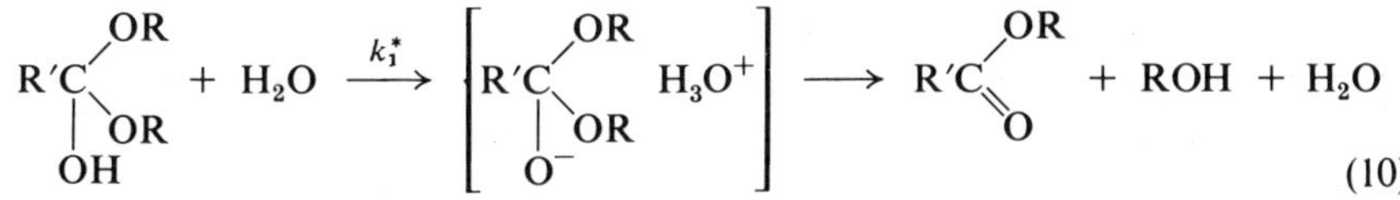

reaction, k_{H_2O}, should be equal to the rate of ionization, k_1^* which may be calculated from the K_a if it is assumed that the rate constant for the recombination of the anion and H_3O^+ has a rate constant of $10^{11}M^{-1}s^{-1}$. The third possibility is that the reaction is concerted either via a non-cyclic (11) or

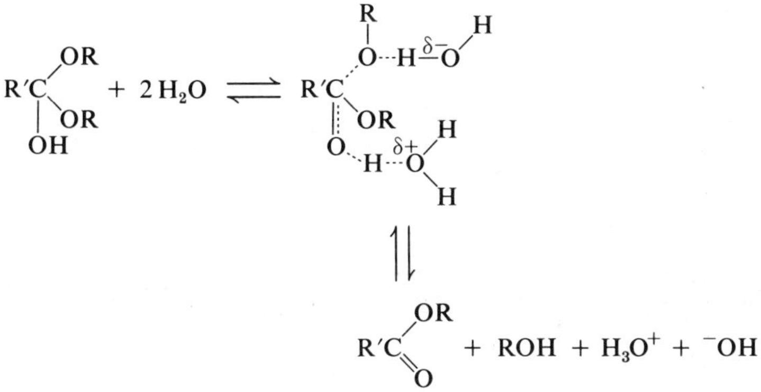

$$R'C\begin{matrix}OR\\OR\\OH\end{matrix} + 2H_2O \rightleftharpoons R'C\begin{matrix}\cdots O\cdots H\overset{\delta-}{-}O\overset{R}{\diagdown}{}^H\\OR\\\cdots O\cdots H\overset{\delta+}{-}O\diagdown_H^H\end{matrix}$$

$$\Updownarrow$$

$$R'C\begin{matrix}OR\\O\end{matrix} + ROH + H_3O^+ + {}^-OH \quad (11)$$

$$R'C\begin{matrix}OR\\OR\\OH\end{matrix} + 2H_2O \rightleftharpoons R'C\begin{matrix}\cdots O\cdots H\diagdown O\diagup H\\OR\\\cdots O\cdots H\diagup \diagdown H\\H\cdots O\diagdown H\end{matrix} \rightleftharpoons RC\begin{matrix}OR\\O\end{matrix} + ROH + 2H_2O \quad (12)$$

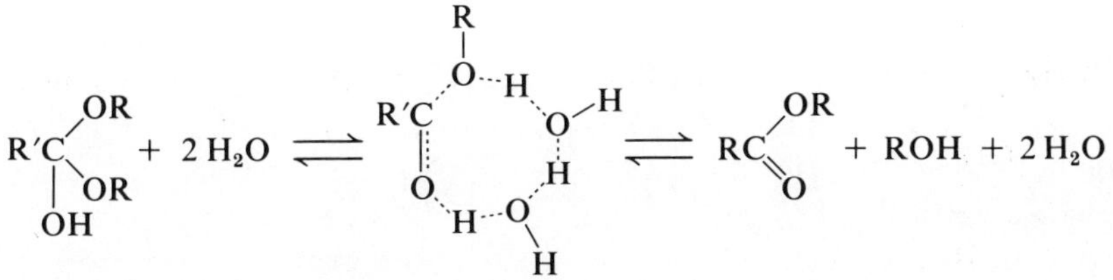

cyclic (12) transition state (McClelland and Patel, 1981b). These mechanisms, unlike the other two, would explain a value of k_{H_2O} greater than the rate constant for ionization of the hydroxyl group k_1^*. Therefore in principle one should be able to decide which mechanism explains the experimental results best by making estimates of k_1^* and k_2^* and finding out how they obey the above criteria. Such estimates are given in Table 16 and, owing to the assumptions involved, are probably only accurate to an order of magnitude. All the

TABLE 16

Extrapolated rate constants for the spontaneous (or water-catalysed) breakdown of hemiorthoesters and hemiacetals in water at 25° and values of k_2^* estimated on the basis of the mechanism of eqn (9) and of k_1^* estimated on the basis of the mechanism of eqn (10)

Compound	k_{H_2O} [a,b]	pK_a [c]	k_1^* [d,b]	k_2^* [e,f]
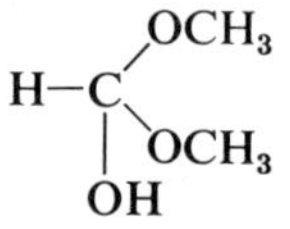 H–C(OCH₃)(OCH₃)(OH)	1.8	11.79	1.6×10^{-1}	1.1×10^{12}
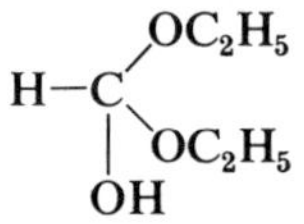 H–C(OC₂H₅)(OC₂H₅)(OH)	3.1	11.48	3.3×10^{-1}	9.4×10^{11}
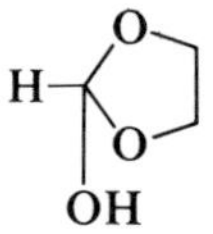 H–(dioxolane)–OH	1.91	11.48	3.3×10^{-1}	5.8×10^{11}
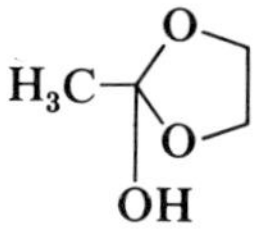 H₃C–(dioxolane)–OH	1.08	12.12	7.6×10^{-2}	1.4×10^{12}
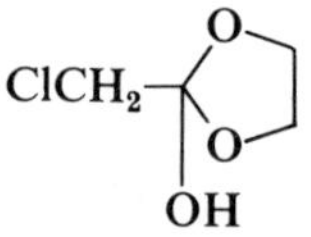 ClCH₂–(dioxolane)–OH	0.63	10.74	1.8	3.5×10^{10}
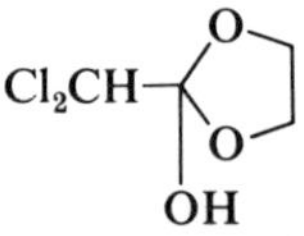 Cl₂CH–(dioxolane)–OH	1.1	9.57	2.6×10^{1}	4.1×10^{9}
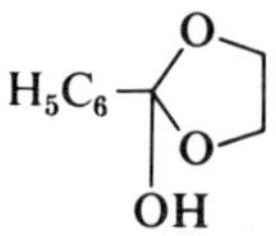 H₅C₆–(dioxolane)–OH	1.5[g]	11.33	4.7×10^{-1}	3.2×10^{11}

TABLE 16 *(continued)*

Compound	k_{H_2O} [a,b]	pK_a [c]	k_1^* [d,b]	k_2^* [e,f]
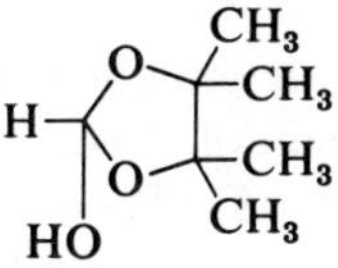 ![] H–C(OH)(O-C(CH₃)₃)(O-C(CH₃)₃) (pinacol-type orthoester, H substituent)	0.48	11.37	4.3×10^{-1}	1.1×10^{11}
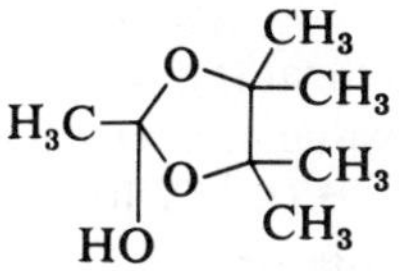 H₃C–C(OH)(O-C(CH₃)₃)(O-C(CH₃)₃)	0.11	12.02	9.5×10^{-2}	1.2×10^{11}
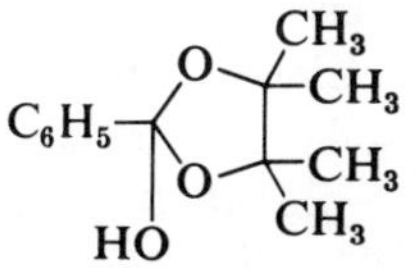 C₆H₅–C(OH)(O-C(CH₃)₃)(O-C(CH₃)₃)	0.074[h]	11.23	5.9×10^{-1}	1.3×10^{10}
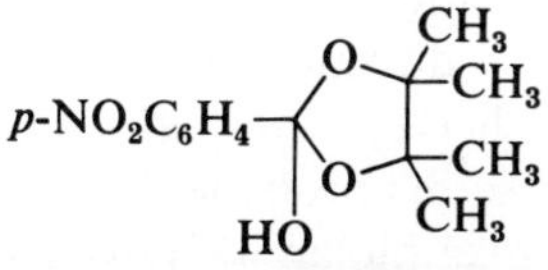 p-NO₂C₆H₄–C(OH)(O-C(CH₃)₃)(O-C(CH₃)₃)	0.15[i]	10.4	4	3.7×10^9
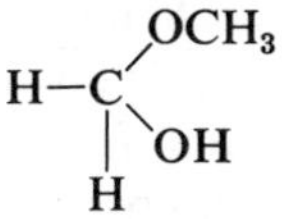 H–CH(OCH₃)(OH)	1.8×10^{-3} [j]	13.20	6.3×10^{-3}	2.8×10^{10}
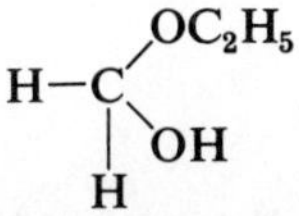 H–CH(OC₂H₅)(OH)	1.6×10^{-3} [j]	13.04	9.1×10^{-3}	1.8×10^{10}

[a] The extrapolated values at 15° from Capon and Sanchez (1983) multiplied by 2.5
[b] Units: s^{-1}
[c] Estimated according to Capon and Ghosh (1981); cf. Capon and Sanchez (1983)
[d] Calculated from the K_a on the assumption that the rate constant for recombination of the anion with H_3O^+ is 10^{11} $M^{-1}s^{-1}$
[e] Estimated on the basis of the mechanism of eqn (9); $k_2^* = k_{H_2O}/K_a$
[f] Units: $M^{-1}s^{-1}$
[g] Experimental value (Ahmad et al., 1979)
[h] Experimental value (McClelland et al., 1979a)
[i] Experimental value (McClelland et al., 1984)
[j] Experimental value (Funderburk et al., 1978)

values of k_2^* lie close to or above the diffusion controlled limit of c. $10^{10} M^{-1} s^{-1}$ so mechanism (9) is probably excluded. Generally k_1^* is within an order of magnitude of k_{H_2O} so on this basis mechanism (10) may be a valid one. However, it is ruled out by the very small effect of substituents on k_{H_2O}. If the rate-limiting step were ionization of the hydroxyl group the rate constants should parallel the K_a-values since the reverse reaction of the anion with H_3O^+ should be diffusion controlled and independent of structure. In fact, k_{H_2O} varies very little with changes in the estimated pK_a (Table 16) so mechanism (9) is also excluded. This leaves the concerted mechanisms (10) and (11) as the most likely. They would explain values of k_{H_2O} greater than the calculated values of k_1^* (if these differences are significant) and the very small effect of substituents on k_{H_2O} as indicated by a ρ-value of near zero for breakdown of dimethyl hemiorthobenzoates (cf. Table 2) (McClelland and Patel, 1981b) and the small effect of chloro substituents on k_{H_2O} for the breakdown of 2-hydroxy-2-methyl-1,3-dioxolan (cf. Tables 12 and 16) (Capon and Dosunmu, 1984). Also, as pointed out by McClelland and Santry (1983) they are consistent with a value of $k_{H_2O}/k_{D_2O} = 4.5$ for breakdown of 2-hydroxy-2-phenyl-1,3-dioxolan and of $\Delta S^{\neq} = -22$ cal $K^{-1} mol^{-1}$ for the breakdown of 2-hydroxy-2-phenyl-1,3-dioxolan. Of these two mechanisms (10) and (11) the one with the cyclic transition state (11) seems more attractive since mechanism (10) in the reverse direction involves a collision between four species at least one of which (H_3O^+ or HO^-) is present in low concentration. A similar mechanism seems probable for the breakdown of 2-hydroxy-2-(4-methoxyphenyl)-1,3-oxathiolan [83] which occurs only with C—S bond cleavage (Santry and McClelland, 1983a). As pointed out by Santry and McClelland C—S bond cleavage is favoured by a "combination of a better leaving group and the production of a more stable product". The better leaving group ability of the S-group also explains the 4.4-fold greater rate of breakdown of [83] compared to the analogous dioxolan.

FORMATION OF OXOCARBOCATIONS; EQUILIBRIUM CONSTANTS FOR FORMATION OF HEMIORTHOESTERS

Cyclic hemiorthoesters [112] besides undergoing breakdown to form esters [111] also undergo fission of the C—OH bond to form oxocarbocations [113] (McClelland et al., 1979a). In acidic solutions this can sometimes be demonstrated by detecting the cation or by demonstrating exchange of the hydroxyl group with water by isotopic labelling. To demonstrate this exchange it is not necessary (or easy) to start with the labelled hemiorthoester, but instead one can use the ester labelled on the carbonyl group. The most thoroughly investigated ester is pinacol monobenzoate, the carbonyl group of which undergoes exchange with water in acidic solutions much faster than the ester undergoes hydrolysis. As proposed by McClelland and his coworkers the most reasonable pathway for exchange is cyclization to form the hemiorthoester which then forms the oxocarbocation [115] much faster than it reverts to ester [114]. A rough estimate for the rate constant,

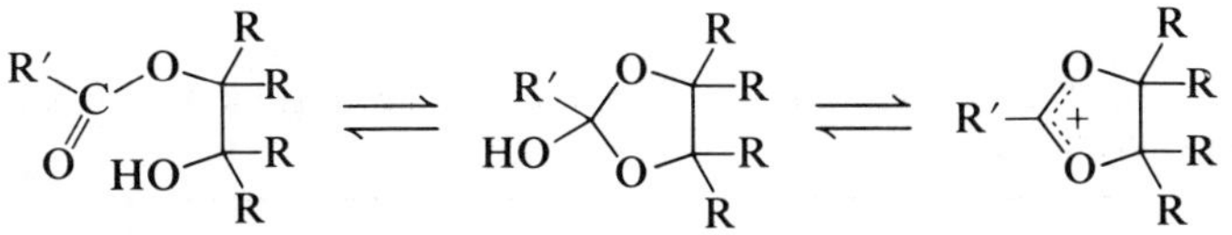

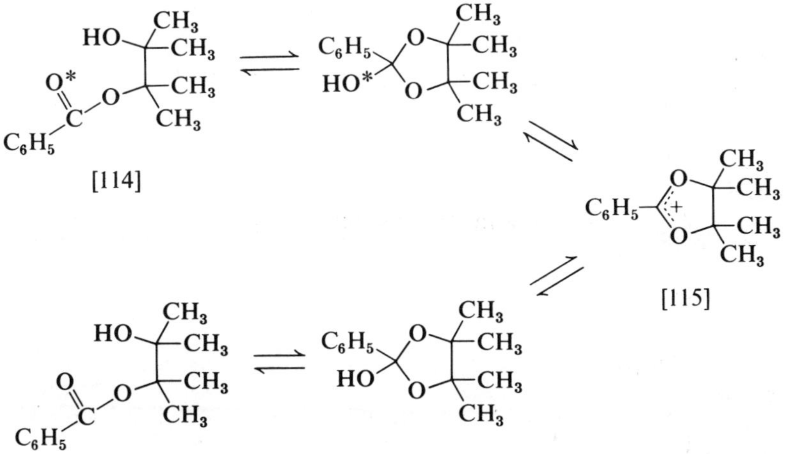

k_{H^+}, for the cleavage of the C—OH bond may be obtained from k_{H^+} for cleavage of the C—OMe bond of the corresponding methyl orthoester, $3.1 \times 10^4 \mathrm{M}^{-1}\mathrm{s}^{-1}$ (25°), which is considerably greater than k_{H^+} for breakdown of the hemiorthoester to ester, $13\,\mathrm{M}^{-1}\mathrm{s}^{-1}$. The rate constant for C—OH bond cleavage may also be measured directly by studying the dependence on acid concentration of the rate of equilibration of cation and hemiorthoester which gives a value of $2 \times 10^4 \mathrm{M}^{-1}\mathrm{s}^{-1}$ (25°), slightly less than that for C—OMe bond cleavage of the methyl orthoester. When starting with labelled ester (e.g. [114]) the rate-limiting step for exchange of the label in acidic solutions (pH < 3.5) is then cyclization and so the rate of exchange is the rate of cyclization. Since the rate of ring-opening has been measured independently, the equilibrium constant for cyclization may be evaluated (Table 17). Similar, but less extensive, studies were made with labelled ethylene glycol monobenzoate (McClelland *et al.*, 1979a) and with the labelled monobenzoate of norbornane-*exo,exo*-2,3-diol (Santry and McClelland, 1983b) (Table 17).

TABLE 17

Rate and equilibrium constants for formation of hemiorthoesters from the corresponding esters at 25°

	k_{H^+}(for)	k_{H^+}(dec)	K	$\Delta G^{\circ\,e}$	Ref.
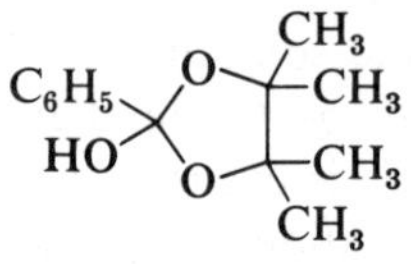 [119] C_6H_5, HO — cyclic orthoester with 4 CH$_3$ groups	$4.8 \times 10^{-4\,a}$	13^a	$3.7 \times 10^{-5\,c}$	6.0	f
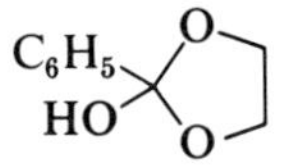 C_6H_5, HO — 1,3-dioxolane	$8.1 \times ^{-7\,a}$	$3 \times 10^{2\,a}$	$2.7 \times 10^{-9\,c}$	11.7	f
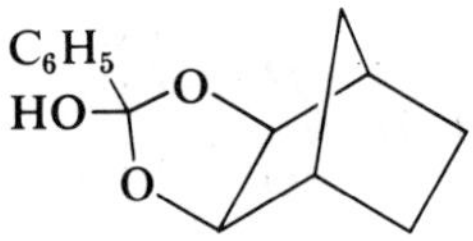 [120] C_6H_5, HO — norbornane-fused dioxolane	$2.3 \times 10^{-3\,a}$	$7 \times 10^{1\,a}$	$3.3 \times 10^{-5\,c}$	6.1	g
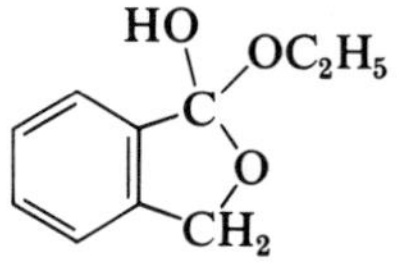 [121] HO, OC$_2$H$_5$ benzo-fused cyclic with CH$_2$	$6 \times 10^{-3\,a}$	$(6\text{–}12) \times 10^{3\,a}$	$5 \times 10^{-7} - 1 \times 10^{-6\,c}$	8.6–8.1	h
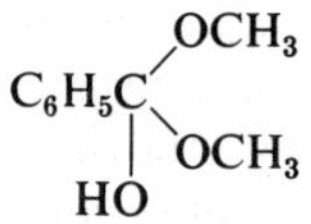 $C_6H_5C(OCH_3)_2$ HO	$4.2 \times 10^{-8\,b}$	$1.9 \times 10^{4\,a}$	$2.2 \times 10^{-12\,d}$	15.9	i, j
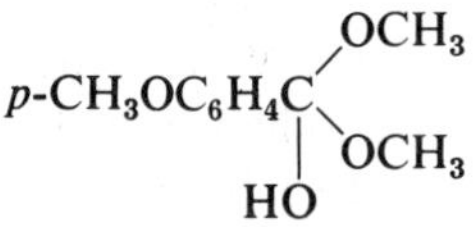 $p\text{-}CH_3OC_6H_4C(OCH_3)_2$ HO	$3.0 \times 10^{-8\,b}$	$4.5 \times 10^{4\,a}$	$6.7 \times 10^{-13\,d}$	16.6	i, j
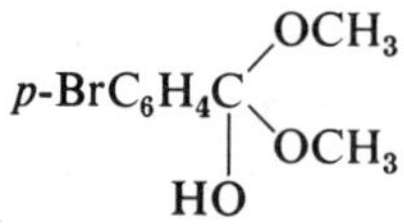 $p\text{-}BrC_6H_4C(OCH_3)_2$ HO	$4.6 \times 10^{-8\,b}$	$6.6 \times 10^{3\,a}$	$7.0 \times 10^{-12\,d}$	15.2	i, j

^{a} Units: $M^{-1}s^{-1}$
^{b} Units: $M^{-2}s^{-1}$

In addition to this method for measuring the rate constants for formation of hemiorthoesters and hence the equilibrium constants, two other methods have been used. The rate of formation of hemiorthoester [117] was determined from the rate of conversion of ester [116] into lactone [118] and the independently determined ratio for the partitioning of hemiorthoester between ester and lactone (McClelland and Alibhai, 1981). Also the rate constants (and hence the equilibrium constants) for formation of methyl hemiorthobenzoates have been determined from the rates of exchange of the trideuteriomethyl esters (Table 17) (McClelland and Patel, 1981b). The values of the equilibrium constants and free energies for the formation of cyclic hemiorthoformates and hemiorthoacetates calculated by Guthrie (1977) (Table 18) are in fairly good agreement with the experimental results since addition to a benzoate ester would be expected to be more difficult than to an acetate or formate ester.

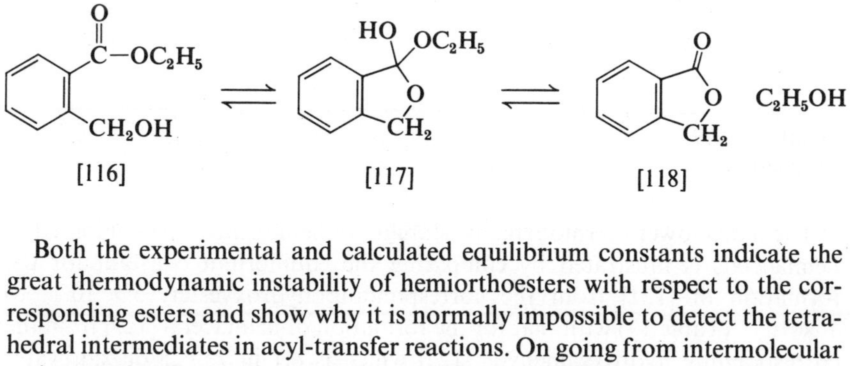

Both the experimental and calculated equilibrium constants indicate the great thermodynamic instability of hemiorthoesters with respect to the corresponding esters and show why it is normally impossible to detect the tetrahedral intermediates in acyl-transfer reactions. On going from intermolecular to intramolecular reactions the tetrahedral intermediate becomes relatively more stable, and if the structure is more rigid (cf. [120], [121] in Table 17) or more sterically crowded (cf. [119]) the tetrahedral intermediate is more stable still. However, it is only with structues as rigid as tetrodotoxin or with the trifluoroacetate of pinacol that the hemiorthoester is more stable than the ester (see Section 1).

---

[c] Dimensionless
[d] Units: M^{-1}
[e] Units kcal mol^{-1}
[f] McClelland *et al.*, 1979a
[g] Santry and McClelland, 1983b
[h] McClelland and Alibhai, 1981
[i] McClelland and Patel, 1981b
[j] In 50% H_2O–dioxan

TABLE 18

Calculated equilibrium constants and free energies for the formation of cyclic hemiorthoesters from the corresponding esters at 25° [a]

	K^b	$\Delta G°$/kcal mol^{-1}
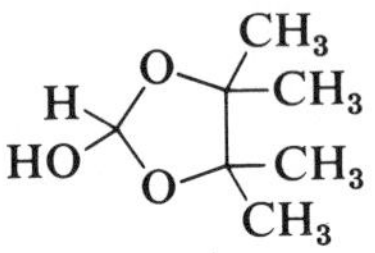 (structure with H, HO, O, O, CH$_3$, CH$_3$, CH$_3$, CH$_3$)	4.4×10^{-3}	3.21
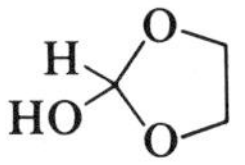 (structure with H, HO, O, O)	2.9×10^{-6}	7.54
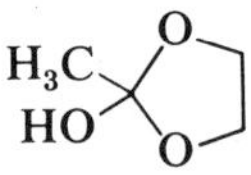 (structure with H$_3$C, HO, O, O)	1.4×10^{-8}	10.72

[a] Guthrie, 1977
[b] Dimensionless

The much lower thermodynamic stability of hemiorthoesters compared to hemiacetals is illustrated by comparing the equilibrium constant for the formation of [121] from the corresponding hydroxyester, 5×10^{-7} to 1×10^{-6} (Table 17) with that for the formation of hemiacetal [123] from the corresponding hydroxyaldehyde [122] which is 8.1 in 75:25 dioxan-water

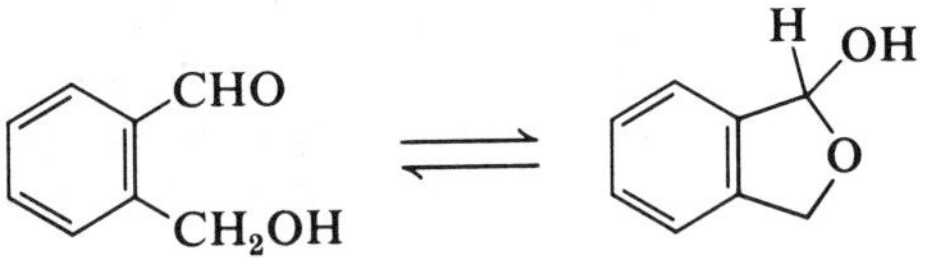

[122]   ⇌   [123]

(Harron et al., 1981). The hemiacetal is therefore relatively more stable than the hemiorthoester by 9–10 kcal mol^{-1}. A similar comparison of the equilibrium constant for the formation of dimethyl hemiorthobenzoate 2.2×10^{-12}M^{-1} with that for formation of benzaldehyde methylhemiacetal, 4×10^{-3}M^{-1}, yields a difference of about 12.5 kcal mol^{-1}. These values are a little less than the value of the delocalization energy of methyl acetate, 16.8 kcal mol^{-1}, calculated by Fastrez (1977).

5 Nitrogen-containing tetrahedral intermediates

There have been kinetic investigations of the breakdown of four nitrogen-containing tetrahedral intermediates [124]–[127] (see Table 19). In addition, compound [28] was studied but the rate constants for individual steps were not evaluated (Rogers and Bruice, 1973, 1974). Compounds [124], [126] and [127] were only studied in basic solutions but [125] was studied over a wider pH range. The values of k_{obs} for the breakdown of [124] and [126] did not increase linearly with increasing concentration of hydroxide ion, but started to level off at high concentrations; this was attributed to ionization of the hydroxyl groups, and the anion of [126] was detected by ^{1}H nmr spectroscopy (Tee et al., 1982). The pK_a-values for these compounds were evaluated from the plots of k_{obs} against [HO$^-$] but that for [125] was estimated by

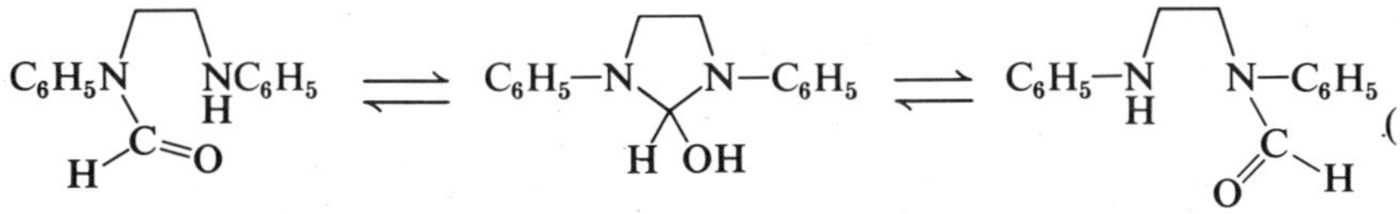

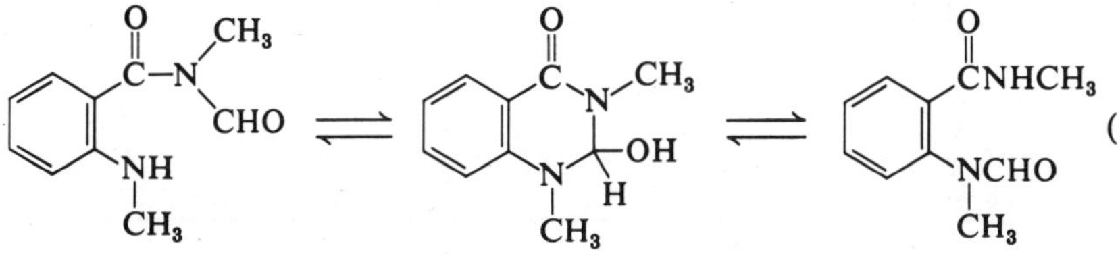

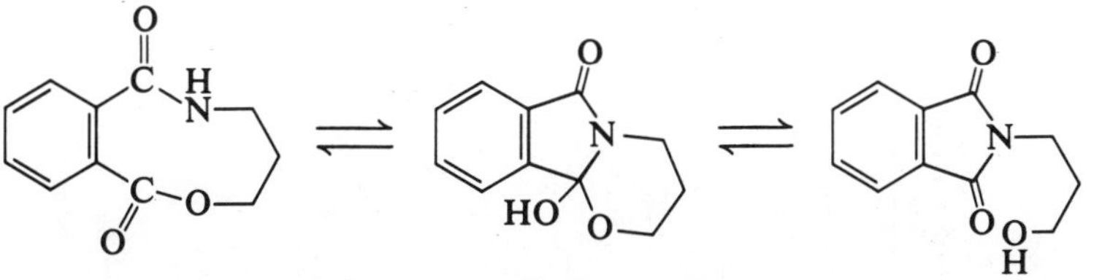

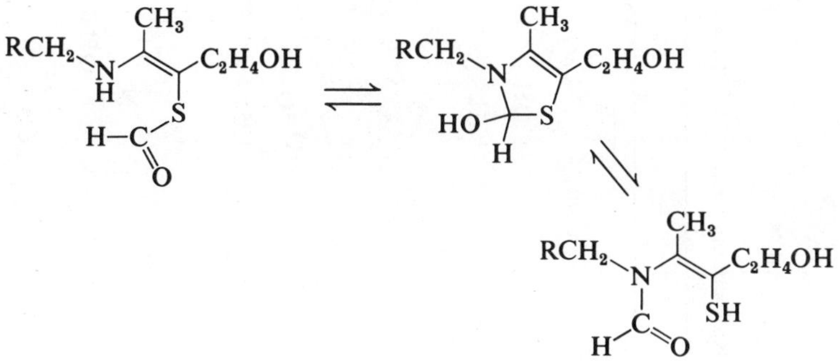

TABLE 19

Rate constants for the breakdown of nitrogen-containing tetrahedral intermediates at 25°[a]

	k_{H^+}[b]	k_{H_2O}[c]	k_{HO^-}[b]	pK_a	k_2[c,d]	Ref.
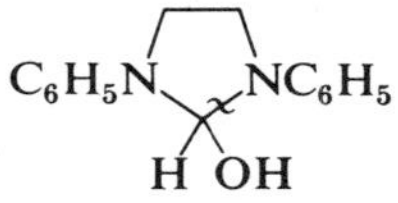 C₆H₅N⟩⟨NC₆H₅ H OH [124]	—	—	3.24×10^3	12.7	180	e
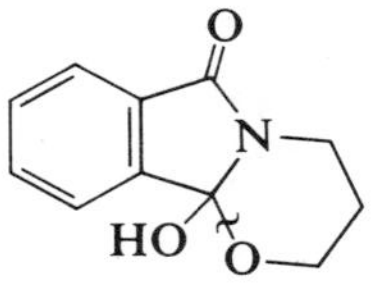 (bicyclic structure with N, O, HO) [125]	10.6	1.84	6.2×10^9	7.5	1.6×10^3	f
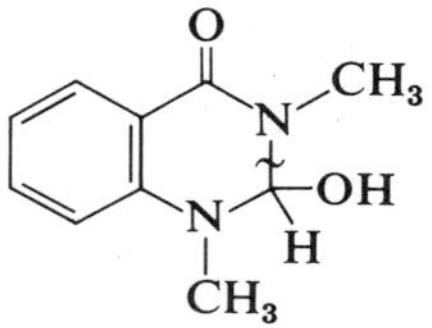 (structure with N-CH₃, OH, N-CH₃) [126]	—	—	5.6×10^4[h] 1.2×10^3[h]	11.06	64[h] 1.42[h]	g

90

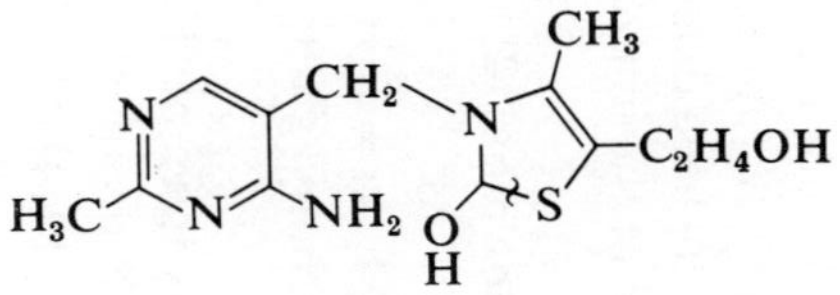

— 6.75 × 10⁴ ʲ — ⁱ

[structure 127, a thiazolidine derivative with pyrimidine substituent]

[127]

a Position of bond fission indicated by wavy line
b Units: $M^{-1} s^{-1}$
c Units: s^{-1}
d Rate constant for breakdown of the monoanion
e Robinson, 1968, 1970
f Gravitz and Jencks, 1974
g Tee et al., 1982
h Rate constants for breakdown into two different rotational isomers measured
i Chahine and Dubois, 1983
j At 5°

assuming that the effect of the "phthalimide nucleus" on the pK_a of the hydroxyl group was the same as on that of an amine (which had been measured independently) and a pK_a of 15.9 for ethanol (Gravitz and Jencks, 1974). Compounds [124] and [126] are the expected intermediates of N,N-acyl transfer reactions (13), (14); [125] is the expected intermediate of the O,N-acyl transfer reaction (15); and [127] is the expected intermediate in the S,N-acyl transfer reaction (16).

The second-order rate constants for the hydroxide-ion catalysed breakdown of compounds [124] and [126] (Table 19) are very much less than those for the breakdown of hemiorthoesters (cf. Tables 13–15) presumably as a result of the difficulty of expelling a nitrogen anion. Nevertheless both compounds break down more rapidly than rotation about the C—N bond of the amide product, since in both reactions rotational isomers of the products have been detected in non-equilibrium proportions by nmr spectroscopy, and their subsequent equilibration has been followed (Capon et al., 1981a; Tee et al., 1982). With [126] the rate constants for breakdown into both rotational isomers were measured, although their structures were uncertain.

Compound [125] breaks down with expulsion of an ionized oxygen group and the second-order constant k_{HO^-} is much larger than for [124] and [126]. This is probably partly due to the greater acidity of the hydroxyl and partly due to the greater ease of expulsion of ionized oxygen compared to ionized nitrogen. The exact proportion depends on the pK_a and Gravitz and Jencks estimated a very low value of 7.5. On this basis the $c.\ 10^5$ greater value of k_{HO^-} for the breakdown of [125] compared to [126] arises from a 3.6×10^3 more favourable pK_a and a 25-fold greater value of k_2, the rate constant for breakdown of the monoanion. If, however, the pK_a value of 7.5 were incorrect then these values would have to be adjusted accordingly.

Unlike with the breakdown of the more reactive hemiorthoesters a mechanism which involves a rapid and reversible ionization followed by a unimolecular breakdown of the monoanion appears to be a valid one for the hydroxide-ion catalysed breakdown of these nitrogen containing tetrahedral intermediates. Not only have the ionized forms of [124] and [126] been detected but the values of k_{-1}, the rate constant for reprotonation of the ionized form calculated as above, is always much greater than k_2.

The rate-constant for the hydronium-ion catalysed breakdown of [125] is $10.6\ M^{-1}s^{-1}$ at 25°, considerably less than the value of $3 \times 10^3\ M^{-1}s^{-1}$ for the breakdown of the six-membered cyclic hemiorthoester [74; Ar=C_6H_5]. Presumably the "push" of a nitrogen atom conjugated to a carbonyl group is less than that of an oxygen atom.

The most reasonable mechanism for the spontaneous breakdown of [125] is probably (17), which involves ionization followed by a hydronium-ion catalysed breakdown of the monoanion (Gravitz and Jencks, 1974). On the basis of a pK_a of 7.5, k_2^* ($=k_{H_2O}/K_a$) would have to be 5.9×10^7 M^{-1}s^{-1}, although if the pK_a were higher it might be necessary to postulate a concerted mechanism.

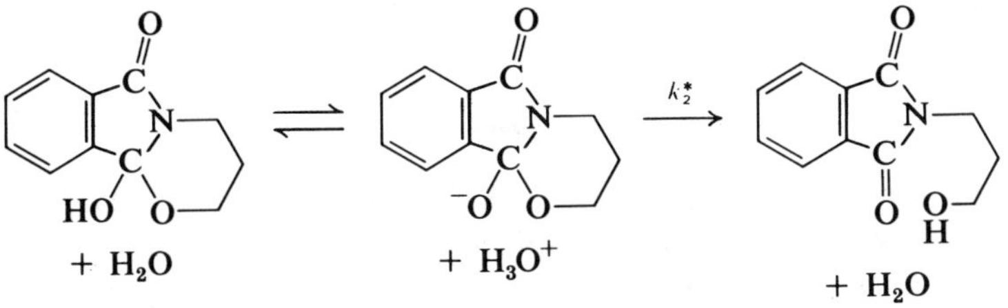

(17)

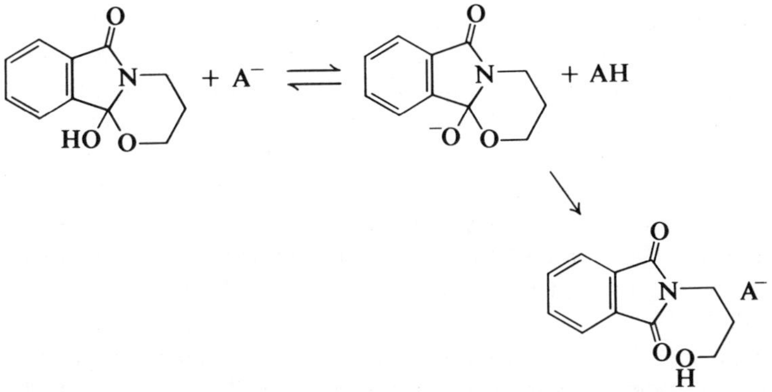

(18)

General-base catalysis was also detected in this reaction and attributed to general-acid catalysis of breakdown of the monoanion (18). The Brønsted β-value was 0.57 and so the α-value for expulsion of the alcohol from the ionized form is 0.43. The point for the water-catalysed reaction lies on the Brønsted plot for catalysis by other bases which supports the view that these reactions proceed by similar mechanisms as formulated in (17) and (18).

6 Future investigations

Although tetrahedral intermediates with the structure of a dialkyl hemiorthoester can now be detected with not too much difficulty there are several other types of tetrahedral intermediate which have yet to be detected. Some of these are shown in Scheme 12. Monoalkyl hemiorthoesters [128] would be the intermediates in the hydrolysis of esters. The calculations of Guthrie (1973) suggest that in acidic solutions they should have a reactivity similar to that of dialkyl hemiorthoesters. However, the problem of releasing the two free hydroxyl groups from a precursor sufficiently quickly so that any intermediate with one free hydroxyl group does not break down faster than the second hydroxyl group is formed has yet to be solved.

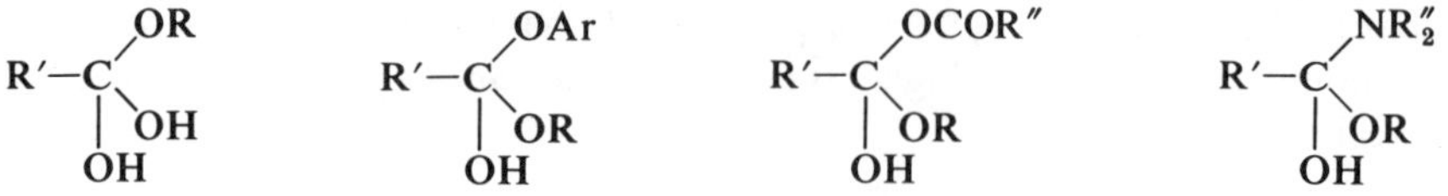

$$\underset{[128]}{\underset{OH}{\overset{OR}{R'-C}}\begin{array}{c}\\OH\end{array}} \qquad \underset{[129]}{\underset{OH}{\overset{OAr}{R'-C}}\begin{array}{c}\\OR\end{array}} \qquad \underset{[130]}{\underset{OH}{\overset{OCOR''}{R'-C}}\begin{array}{c}\\OR\end{array}} \qquad \underset{[131]}{\underset{OH}{\overset{NR''_2}{R'-C}}\begin{array}{c}\\OR\end{array}}$$

Scheme 12

As the leaving group of a tetrahedral intermediate becomes better the question arises as to whether it will become too unstable to exist. Compounds [129] and [130] which have aryloxy and acyloxy leaving groups would be the tetrahedral intermediates in the hydrolysis (R = H) or alcoholysis (R = alkyl) of aryl esters and anhydrides. In view of the very high reactivity of dialkyl hemiorthoesters in basic solution there must be some doubt as to whether [129] and especially [130], which have better leaving groups, are capable of existence under these conditions. It is perhaps significant that methyl vinyl hemiorthoformate which has a vinyloxy leaving group could not be detected in the hydrolysis of methoxyvinyloxymethyl acetate or chloroacetate (Capon et al., 1981b) (see Section 1).

Compound [131] would be the tetrahedral intermediate in the hydrolysis (R = H) or alcoholysis (R = alkyl) of an amide or the aminolysis of an ester (R = alkyl). As discussed earlier (Sections 1 and 5) structures of this type are known when R″ is a group capable of conjugating with the nitrogen and thus reducing its ability to expel RO⁻. However, when R″ is a simple alkyl group such as methyl, the calculations of Guthrie (1974b) suggest that in basic solutions the energy of activation for its breakdown would be very low and so it may be very difficult or impossible to detect this class of intermediate

References

Ahmad, M., Bergstrom, R. G., Cashen, M. J., Kresge, A. J., McClelland, R. A. and Powell, M. F. (1977). *J. Amer. Chem. Soc.* **99**, 4827

Ahmad, M., Bergstrom, R. G., Cashen, M. J., Chiang, Y., Kresge, A. J., McClelland, R. A. and Powell, M. F. (1979). *J. Amer. Che. Soc.* **101**, 2669

Ahmad, M., Bergstrom, R. G., Cashen, M. J., Chiang, Y., Kresge, A. J., McClelland, R. A. and Powell, M. F. (1982). *J. Amer. Chem. Soc.* **104**, 1156

Antonov, V. K., Shkrob, A. M. and Shemyakin, M. M. (1963). *Tetrahedron Lett.* 439

Antonov, V. K., Shkrob, A. M. and Shemyakin, M. M. (1965). *Zh. Obshch. Khim.* **35**, 1380; *J. Gen. Chem. USSR* **35**, 1385

Barnett, R. and Jencks, W. P. (1969). *J. Org. Chem.* **34**, 2777

Bender, M. L. (1951). *J. Amer. Chem. Soc.* **73**, 1626

Bender, M. L. (1953). *J. Amer. Chem. Soc.* **75**, 5986

Bender, M. L. (1960). *Chem. Rev.* **60**, 53

Bender, M. L. and Heck, H. d'A. (1967). *J. Amer. Chem. Soc.* **89**, 1211

Blackburn, G. M. and Jencks, W. P. (1968). *J. Amer. Chem. Soc.* **90**, 2638

Bladon, P. and Forrest, G. C. (1966). *Chem. Commun.* 481
Bobrański, B. and Śladowska, M. (1972). *Rocznicki Chem.* **46**, 451
Bordner, J., Thiessen, W. E., Bates, H. A. and Rapoport, H. (1975). *J. Amer. Chem. Soc.* **97**, 6008
Bull, H. G., Koehler, K., Pletcher, T. C., Ortiz, J. J. and Cordes, E. H. (1971). *J. Amer. Chem. Soc.* **93**, 3002
Burt, R. A., Chiang, Y., Kresge, A. J. and McKinney, M. A. (1982). *J. Amer. Chem. Soc.* **104**, 3685
Capon, B. and Dosunmu, M. I. (1984). *Tetrahedron* **40**, 3625
Capon, B. and Ghosh, A. K. (1981). *J. Amer. Chem. Soc.* **103**, 1765
Capon, B. and Grieve, D. McL. A. (1980). *J.C.S. Perkin II*, 300
Capon, B. and Nimmo, K. (1975). *J.C.S. Perkin II*, 1113
Capon, B. and Sanchez, M. de N. de M. (1983). *Tetrahedron* **39**, 4143
Capon, B., Nimmo, K. and Reid, G. P. (1976a). *J.C.S. Chem. Commun.* 871
Capon, B., Gall, J. H. and Grieve, D. McL. A. (1976b). *J.C.S. Chem. Commun.* 1034
Capon, B., Labbé, C. and Rycroft, D. S. (1979). *Can. J. Chem.* **57**, 2978
Capon, B., Ghosh, A. K. and Grieve, D. McL. A. (1981a). *Accounts Chem. Res.* **14**, 306
Capon, B., Rycroft, D. S., Watson, T. W. and Zucco, C. (1981b). *J. Amer. Chem. Soc.* **103**, 1761
Cerrini, S., Fedeli, W. and Mazza, F. (1971). *Chem. Commun.* 1607
Chahine, J. M. El H. and Dubois, J. E. (1983). *J. Amer. Chem. Soc.* **105**, 2335
Challand, S. R., Herbert, R. B. and Holliman, F. G. (1970). *Chem. Commun.* 1423
Chatterjee, A., Bose, S. and Ghosh, C. (1959). *Tetrahedron* **7**, 257
Chaturvedi, R. K. and Schmir, G. L., (1968). *J. Amer. Chem. Soc.* **90**, 4413
Chaturvedi, R. K., MacMahon, A. E. and Schmir, G. L. (1967). *J. Amer. Chem. Soc.* **89**, 6984
Chiang, Y., Kresge, A. J., Salomaa, P. and Young, C. I. (1974). *J. Amer. Chem. Soc.* **96**, 4494
Chiang, Y., Kresge, A. J., Lahti, M. O. and Weeks, D. P. (1983). *J. Amer. Chem. Soc.* **105**, 6852
Cipiciani, A., Linda, P. and Savelli, G. (1977). *J.C.S. Chem. Commun.* 857
Cipiciani, A., Linda, P., Savelli, G. and Bunton, C. A. (1981). *J. Amer. Chem. Soc.* **103**, 4874
Collet, H., Germain, A. and Commeyras, A. (1974). *Bull. Soc. Chim. Fr.* 279
Dahlqvist, K. I. and Forsén, S. (1969). *J. Phys. Chem.* **73**, 4124
Day, J. N. E. and Ingold, C. K. (1941). *Trans. Faraday Soc.* **37**, 686
Deslongchamps, P. (1975). *Tetrahedron* **31**, 2463
Deslongchamps, P., Cheriyan, U. O., Guida, A. and Taillefer, R. J. (1977). *Nouveau J. Chem.* **1**, 235
Deslongchamps, P., Barlet, R. and Taillefer, R. J. (1980). *Can. J. Chem.* **58**, 2167
DeWolfe, R. H. (1970). "Carboxylic Ortho Acid Derivatives", p. 440. Academic Press, New York
DeWolfe, R. H. and Jensen, J. L. (1963). *J. Amer. Chem. Soc.* **85**, 3264
Fastrez, J. (1977). *J. Amer. Chem. Soc.* **99**, 7004
Fedor, L. R. and Bruice, T. C. (1965). *J. Amer. Chem. Soc.* **87**, 4138
Fife, T. H. and Jao, L. K. (1965). *J. Org. Chem.* **30**, 1492
Fodor, G., Letourneau, F. and Mandava, N. (1970). *Can. J. Chem.* **48**, 1465
Fraenkel, G. and Watson, D (1975). *J. Amer. Chem. Soc.* **97**, 231
Funderburk, L. H., Aldwin, L. and Jencks, W. P. (1978). *J. Amer. Chem. Soc.* **100**, 5444

Gensmantel, N. P. and Page, M. I. (1979). *J.C.S. Perkin II*, 137
Goto, T., Kishi, Y., Takahashi, S. and Hirata, Y. (1965). *Tetrahedron* **21**, 2059
Grace, M. E., Loosemore, M. J., Semmel, M. L. and Pratt, R. F. (1980). *J. Amer. Chem. Soc.* **102**, 6784
Gravitz, N. and Jencks, W. P. (1974). *J. Amer. Chem. Soc.* **96**, 489
Grieve, D. McL. A. (1978). Ph.D. Thesis, p. 235. University of Glasgow
Griot, R. G. and Frey, A. J. (1963). *Tetrahedron* **19**, 1661
Guthrie, J. P. (1973). *J. Amer. Chem. Soc.* **95**, 6999
Guthrie, J. P. (1974a). *J. Amer. Chem. Soc.* **96**, 588
Guthrie, J. P. (1974b). *J. Amer. Chem. Soc.* **96**, 3608
Guthrie, J. P. (1976). *Can. J. Chem.* **54**, 202
Guthrie, J. P. (1977). *Can. J. Chem.* **55**, 3562
Guthrie, J. P. and Cullimore, P. A. (1980). *Can. J. Chem.* **58**, 1281
Hallas, M. D., Reed, P. B. and Martin, R. B. (1971). *Chem. Commun.* 1506
Hammett, L. P. (1940). "Physical Organic Chemistry", p. 357. McGraw-Hill, New York and London
Harron, J., McClelland, R. A., Thankachan, C. and Tidwell, T. T. (1981). *J Org. Chem.* **46**, 903
Hershfield, R. and Schmir, G. L. (1972a). *J. Amer. Chem. Soc.* **94**, 1263
Hershfield, R. and Schmir, G. L. (1972b). *J. Amer. Chem. Soc.* **94**, 6788
Hershfield, R. and Schmir, G. L. (1973). *J. Amer. Chem. Soc.* **95**, 8032
Hershfield, R., Yeager, M. J. and Schmir, G. L. (1975). *J. Org. Chem.* **40**, 2940
Hibbert, F. and Satchell, D. P. N. (1967) *J. Chem. Soc. (B)* 653
Hine, J., Ricard, D. and Perz, R. (1973). *J. Org. Chem.* **38**, 110
Hinshelwood, C. N. and Legard, A. R. (1935). *J. Chem. Soc.* 587
Jencks, W. P. (1964). *Prog. Phys. Org. Chem.* **2**, 104
Jencks, W. P. (1969). "Catalysis in Chemistry and Enzymology", pp. 465–90. McGraw-Hill, New York
Jencks, W. P. and Gilchrist, M. (1964). *J. Amer. Chem. Soc.* **86**, 5616
Jencks, W. P. and Gilchrist, M. (1968). *J. Amer. Chem. Soc.* **90**, 2622
Johnson, S. L. (1967). *Adv. Phys. Org. Chem.* **5**, 237
Jones, D S., Kenner, G. W. and Sheppard, R. C. (1963). *Experientia* **19**, 136
Kanchuger, M. S. and Byers, L. D. (1979). *J. Amer. Chem. Soc.* **101**, 3005
Kankaanperä, A. and Tuominen, H. (1967). *Suomen Kem.* **B40**, 271
Khouri, F. and Kaloustian, M. K. (1979). *J. Amer. Chem. Soc.* **101**, 2249
Kirby, A. J. (1972). *In* "Comprehensive Chemical Kinetics"(Bamford, C. H. and Tipper, C. F. H., eds) Vol. 10, Ester formation and hydrolysis and related reactions, p. 104. Elsevier, Amsterdam, Oxford and New York
Kresge, A. J. and Straub, T. S. (1983). *J. Amer. Chem. Soc.* **105**, 3957
Kuhn, R. and Weiser, D. (1957). *Angew. Chem.* **69**, 371
Kwart, H. and Price M. B. (1960). *J. Amer. Chem. Soc.* **82**, 5123
Lane, C. A., Cheung, M. F. and Dorsey, G. E. (1968). *J. Amer. Chem. Soc.* **90**, 6492
Levesque, G., Mahjoub, A. and Thuillier, A. (1978). *Tetrahedron Lett.* 3847
Lienhard, G. E. and Jencks, W. P. (1966). *J. Amer. Chem. Soc.* **88**, 3982
Lowry, T. M. (1926). *2ième Cons. Chim. Inst. Intern. Chim. Solvay*, 135
Lucente, G. and Romeo, A. (1971). *Chem. Commun.* 1605
McCapra, F. and Leeson, P. (1976). *J.C.S. Chem. Commun.* 1037
McCapra, F. and Leeson, P. (1979). *J.C.S. Chem. Commun.* 114
McClelland, R. A. (1975). *J. Amer. Chem. Soc.* **97**, 5281
McClelland, R. A. (1976). *J. Org. Chem.* **41**, 2776
McClelland, R. A. (1978). *J. Amer. Chem. Soc.* **100**, 1844

McClelland, R. A. and Ahmad, M. (1979). *J. Org. Chem.* **44**, 1855
McClelland, R. A. and Alibhai, M. (1981). *Can. J. Chem.* **59**, 1169
McClelland, R. A. and Patel, G. (1981a). *J. Amer. Chem. Soc.* **103**, 6908
McClelland, R. A. and Patel, G. (1981b). *J. Amer. Chem. Soc.* **103**, 6912
McClelland, R. A. and Santry, L. J. (1983). *Accounts Chem. Res.* **16**, 394
McClelland, R. A. and Somani, R. (1979). *J.C.S. Chem. Commun.* 407
McClelland, R. A., Ahmad, M., Bohonek, J. and Gedge, S. (1979a). *Can. J. Chem.* **57**, 1531
McClelland, R. A., Somani, R. and Kresge, A. J. (1979b). *Can. J. Chem.* **57**, 2260
McClelland, R. A., Gedge, S. and Bohonek, J. (1981). *J. Org. Chem.* **46**, 886
McClelland, R. A., Seaman, N. E. and Cramm, D. (1984). *J. Amer. Chem. Soc.* **106**, 4511
Mager, H. I. X. and Addink, R. (1979). *Tetrahedron Lett.* 3545
Milstein, S. and Cohen, L. A. (1969). *J. Amer. Chem. Soc.* **91**, 4585
Milstein, S. and Cohen, L. A. (1970). *J. Amer. Chem. Soc.* **92**, 4377
Moffat, A. and Hunt, H. (1959). *J. Amer. Chem. Soc.* **81**, 2082
Nakamura, S., Maeda, K. and Umezawa, H. (1964). *J. Antibiotics (A)*, **17**, 33
Okada, K., Sakuma, H. and Inoue, S (1979). *Chem. Lett.* **1979**, 131
Olsson, K., Adolfsson, L. and Anderson, R. (1976). *Chem. Scripta* **10**, 122
Pachter, I. J. and Suld, G. (1960). *J. Org. Chem.* **25**, 1680
Patchornik, A., Berger, A., Katchalski, E. (1957). *J. Amer. Chem. Soc.* **79**, 6416
Pocker, Y. and Green, E. (1974). *J. Amer. Chem. Soc.* **96**, 166
Przystas, T. J. and Fife, T. H. (1981). *J. Amer. Chem. Soc.* **103**, 4884
Robinson, D. R. (1968). *Tetrahedron Lett.* 5007
Robinson, D. R. (1970). *J. Amer. Chem. Soc.* **92**, 3138
Robinson, R. A. and Stokes, R. H. (1959). "Electrolyte Solutions", 2nd edn, p. 544. Butterworths, London
Rogers, G. A. and Bruice, T. C. (1973). *J. Amer. Chem. Soc.* **95**, 4452
Rogers, G. A. and Bruice, T. C. (1974). *J. Amer. Chem. Soc.* **96**, 2473, 2481
Rolfe, A. C. and Hinshelwood, C. N. (1934). *Trans. Faraday Soc.* **30**, 935
Rothe, M. and Steinberger, R. (1968). *Angew. Chem. Int. Ed. Engl.* **7**, 884
Rothe, M. and Steinberger, R. (1970a). *Tetrahedron Lett.* 649
Rothe, M. and Steinberger, R. (1970b). *Tetrahedron Lett.* 2467
Samuel, D. and Silver, B. L. (1965). *Adv. Phys. Org. Chem.* **3**, 123
Santry, L. J. and McClelland, R. A. (1983a). *J. Amer. Chem. Soc.* **105**, 3167
Santry, L. J. and McClelland, R. A. (1983b). *J. Amer. Chem. Soc.* **105**, 6138
Schantz, E. J., Ghazarossian, V. E., Schnoes, H. K., Strong, F. M., Springer, J. P., Pezzanite, J. O. and Clardy, J. (1975). *J. Amer. Chem. Soc.* **97**, 1238
Scherowsky, G. and Matloubi, H. (1978). *Annalen* 98
Scheuer, P. J. (1977). *Accounts Chem. Res.* **10**, 33
Schmir, G. L. (1968). *J. Amer. Chem. Soc.* **90**, 3478
Schmir, G. L., and Cunningham, B. A. (1965). *J. Amer. Chem. Soc.* **87**, 5692
Schmir, G. L., and Cunningham, B. A. (1966). *J. Amer. Chem. Soc.* **88**, 551
Schowen, R. L., Jayaraman, H. and Kershner, L. (1966). *J. Amer. Chem. Soc.* **88**, 3373
Shain, S. A. and Kirsch, J. F. (1968). *J. Amer. Chem. Soc.* **90**, 5848
Shemyakin, M. M., Antonov, V. K., Shkrob, A. M., Sheinker, Y. N. and Senyavina, L. B. (1962). *Tetrahedron Lett.* 701
Shemyakin, M. M., Antonov, V. K., Shkrob, A. M., Shchelokov, V. I. and Agadzhanyan, Z. E. (1965). *Tetrahedron* **21**, 3537
Sheppard, R. C. (1963). *Experientia* **19**, 125

Smith, S. G. and Feldt, R. J. (1968). *J. Org. Chem.* **33**, 1022
Stauffer, C. E. (1972). *J. Amer. Chem. Soc.* **94**, 7887
Stauffer, C. E. (1974). *J. Amer. Chem. Soc.* **96**, 2489
Straub, T. S. (1970). Ph.D. Thesis. Illinois Institute of Technology, *Diss. Abs. Int. B.* **30**, 5439
Takagi, M., Ishihara, R. and Matsudu, T. (1977). *Bull. Chem. Soc. Jpn.* **50**, 2193
Taylor, P. J. (1968). *Chem. Commun.* 968
Tee, O. S., Trani, M., McClelland, R. A. and Seaman, N. E. (1982). *J. Amer. Chem. Soc.* **104**, 7219
Temple, C., McKee, R. L. and Montgomery, J. A. (1965). *J. Org. Chem.* **30**, 829
Thornton, E. R. (1967). *J. Amer. Chem. Soc.* **89**, 2915
Tomoto, N., Boyle, W. J. and Bunnett, J. F. (1972). *J. Org. Chem.* **37**, 4315
Vinnik, M. I. and Moiseyev, Y. V. (1963). *Tetrahedron* **19**, 1441
Weeks, D. P. (1965). quoted by Bunnett, J. F. and Hauser, C. F. *J. Amer. Chem. Soc.* **87**, 2214 (1965)
Wong, J. L., Oesterlin, R. and Rapoport, H. (1971). *J. Amer. Chem. Soc.* **93**, 7344
Woodward, R. B. (1964). *Pure Appl. Chem.* **9**, 49
Woodward, R. B. and Gougoutas, J. Z. (1964). *J. Amer. Chem. Soc.* **86**, 5030

A General Approach to Organic Reactivity: The Configuration Mixing Model

ADDY PROSS

Department of Chemistry, Ben-Gurion University of the Negev, Beer Sheva, Israel

1 Introduction 99
2 Theoretical background 102
    VB configurations 103
    MO configurations 107
    MO–VB correspondence. Two-Electron bonds 110
    Reaction profiles 112
3 Applications 139
    Photochemical processes 139
    Nucleophilic substitution 145
    Elimination 161
    Proton transfer 167
    Pericyclic reactions 173
4 General consequences 177
    The significance of the Brønsted parameter 177
    Relation to Marcus theory 182
    The relationship between geometric and charge progression along the reaction co-ordinate 186
5 Conclusion 190
Acknowledgements 191
References 191

1 Introduction

The question of organic reactivity and the factors that control it remain poorly understood. As a consequence, a significant body of organic chemistry remains within the realm of empirical knowledge. What is quite paradoxical, however, is the fact that this lack of understanding exists, even though the underlying quantum mechanical principles governing all of chemical behaviour have been remarkably well understood for at least half a century. The reason for this apparent anomaly is that the mathematical complexity associated with the solution of the Schrödinger equation has meant that, until about 1970, solutions to practical problems were out of reach. More recently, though, modern computer technology and sophisticated programming has provided theoreticians with a unique opportunity to test the validity of the quantum-mechanical principles by comparing the computational prediction with experiment. This meeting of experiment and theory can only be described as a resounding success: both questions of stucture and energy, the pillars of modern chemistry, were shown to be tractable computationally so that in many cases theoretical calculations on new systems are in themselves considered "experimental" data, and often stand shoulder-to-shoulder in the literature with genuine experimental results. Indeed, cases exist where theory has been known to improve on or even invalidate experimental data (at least incorrectly obtained data) so that today the standing of computational procedures, in exploring new frontiers in chemistry, is rightfully high. However, this obvious achievement has failed to resolve the basic problem; the success of the computational procedure has not, in itself, satisfied the chemist's need to understand. The calculation on its own does not provide any insight as to the reason behind the result. The very complexity of the sophisticated computational procedures detaches much of the mathematical process from human comprehension so that while the computation on its own may provide an answer to a "what" question it does not necessarily provide one to a "why" question. As Hoffmann (Libit and Hoffmann, 1974) has noted, "the problem is understanding why the calculation came out the way it did." This important task must be performed by qualitative theory.

Modern organic chemistry abounds with qualitative theories. These range from simple rules, such as "electronegative substituents generally increase the acidity of a substrate," to more theoretically based models such as the Frontier Molecular Orbital (FMO) theory [for a review see, Fleming (1976)], which has revolutionized current chemical thinking. Yet, existing models fail to clarify the mechanism of formation of activation barriers, and for this reason must be limited in their ability to tackle general problems of reactivity.

The purpose of this review is to present in some detail a simple, qualitative framework for understanding the factors which go into the creation of a reaction profile. Since the question raised is so fundamental – "What determines the barrier in any chemical reaction?" – the model encompasses within its single structure reactions as different as electron-transfer reactions, e.g., (1) and (2), nucleophilic substitution (3), cycloadditions (4), proton transfer (5), nucleophile–electrophile combination (6), elimination reactions (7), to mention a few. Thus, to quote just one example, the very same elements are utilized to explain why [2 + 2] cycloadditions are forbidden and [4 + 2] allowed, and why the Brønsted parameter for nitroalkane deprotonation is anomalous. In addition it brings under one roof both thermal *and* photochemical processes. The theory which we have termed the Configuration Mixing (CM) model, or when valence-bond configurations are solely utilized, the Valence-Bond Configuration Mixing (VBCM) model (Pross and Shaik, 1983) has been applied over recent years to many of the above mentioned reaction types (Pross and Shaik, 1981, 1982a,b; McLennan and Pross, 1984; Shaik, 1983). Others will be described here.

$$Na + Cl \rightarrow Na^+ + Cl^- \qquad (1)$$

$$Fe^{2+} + Fe^{3+} \rightarrow Fe^{3+} + Fe^{2+} \qquad (2)$$

$$I^- + CH_3Br \rightarrow ICH_3 + Br^- \qquad (3)$$

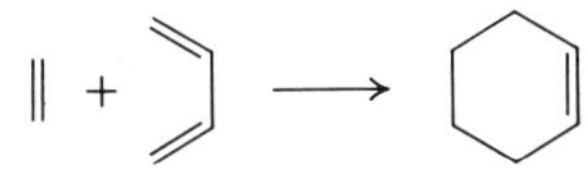

$$\| + \big\rangle \longrightarrow \bigcirc \qquad (4)$$

$$HO^- + CH_3NO_2 \rightarrow HOH + CH_2NO_2^- \qquad (5)$$

$$(Ph)_3C^+ + Br^- \rightarrow (Ph)_3C\text{—Br} \qquad (6)$$

$$HO^- + PhCH_2CH_2Cl \rightarrow HOH + PhCH=CH_2 + Cl^- \qquad (7)$$

The essence of the model is to build up any reaction profile through the qualitative mixing of appropriate electronic configurations. Thus, just as the properties of molecular orbitals (energy, geometry) may be readily appreciated by analysis of the atomic orbitals from which they are derived, the essence of a reaction profile may be more readily comprehended by understanding the building blocks from which it is constructed.

In order to justify any new theory it must be able to explain existing facts and be able to make predictions that conventional theories are unable to make. In this endeavour we feel the VBCM model is successful. The theory provides a basis for understanding a variety of general problems in physical organic chemistry as well as providing specific answers to specific questions. Pleasingly, the model overlaps with other existing theories, primarily the Marcus relation (Marcus, 1964, 1977) and the potential energy surface models popularized by More O'Ferrall (1970), Thornton (1967), and Jencks (1972). The model, in addition, however, tackles a variety of questions not directly addressed by these theories.

Some typical questions of a general nature that the VBCM model attempts to resolve include:

1   When do rates and equilibria correlate in organic chemistry, and why do rate–equilibrium relationships break down?
2   What is the significance of anomalous Brønsted parameters?
3   How applicable is the Marcus rate equation to organic chemistry?

4   What is the relationship between charge and geometric progression in a reaction transition state?
5   Why are some reactions concerted while others proceed through an intermediate?
6   What is the molecular significance of LFER parameters such as the Brønsted parameter, α, or the Hammett ρ parameter?
7   What factors govern the heterolytic versus homolytic dissociation of a bond?
8   Are intermediates permissible in reactions allowed according to the Woodward–Hoffmann rules?

In the course of this review, these questions and others will be discussed, and, hopefully, will lend substance to our view that CM theory is a useful conceptual framework on which the organic chemist may relate much of the large body of existing empirical knowledge.

In view of the fundamental role played by quantum theory in modern chemistry it is not surprising that the theoretical basis of the CM model goes back to the earliest days of quantum chemistry. In fact the basic concept of CM theory may be considered a theoretical precursor of resonance theory. Not surprisingly, we are of the view that resonance theory plays and will continue to play a dominant role in chemical thinking. To quote Salem (1982) in the opening paragraph of his monograph: "Resonance is the conceptual heart of chemistry. The notion that a molecule 'hesitates' between different structures, borrows its characteristics from all of these and finally adopts a structure that is somewhat intermediate between them, is central to understanding electronic behaviour." This author fully concurs. It is from this single yet powerful principle that CM theory borrows its utility and generality.

2   Theoretical background

In order to develop a qualitative model capable of tackling a wide range of organic reactivity problems, we require the means of generating a simple, qualitative description of any organic reaction profile. The basic principles are well-established since they go back to the early days of quantum mechanics. Early contributions are due to Ogg and Polanyi (1935), Evans and Warhurst (1938), Evans and Polanyi (1938), Evans (1939), Baughan and Polanyi (1941) and Laidler and Shuler (1951). In fact, one cannot but marvel at the extraordinary chemical insight shown by these early workers at a stage in which much experimental chemistry was still to be uncovered.

More recent work which has strongly influenced the present ideas includes that of Mulliken and Person (1969), Nagakura (1963), Longuet-Higgins and Abrahamson (1965), Woodward and Hoffmann (1970), van der Lugt and

Oosterhoff (1969), Michl (1974), Dauben *et al.* (1975), Salem *et al.* (1975), Fukui (1975), Epiotis (1978) and Warshel and Weiss (1980) though this list is no more than a selection of many relevant papers. The essence of the approach is to use electronic configurations as the building blocks from which the entire reaction profile, including ground states, intermediates and transition states, may be generated.

Let us begin by briefly examining two relatively accessible points on the reaction profile, the reactants and the products. As these points represent ground states on the reaction surface, the treatment is well known. We will see, however, that the same basic principles carry over into generating less accessible points on the reaction profile, such as the reaction transition state.

At any given geometry a molecule, or group of molecules, may exist in a number of predetermined *states*. Lowest in energy is the *ground state* and above the ground state lie a series of *excited states* which are often accessible through photochemical excitation. Any given state may be approximately described by a particular electronic *configuration*. A configuration is characterized by the specified distribution of available electrons to either atomic orbitals (for VB configurations) or molecular orbitals (for MO configurations). Indeed, the CM model may be employed using either VB or MO configurations. Let us illustrate these in turn.

VB CONFIGURATIONS

The state of a given system may be conveniently represented by the major VB configuration which describes that state. For example, a VB configuration which gives a rough representation of the ground state of the R—X bond is (8). This is just the well-known Heitler–London VB wave-function (Heitler

$$\psi_{cov} = (1/\sqrt{2})\{|\varphi_R(1)\bar{\varphi}_X(2)| - |\bar{\varphi}_R(1)\varphi_X(2)|\} \qquad (8)$$

and London, 1927) which allocates one electron to an atomic (or hybrid) orbital on R, φ_R, and one to an orbital on X, φ_X. The bar over the wave function signifies electron spin α; absence of a bar signifies electron spin β.

This configurational description of the R—X bond is, of course, only a crude one – even the polarity of such a bond is not described by ψ_{cov} – and the R—X bond strength will be greatly underestimated by such a wave-function. A superior description of the R—X bond may be obtained by the mixing in of additional configurations. These additional configurations are simply obtained by redistributing the relevant valence electrons in the available orbitals, φ_R and φ_X. In this way two higher energy ionic configurations, $\psi_{ion(1)}$ (9) and $\psi_{ion(2)}$ (10), are generated. Thus a better description of the

$$\psi_{ion(1)} = |\varphi_X(1)\bar{\varphi}_X(2)| \quad (9)$$

$$\psi_{ion(2)} = |\varphi_R(1)\bar{\varphi}_R(2)| \quad (10)$$

R—X bond is obtained through a linear combination of ψ_{cov}, $\psi_{ion(1)}$ and $\psi_{ion(2)}$ as indicated in (11), where a, b and c are coefficients.

$$\psi_{R-X} = a\psi_{cov} + b\psi_{ion(1)} + c\psi_{ion(2)} \quad (11)$$

This mixing of configurations is commonly termed *configuration interaction*. Since application of the configuration mixing model is non-mathematical, the cumbersome mathematical description of various wave-functions in (8)–(11) may be expressed more simply using structural formulae and placing the valence electrons adjacent to the appropriate atom. The result is a pictorial and chemically meaningful representation. Thus (8), (9) and (10) now become (12), (13) and (14).

$$\psi_{cov} \equiv R\cdot\uparrow\downarrow\cdot X \longleftrightarrow R\cdot\downarrow\uparrow\cdot X, \text{ or more simply } R\cdot\ \cdot X \quad (12)$$

$$\psi_{ion(1)} \equiv R^+ :X^- \quad (13)$$

$$\psi_{ion(2)} \equiv R:^- X^+ \quad (14)$$

Similarly ψ_{R-X} (11) becomes:

$$\psi_{R-X} = a(R\cdot\ \cdot X) + b(R^+ :X^-) + c(R:^- X^+) \quad (15)$$

In the course of this review, wave-functions will almost entirely be described using this chemically lucid method.

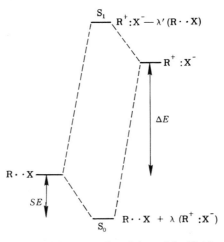

FIG. 1 Energy diagram that illustrates the mixing of the Heitler–London configuration, R· ·X, with the zwitterionic configuration, $R^+ :X^-$, to generate ground (S_0) and excited (S_1) states of the R—X bond

CONFIGURATION MIXING MODEL

The same configurations that are utilized to generate the *ground* state may also be utilized to generate the various possible *excited* states. Let us see how this is done for the R—X bond using, for simplicity, two configurations, R· ·X, the covalent configuration, and R$^+$:X$^-$, the lower energy ionic configuration.

The interaction of two configurations, R· ·X and R$^+$:X$^-$, may be illustrated using a simple interaction diagram. The diagram is similar to that used to generate molecular orbitals from atomic orbitals and is shown in Fig. 1. The high energy ionic configuration R$^+$:X$^-$ mixes into the low energy covalent configuration R· ·X in a *stabilizing* manner to generate the ground state of the R—X bond, S_0, while the low energy covalent form mixes into the high energy ionic form in a *destabilizing* manner to generate the first excited state of the R—X bond, S_1. S_0 and S_1, based solely on the mixing of two configurations, are represented by (16) and (17). We see therefore that

$$(R\text{—}X)_{S_0} \sim (R\cdot \cdot X) + \lambda(R^+ :X^-) \qquad (16)$$

$$(R\text{—}X)_{S_1} \sim (R^+ :X^-) - \lambda'(R\cdot \cdot X) \qquad (17)$$

the mixing of resonance forms can take place in an *in-phase* manner to generate a stabilized, ground state and in an *out-of-phase* manner so as to generate an excited state. So, just as the interaction of two atomic orbitals generates two molecular orbitals, the interaction of two configurations generates two *states*. Of course, mixing in of the third configuration, i.e. R:$^-$ X$^+$, will

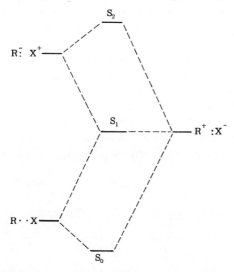

FIG. 2 Energy diagram showing the mixing of three configurations of RX, R· ·X, R$^+$:X$^-$, and R:$^-$ X$^+$, to generate ground (S_0), first excited singlet (S_1), and second excited singlet (S_2), states of the R—X bond

result in a more complex interaction diagram (Fig. 2) in which, not one, but two excited states are generated, again in analogy with the interaction of orbitals. The number of states generated will equal the number of configurations which are allowed to interact.

The equations that govern the way configurations interact are essentially the same as those used for interacting orbitals (Libit and Hoffmann, 1974). The stabilization energy *(SE)*, induced in R· ·X through interaction with R$^+$:X$^-$ (Fig. 1) is given by (18), where β is the matrix element defined by (19) and ΔE is the energy difference between the interacting configurations.

$$SE = \beta^2/\Delta E \tag{18}$$

$$\beta = <(R\cdot\ \cdot X)|H|(R^+ :X^-)> \tag{19}$$

The form of β may be greatly simplified since the matrix element (19) is roughly equal to the matrix element of those orbitals within the two interacting configurations that differ in one-electron occupancy (Salem *et al.*, 1975; Epiotis, 1978; Epiotis and Shaik, 1978a). In the present case the relevant orbitals are φ_R and φ_X, the hybrid orbitals on R and X respectively. Thus the new relation for β becomes (20). Since this term is strongly dependent on the

$$\beta = <\varphi_R|H|\varphi_X> \tag{20}$$

overlap between φ_R and φ_X, it is often approximated by (21), where S_{RX} is the overlap integral between φ_R and φ_X and K is an energy constant.

$$<\varphi_R|H|\varphi_X> = KS_{RX} \tag{21}$$

In words, the above relations merely state that the stabilization of the lower energy configuration, *SE*, is roughly proportional to the square of the overlap integral of the relevant atomic orbitals within the two configurations that differ in one-electron occupancy, and is inversely proportional to the energy gap between the interacting configurations.

The mixing parameter, λ, that measures the extent to which the high energy configuration, R$^+$:X$^-$, mixes into the low energy configuration,

$$\lambda = \beta/\Delta E \tag{22}$$

R· ·X, is given by (22). On the basis of (20) and (21) this becomes (23)

$$\lambda = KS_{RX}/\Delta E \tag{23}$$

which, expressed in words, states that the extent to which the high-energy configuration mixes into the low-energy one is proportional to the overlap and inversely proportional to the energy gap.

The above perturbation equations are all based on the assumption of a large energy gap between the interacting configurations. When the two configurations become degenerate, (18) and (22) are no longer applicable. In that case SE is just equal to β (24), and the two states, ψ_G and ψ_E, are described by

$$SE = \beta \qquad (24)$$

$$\psi_G = (1/\sqrt{2})(\chi_A + \chi_B) \qquad (25)$$

$$\psi_E = (1/\sqrt{2})(\chi_A - \chi_B) \qquad (26)$$

(25) and (26), where χ_A and χ_B represent the interacting configurations. The corresponding interaction diagram is illustrated in Fig. 3.

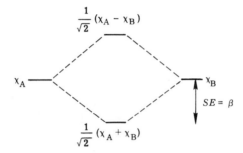

FIG. 3 Energy diagram that illustrates the mixing of two degenerate configurations, χ_A and χ_B, to give ground $[(1/\sqrt{2})(\chi_A + \chi_B)]$, and excited $[(1/\sqrt{2})(\chi_A - \chi_B)]$, states

It is worth noting that, while the various states of a given system may be generated through a linear combination of configurations, for many qualitative applications any given state may be approximated by the major configuration contributing to it. Thus, returning to our example of the R—X bond, the ground state may be approximated by the covalent configuration, R· ·X, while R$^+$:X$^-$ is a crude approximation for the first excited singlet state, and R:$^-$ X$^+$, that for the second excited singlet state.

MO CONFIGURATIONS

In the previous section we saw how a linear combination of VB configurations may be utilized in order to generate states. The same procedure may also be utilized with MO configurations. The theoretical equivalence of these separate routes will be illustrated in the next section.

The starting point for the MO procedure is the donor (D)–acceptor (A) model originally proposed by Mulliken (Mulliken, 1952a,b; Mulliken and Person, 1969). Let us consider the possible electronic MO configurations of a donor and acceptor pair. The donor molecule is represented by an electron pair in the HOMO orbital, n, and the acceptor molecule by a two orbital system, an occupied σ orbital and an unoccupied σ* orbital. The choice of a σ-system here is arbitrary; a π-system could serve equally well. Thus the ground state of the D–A system as a whole may be represented by the DA configuration as indicated in (27) and Scheme 1. Additional configurations

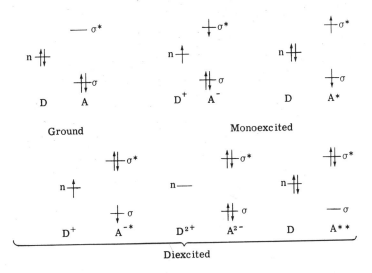

Scheme 1: Ground and excited configurations for a donor (D)–acceptor (A) system

may be obtained through the reorganization of the four valence electrons among the three available orbitals (n, σ, and σ*). In this way a charge-transfer configuration, D^+A^-, is obtained by the transfer of an electron from n to σ* (28), and a locally excited acceptor configuration, DA^*, by the internal excitation of a σ electron to σ* (29). These are the two monoexcited configurations. Diexcited configurations, D^+A^{-*}, $D^{2+}A^{2-}$ and DA^{**}, may be obtained by the excitation of two electrons. Thus the six MO configurations may be represented by (27)–(32), and these are illustrated in Scheme 1.

$$DA \equiv n^2\sigma^2 \quad (27)$$

$$D^+A^- \equiv n^1\sigma^2\sigma^{*1} \quad (28)$$

$$DA^* \equiv n^2\sigma^1\sigma^{*1} \quad (29)$$

$$D^+A^{-*} \equiv n^1\sigma^1\sigma^{*2} \quad (30)$$

$$D^{2+}A^{2-} \equiv n^0\sigma^2\sigma^{*2} \quad (31)$$

$$DA^{**} \equiv n^2\sigma^0\sigma^{*2} \quad (32)$$

The actual ground *state* of the A molecule (and hence the DA moiety as well) is only approximately represented by the σ^2 (or $n^2\sigma^2$ for DA) configuration. This is directly analogous to the situation in which the R· ·X configuration is an approximate description of the R—X bond [see (15)]. In order to obtain a better description of the ground state of the A molecule, excited configurations need to be mixed into the ground configuration. Thus for the case in which σ represents the R—X bond the ground state will be more accurately described by (33).

$$R—X = \sigma^2 - \lambda\sigma^{*2} \qquad (33)$$

In other words, mixing in of the diexcited configuration, σ^{*2} into the ground configuration provides a more accurate wave-function that describes the R—X bond. The way in which (33) is obtained will be described in the next Section (p. 110).

The most common application of MO configuration mixing is that used to describe charge transfer complexes. If a donor molecule, D, possessing a high energy HOMO (low ionization potential, I_D) interacts with an acceptor molecule, A, possessing a low-energy LUMO (high electron affinity, A_A) then these two molecules are likely to generate a charge transfer complex, as illustrated in Fig. 4. The complex comes about through interaction of the ground (or no-bond) configuration, DA, with the charge transfer configuration, D^+A^-.

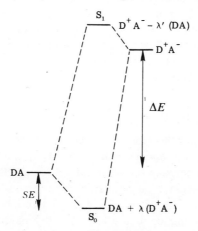

FIG. 4 Energy diagram showing the mixing of a ground configuration, DA, with a charge transfer configuration, D^+A^-, to generate ground, S_0, and first excited, S_1, states of the charge transfer complex. The stabilization energy, SE, of the complex is governed by (18).

It is only through configuration mixing that the stability of the complex with respect to separated donor and acceptor may be readily understood

(Mulliken and Person, 1969). The stabilization energy, SE, is given by (18) where β is now given by (34). The energy gap, ΔE, is given by $I_D - A_A$.

$$\beta = <DA|H|D^+A^-> \tag{34}$$

Since DA and D^+A^- are related by a single electron transfer, (34) may be simplified to (35) in analogy to (21). Thus the stability of charge transfer

$$\beta = KS_{n,\sigma^*} \tag{35}$$

complexes is enhanced by large donor HOMO-acceptor LUMO overlap (the β term) and a small HOMO–LUMO energy gap (the ΔE term).

The configuration mixing procedure means that the ground *state* of the complex is described by (36) while the excited state is described by (37). The

$$S_0 = DA + \lambda(D^+A^-) \tag{36}$$

$$S_1 = D^+A^- - \lambda'(DA) \tag{37}$$

charge transfer band that characterizes charge transfer complexes comes about from the excitation of the S_0 state to the S_1 state.

Note that while the ground state of the complex may be approximately described by the DA configuration, this is not an accurate description since the complex does have some charge transfer character mixed into the ground state. In similar fashion the excited charge transfer state is largely but not entirely described by the charge transfer configuration, D^+A^-. We see therefore that the subject of charge transfer complexes may be viewed as an exercise in configuration mixing.

MO–VB CORRESPONDENCE. TWO-ELECTRON BONDS

In the previous section we have indicated that both MO and VB configurations may be used to generate states. In this section we demonstrate the precise relationship between the MO and VB forms (Shaik, 1981; Pross and Shaik, 1981). We begin by considering the relationship between the MO description of the R—X bond, σ^2, and the VB description, R· ·X.

The acceptor ground configuration, A, may be expressed in terms of MO's [in analogy with (27)] as (38). The σ orbital may be described by its constituent hybrid orbitals on R and X, φ_R and φ_X respectively, as (39).

$$A \equiv \sigma^2 \tag{38}$$

$$\sigma = (a\varphi_R + b\varphi_X) \qquad (b > a) \tag{39}$$

Substituting (39) into (38) we may write (40).

CONFIGURATION MIXING MODEL

$$A \equiv \sigma^2 = |(a\varphi_R + b\varphi_X)(1)\overline{(a\varphi_R + b\varphi_X)}(2)| \tag{40}$$

Multiplying this out,

$$\sigma^2 = a^2|\varphi_R(1)\overline{\varphi_R}(2)| + b^2|\varphi_X(1)\overline{\varphi_X}(2)| + ab\{|\varphi_R(1)\overline{\varphi_X}2)| - |\overline{\varphi_R}(1)\varphi_X(2)|\} \tag{41}$$

Since

$$|\varphi_R(1)\overline{\varphi_R}(2)| \equiv R{:}^- \ X^+ \tag{42}$$

$$|\varphi_X(1)\overline{\varphi_X}(2)| \equiv R^+ {:}X^- \tag{43}$$

and

$$\frac{1}{\sqrt{2}}\{|\varphi_R(1)\overline{\varphi_X}(2)| - |\overline{\varphi_R}(1)\varphi_X(2)|\} \equiv R{\cdot}\ {\cdot}X \tag{44}$$

σ^2 can now be expressed as (45).

$$A \equiv \sigma^2 = \sqrt{2}ab(R{\cdot}\ {\cdot}X) + b^2(R^+{:}X^-) + a^2(R{:}^-\ X^+) \tag{45}$$

Thus we see that the MO configuration, σ^2, has a precise VB equivalent. What is more, *any* MO representation may be converted to its VB analogue, and vice versa, by simply describing the MO configuration in terms of the atomic or hybrid orbitals from which it is composed. It follows, therefore, that just as $R{\cdot}\ {\cdot}X$ is a poor VB representation of the R—X bond because it does not take into account ionic contributions, σ^2 as represented by (41), is also seen to be an unsatisfactory MO wave-function since it places an excessive emphasis on ionic configurations where the two electrons lie in the same hybrid orbital, and hence repel each other.

Allowing MO configuration mixing of the A** configuration (i.e. σ^{*2}) leads to an *improved* MO wave-function. Since $\sigma^* = (-b\varphi_R + a\varphi_X)$, then in the same way that (41) was derived for σ^2 we may derive expression (46) for $\sigma^2 - \lambda\sigma^{*2}$. We see in (46) that the coefficient for the covalent configuration,

$$\sigma^2 - \lambda\sigma^{*2} = \sqrt{2}ab(1 + \lambda)(R{\cdot}\ {\cdot}X) + (b^2 - \lambda a^2)(R^+{:}X^-) + (a^2 - \lambda b)(R{:}^-\ X^+) \tag{46}$$

$R{\cdot}\ {\cdot}X$, has been *increased*, while those for the ionic configurations, $R^+ {:}X^-$ and $R{:}^- \ X^+$ have been *decreased*, leading to a more satisfactory wavefunction. The magnitude of λ, the mixing parameter, may be optimized in order to minimize the energy.

The VB analogue of the first singlet excited state, A*, may likewise be obtained by expansion of (47), giving (48).

$$A^* \equiv \sigma^1\sigma^{*1} = |(a\varphi_R + b\varphi_X)(1)(-b\overline{\varphi_R} + a\overline{\varphi_X})(2)| \tag{47}$$

$$A^* \equiv \sigma^1\sigma^{*1} = (a^2 - b^2)(R{\cdot}\ {\cdot}X) - \sqrt{2}ab[(R{:}^-\ X^+) - (R^+{:}X^-)] \tag{48}$$

According to (48), A* is mainly an equal mix of the two ionic forms $R^+:X^-$ and $R:^-X^+$. In actual fact, mixing in of DA** and DA to obtain an improved wave-function leads to a dominant contribution of $R^+:X^-$. Thus $R^+:X^-$ is the simplest VB description of the first excited singlet state of R—X, in agreement with the earlier VB analysis.

This section illustrates therefore that both ground and excited states of the R—X bond may be simply described by either VB or MO configurations. These configurations are mutually related and an improved wave-function may be obtained through configuration mixing.

REACTION PROFILES

Dissociation of the R—X Bond

In the previous sections we summarized the rules which govern the interaction of VB and MO configurations and illustrated how the mixing of configurations generates improved descriptions of the possible molecular states. All interactions were at that stage conducted at a fixed nuclear geometry, and we now consider the interaction of configurations *at each point along a given reaction co-ordinate*. If we know the energy of a given configuration as a function of geometry then we have, through the configuration mixing process, a means of generating an entire reaction profile. Let us again turn to the R—X bond to illustrate the approach. By plotting the energies of the two principal configurations, $R\cdot\ \cdot X$, and $R^+:X^-$, as a function of the reaction co-ordinate and then allowing the configurations to interact we can generate an energy profile for the dissociation of the R—X bond.

A plot of the energies of the two configurations, $R\cdot\ \cdot X$ and $R^+:X^-$, in the gas phase as a function of the reaction co-ordinate is illustrated by the bold lines in Fig. 5. Each curve shows a minimum, the $R\cdot\ \cdot X$ configuration because of the out-of-phase resonance, $R\cdot\downarrow\uparrow\cdot X \longleftrightarrow R\cdot\uparrow\downarrow\cdot X$ (the in-phase resonance describing one of the triplet forms of the R—X bond), while $R^+:X^-$ shows a minimum due to the electrostatic attraction between cation and anion.

The *state* curves, S_0 and S_1 (dotted lines) are generated through the mixing of the two configurations at each point along the R—X dissociation axis. At infinite separation the two configurations fail to mix and the two states are described accurately by the two configurations. There is no mixing [i.e. $\lambda = 0$, eqns (22) and (23)] because at infinity S_{RX} is zero. As R and X approach one another, mixing takes place and the two configuration curves respond as if to repel each other, in analogy with the interaction diagram of Fig. 1. S_0 is *stabilized* with respect to $R\cdot\ \cdot X$, while S_1 is *destabilized* with respect to $R^+:X^-$. Even at this simple level of theory, the state diagram for an R—X bond dissociation provides interesting information concerning

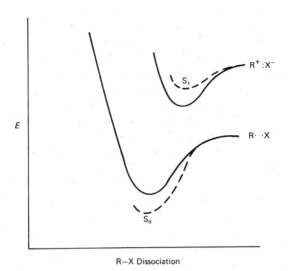

FIG. 5 Energy diagram of Heitler–London, R· ·X, and zwitterionic, R$^+$:X$^-$ configurations (bold lines) as a function of R——X distance. State curves, S_0 and S_1 (dotted lines) are obtained by configuration mixing

heterolytic vs homolytic bond-breaking. If the R—X bond is gradually stretched then two radicals, R· and X· are generated (motion along S_0) while if the R—X bond were to be excited to S_1, then bond-stretching is predicted to lead to the generation of two ions, R$^+$ and :X$^-$.

Dissociation of three-electron bonds

In this section we describe the MO and VB make-up for *three-electron bonds* and the corresponding energy profile for dissociation.

VB description. Just as R—X is a representation of the two-electron bond, (R∴X)$^-$ is the representation we will use for the three-electron bond between R and X. In VB terms such a bond may be described by a linear combination of the configurations, R· :X$^-$ and R:$^-$ ·X. Both of these configurations are repulsive with respect to R---X approach due to *exchange repulsion*. The exchange repulsion comes about from the overlap of two electrons, one on R and the other on X, possessing the same spin (Pauling and Wilson, 1935). The linear combination of the two VB forms may, however, lead to the formation of a stable three-electron bond. Let us see how this comes about.

If the energy gap between the configuration curves R· :X$^-$ and R:$^-$ ·X is large (bold lines, Fig. 6a) then both the ground and excited state curves (dotted lines, Fig. 6a) are repulsive at all R--X distances. This is because mixing

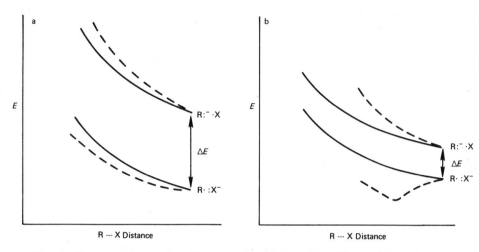

FIG. 6 Energy diagram that illustrates the mixing of R· :X⁻ and R:⁻ ·X configurations (solid lines) as a function of the R---X distance for (a) a large energy gap, ΔE, between the interacting configurations, and (b) a small energy gap, ΔE. Ground and excited state curves are indicated by dotted lines. For a small energy gap, as in (b), a stable three-electron bond is formed in the ground state

is slight [large ΔE in (18)] and the ground state is described predominantly by R· :X⁻. Note that the energy gap, ΔE, is given by $I_{X:^-} - A_R$. If, however, the energy gap is small, then mixing of the high energy R:⁻ ·X form into the R· :X⁻ form is increased and a *local minimum* may result, i.e. a stable three-electron bond is formed. This is illustrated in the state curves of Fig. 6b. A local minimum may result due to the enhanced stabilization energy *(SE)* as the R...X distance decreases. The enhanced stabilization energy is a direct result of the increased overlap (18) between the interacting configurations. Thus the ground state of a three-electron bond (R∴X)⁻ may be simply described by (49) where λ is given by (23) and is less than one.

$$(R \dot{-} X)^- = (R\cdot :X^-) + \lambda(R:^- \cdot X) \tag{49}$$

The most stable three-electron bonds will occur for the case in which R = X since then the two interacting configurations are isoenergetic and contribute equally to the wave-function, so that stabilization is at a maximum.

These ideas may be readily illustrated by a number of examples. Homodiatomic species exhibiting three-electron bonds are common, since this corresponds to the situation in which the two VB forms of (49) contribute equally, leading to maximum stabilization. Examples include He_2^+ (Pauling and Wilson, 1935), and O_2 (Goddard *et al.*, 1973). In fact contrary to the conventional view, three-electron VB structures, [1] and [2] explain

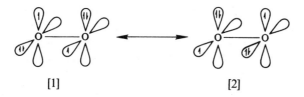

[1]　　　　　　　　　　[2]

the triplet ground state of O$_2$ satisfactorily. In the electronic configuration indicated in [1] and [2], three-electron bonds, each estimated to be c. 30 kcal mol^{-1} (Goddard et al., 1973) are present, whereas in the singlet structure [3] the regular π-bond, generated in the plane of the paper, is counteracted by four-electron repulsion in the perpendicular plane.

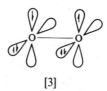

[3]

If the two VB forms describing the three-electron bond are very different in energy then, as indicated in Fig. 6a the bond is expected to break up spontaneously. This is indeed the case for [CH$_3$∴Cl]$^-$ which decomposes spontaneously to CH$_3$ and Cl$^-$ (Wang and Williams, 1980). As the R group is made more electronegative the relative contribution of R:$^-$ ·X increases so that a stable three-electron bond is generated. This is observed for [CF$_3$∴Cl]$^-$ and [(CF$_3$)$_3$C∴I]$^-$, which both exhibit measurable lifetimes (Wang and Williams, 1980).

VB–MO correspondence for three-electron bonds. The MO description of the ground state of the three-electron bonded species, (R∴X)$^-$, is $\sigma^2\sigma^{*1}$, i.e. a doubly occupied σ-orbital and a singly occupied σ* orbital. Substituting the hybrid orbital description of σ and σ* into the MO description of the three-electron bond, we obtain (50).

$$\sigma^2\sigma^{*1} = |(a\varphi_R + b\varphi_X)(1)\overline{(a\varphi_R + b\varphi_X)}(2)(-b\varphi_R + a\varphi_X)(3)| \qquad (50)$$

Multiplying the terms out, eliminating zero determinants, and substituting the chemical representations R· :X$^-$ and R:$^-$ ·X, simplifies (50) to the simple two-term expression (51).

$$\sigma^2\sigma^{*1} = b(R· :X^-) - a(R:^- ·X) \quad (b > a) \qquad (51)$$

Thus we see that the simple VB and MO descriptions of the (R∴X)$^-$ bond are identical! This is in contrast to the situation for two electron bonds where simple MO and VB descriptions were not equivalent for any given case, though they could be related through configuration mixing.

In similar fashion the first excited state of $(R\stackrel{.}{-}X)^-$, described by the MO description, $\sigma^1\sigma^{*2}$, may be shown to have the VB equivalent indicated in (52).

$$\sigma^1\sigma^{*2} = -b(R:^- \cdot X) - a(R\cdot :X^-) \quad (b > a) \tag{52}$$

Dissociation of the NaCl bond

The energy diagram describing NaCl dissociation is quite different to that observed for R—X and $[R\stackrel{.}{-}X]^-$ dissociation. This is because for NaCl the two relevant configuration curves *cross* (Herzberg, 1950; Levine and Bernstein, 1974), in contrast to the situation for R—X and $(R\stackrel{.}{-}X)^-$ where the two curves are separated. The well-known NaCl curve crossing is illustrated in Fig. 7. At infinite separation Na· ·Cl is more stable than

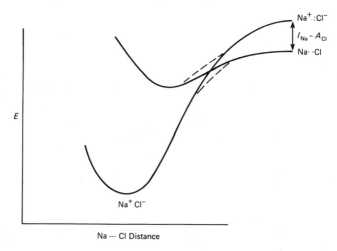

FIG. 7 Energy diagram showing the avoided crossing of Na· ·Cl and Na$^+$Cl$^-$ configuration curves. State curves resulting from configuration mixing are indicated by dotted lines

Na$^+$Cl$^-$ by c. 1.5 eV, the value of $I_{Na} - A_{Cl}$. At the equilibrium bond distance, however, it is the *ionic* configuration, Na$^+$Cl$^-$, which is the more stable one. This stems from the relatively low value of $I_{Na} - A_{Cl}$, as well as the shallow energy minimum in the Na· ·Cl configuration. The state curves, obtained *after* configuration mixing, however, do not cross. Mixing of the two configurations generates what is termed an *intended,* or *avoided* crossing (Salem *et al.,* 1975), and is indicated by the dotted lines. This is because in the region of the crossing point the mixing of the two configurations leads to a low- and a high-energy combination (as illustrated in Fig. 7). The state crossing is thus *avoided*. Stretching of the NaCl bond is predicted, on this basis to convert Na^+Cl^- to Na· and Cl· (and not to Na^+ and Cl^-!). The electron transfer takes place in the region of the avoided crossing and comes about due to the configuration switch in the description of the ground state The actual point of electron transfer may be approximated by (53) which

$$I_{Na} - A_{Cl} \approx e^2/r \quad (53)$$

states that the electron jump will occur as soon as it is energetically profitable. In other words, as soon as the energy expenditure required to transfer the electron, $I_{Na} - A_{Cl}$, is provided by the electrostatic stabilization between a cation and an anion, (e^2/r), the electron will jump.

A number of general points emerge here. Firstly, the better the donor–acceptor pair (i.e. the smaller the covalent–ionic configuration gap), the *earlier* the electron jump. Thus for NaCl $(I - A \approx 1.5 \text{ eV})$ r is large, while for HCl $(I - A \approx 10 \text{ eV})$, r is small. Secondly, the charge redistribution is localized in the region of the avoided crossing. These simple ideas will subsequently be applied in Section 4 (p. 186) to the question of charge development along the reaction co-ordinate. We will see, however, that despite the fact that the above concepts are rooted in the earliest days of quantum chemistry, many simple consequences of these ideas challenge a number of well-established principles of modern organic chemistry.

Generalized reaction profiles

The avoided crossing example of NaCl is extremely general and in many ways typifies that observed for many organic reactions. The ground state of the reactants correlates with the excited state of the products (e.g. for NaCl dissociation, Na^+Cl^- correlates with $Na^+ + Cl^-$) while the excited state of the reactants correlates with the ground state of the products (e.g. for NaCl, Na–Cl correlates with Na· + Cl·). In this section we outline in general terms the way in which reaction profiles may be generated by the avoided crossing of two or more configurations. A similar analysis using a rather different terminology has been presented by Metiu *et al.* (1979). Specific applications of the approach appear in Section 3 (p. 139).

Two-configuration reactions. Since for most organic reactions one or more covalent bonds are broken while one or more new bonds are formed a reaction profile may be schematically generated as indicated in Fig. 8. For the case in which at least one reactant is a closed shell species, two configuration curves, φ_R representing the reactants, and φ_P representing the products, are necessary in order to create a reaction barrier. The reactant configuration, φ_R, contains within it a description of the covalent bond to be

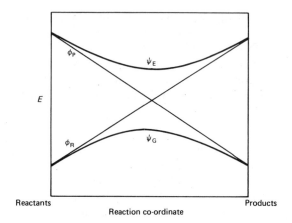

FIG. 8 Energy diagram for a general reaction that illustrates the formation of ground (ψ_G) and excited (ψ_E) state curves from the mixing of two configurations, φ_R, the reactant configuration, and φ_P, the product configuration

broken, while the product configuration, φ_P, contains within it a description of the new bond which is formed. For example in the S_N2 reaction, $N:^- + R—X \rightarrow N—R + :X^-$, the reactant configuration will be described by $N:^-$ $R\cdot$ $\cdot X$ and contains the Heitler–London form, $R\cdot$ $\cdot X$, of the bond to be broken, while the product configuration is $N\cdot$ $\cdot R:X^-$ and contains the Heitler–London form, $N\cdot$ $\cdot R$, of the bond which is formed. The reactant configuration, φ_R, climbs steeply in energy as a function of the reaction co-ordinate. This is because in the product geometry, φ_R represents an excited configuration. For example in the S_N2 reaction, the R—X bond in the reactant geometry, has been broken and replaced by a high-energy $^-N:$ $\cdot R$ three-electron interaction in the product geometry. Symmetrically, the product configuration, φ_P, starts out as an excited configuration of high energy, and drops in energy along the reaction co-ordinate so as to end up as a ground configuration. The ground state reaction complex at any point along the reaction co-ordinate may be described by ψ_G (G = ground) as indicated in (54) and Fig. 8 (lower bold line). At the beginning of the reaction where φ_R

$$\psi_G = a\varphi_R + b\varphi_P \tag{54}$$

is lower in energy than φ_P the wave-function describing the reaction complex, ψ_G, is dominated by φ_R ($a \gg b$). Further along the reaction co-ordinate a decreases while b increases since the energy gap between φ_R and φ_P decreases [see (22)]. At the intended crossing ψ_G is given by (55) since φ_R and φ_P are now degenerate. This is likely to occur in the transition state region.

$$\psi_G = (1/\sqrt{2})(\varphi_R + \varphi_P) \tag{55}$$

As the reaction proceeds toward completion it is now φ_R which mixes slightly into φ_P, (i.e. $b \gg a$) since past the crossing point it is φ_P which is lower in energy. We see, therefore, that there is a continuous change in character from reactant-like to product-like along the reaction co-ordinate with most of the switch-over in character, centred in the avoided crossing region where the wave-function of the reaction complex contains large contributions of both φ_R and φ_P. It is clear then that *breaking of a bond during a reaction must be attended by the disappearance of the specific Heitler–London form of that bond from the wave-function of the reaction complex. Symmetrically, formation of a new bond must be attended by the appearance of that specific Heitler–London form of the new bond within the wave-function of the reaction complex.* Note, also, that the mixing of two configurations (Fig. 8) generates, in addition to the ground state profile, an *excited* state surface, ψ_E, as well (Fig. 8, upper bold line). This upper surface comes about through the out-of-phase mixing of φ_R and φ_P. Thus just as mixing of two atomic orbitals generates bonding and anti-bonding molecular orbitals, mixing of configuration curves generates ground and excited state curves. Thus a reaction may in principle take place in two discrete ways (Fig. 9). In the thermal process (Route T, Fig. 9), the reaction complex "climbs" up the ground state curve, ψ_G, primarily described by the reactant configuration, φ_R, passes through the transition state in the avoided crossing region, and down to products (now described primarily by the excited configuration, φ_P). In the photochemical process (Route P, Fig. 9), the reactants are excited up to the excited state surface, ψ_E, "slide" down this surface toward the minimum, then funnel down onto the ground surface, and down to products (Turro *et al.*, 1979). It

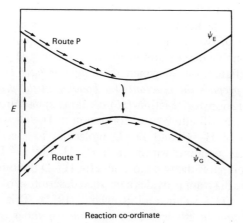

FIG. 9 Energy diagram for a general reaction showing the relationship between the thermal route, T, and the photochemical route, P

is clear then that a qualitative understanding of the nature of the configuration curves which go into building both ground and excited surfaces is the key to a more detailed understanding of both thermal and photochemical processes. Photochemical applications of configuration mixing are discussed in Section 3 (p. 139).

This avoided crossing model for describing a reaction barrier is striking in its generality. It has been applied in a detailed theoretical analysis to reaction (56) (Shaik, 1981) and reaction (57) (Michl, 1975, 1977; Gerhartz et al.,

$$H^- + D\!-\!D \rightarrow H\!-\!D + D^- \tag{56}$$

$$\begin{array}{ccc} H & D & H\!-\!D \\ | & + \; | & \rightarrow \\ H & D & H\!-\!D \end{array} \tag{57}$$

1976). In VB terms, reaction (56) involves the avoided crossing of H:⁻ D· ·D, the reactant configuration, and H· ·D :D⁻, the product configuration, while in MO terms it is the avoided crossing of the DA configuration with the charge transfer D^+A^- configuration (Scheme 1, p. 108). The bond-exchange reaction (57) may be described in VB terms by the avoided crossing of [4] and [5], i.e., the H—H/D—D ground configuration with the

$$\begin{array}{cccc} H\!\uparrow & \cdot D & H\!\uparrow & \uparrow D \\ H\!\downarrow & \cdot D & H\!\downarrow & \downarrow D \\ [4] & & [5] & \end{array}$$

³H—H*/³D—D* ditriplet configuration. In MO terms it is the crossing of DA with ³D* ³A*. The key point is that the barrier comes about through the avoided crossing of two configurations, the reactant configuration, which provides a reasonable description of the reactants in the ground state, and the product configuration, in which *the reactants have been electronically excited so as to prepare the reactants for bonding in the products*. Thus the product configuration for reaction (56) contains spin-paired odd electrons on H and D which will allow bond formation in H—D, while for reaction (57), promotion of H—H and D—D molecules to the ditriplet (overall singlet) eliminates bonding within each H—H and D—D molecule and simultaneously prepares the two molecules for H—D bonding.

The same basic diagram provides a detailed description of the twisting of a π-bond (Salem, 1973; Salem and Rowland, 1971). This is illustrated in Fig. 10. The π-bond in ethylene may exist in a number of predetermined states, each of which may be approximated by the appropriate MO configuration. The possible states are S_0, the ground state, (π^2); T_1, the first

CONFIGURATION MIXING MODEL

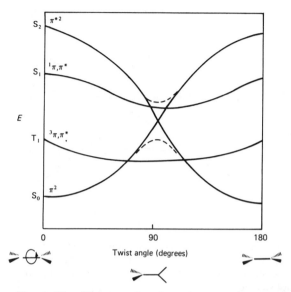

FIG. 10 Energy diagram that illustrates the effect of rotation about the C=C bond of ethylene on the MO-configuration energies. The avoided crossing of π^2 and π^{*2} configurations is indicated by the dotted lines

excited triplet state, $(^3\pi,\pi^*)$; S_1, the first excited singlet state, $(^1\pi,\pi^*)$; and S_2, the second excited singlet state, (π^{*2}). Note that for rotation of 180° about the C—C axis, π^{*2} is the product configuration. This is because the two electrons now occupy the orbital which in the product geometry (180° rotation) enables π-bonding to occur. Rotation by 180° converts an antibonding π^* interaction into a bonding π-interaction, so that S_2 in the reactant geometry (0°) correlates with S_0 in the product geometry (180°). For this reason the barrier to rotation about the C=C bond may be attributed to the avoided crossing of the ground configuration, π^2, with the diexcited configuration, π^{*2}.

In order to complete the configurational description of the rotational process the monoexcited configurations, S_1 and T_1 need to be incorporated, and are included in Fig. 10. Indeed, there are a large number of reactions which are not adequately described by the simple two-configuration diagram of Fig. 8. Such reactions may be classified as multiconfiguration reactions and are discussed subsequently.

Multiconfiguration reactions. There are a large number of reactions for which at least one additional configuration is necessary for a proper description of the reaction complex. Such a reaction is schematically illustrated in Fig. 11. As in the two-configuration case, we have a reactant configuration,

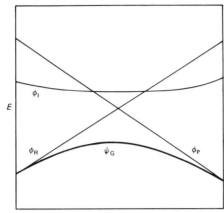

FIG. 11 Energy diagram for a general reaction showing the formation of a ground state curve (ψ_G) by the mixing of three configurations: φ_R, the reactant configuration, φ_P, the product configuration, and φ_I, the intermediate configuration. Excited state curves are not shown

φ_R, starting off as a ground configuration and ending up as an excited configuration, a product configuration, φ_P, which does the reverse; it begins initially as an excited configuration and ends up as the predominant descriptor of the products. However, in addition, we have an intermediate configuration, φ_I, which at the reaction beginning represents a further excited configuration (in addition to φ_P), but at the reaction end *remains* an excited configuration. So by comparison with φ_R and φ_P, φ_I tends to be relatively flat, being an excited configuration at both reaction beginning and end. This was indeed observed in Fig. 10 for the monoexcited configurations, $^1\pi,\pi^*$ and $^3\pi,\pi^*$.

In general, possible intermediate configurations may be obtained by rearranging the available valence electrons in energetically sensible ways. Returning to our example of an S_N2 reaction, the likely intermediate configurations may be obtained by rearranging the four valence electrons (lone pair on N, R—X bonding pair) amongst the three reacting atoms, N, R and X. Two such configurations are $N:^- \ R^+ :X^-$, the carbocation configuration, and $N\cdot \ R:^- \ \cdot X$, the carbanion configuration. The relative flatness of these configurations is also readily understood. For the carbocation configuration, $N:^- \ R^+ :X^-$, at reaction beginning the $R^+ :X^-$ description of the RX linkage is just the VB representation of the first singlet excited state of R—X (Section 2, p. 104). At reaction end the $N:^- \ R^+$ description is just the VB representation of the first excited singlet state of the N—R bond. Thus, at both reactant *and* product ends this configuration represents a monoexcited description of reactants or products and hence is flat in comparison to reactant or product configurations.

For any reaction described by three configurations the reaction complex at any point on the reaction profile is described by ψ_G (Fig. 11, bold line) where ψ_G is given by (58). At the reaction ends the coefficient of φ_I is small since the

$$\psi_G = a\varphi_R + b\varphi_P + c\varphi_I \tag{58}$$

reaction is dominated by either φ_R or φ_P. However, in the transition state region φ_I may be of similar energy to φ_R and φ_P, and hence play an important role in describing the reaction complex. Thus both the energy and the nature of the transition state will be strongly influenced by any low-lying intermediate configurations that may be present. Many aspects of rate–equilibrium relationships, including anomalous α values, may be understood in a straightforward manner on the basis of Fig. 11 (see Section 4, p. 177).

MO vs VB configurations. Up to now we have indicated that in principle both MO and VB configurations may be used in order to build up a reaction profile. The question now arises as to which of these two is more useful in tackling a range of organic reactivity problems. MO configurations have been utilized by Mulliken (1952a,b), Nagakura (1963) and Epiotis *et al.* (Epiotis and Shaik, 1977a,b, 1978a,b,c; Epiotis *et al.*, 1980; Epiotis, 1978). VB configurations have their origin in the early days of quantum chemistry and were utilized by Evans *et al.* (Evans and Polanyi, 1938; Evans and Warhurst, 1938; Evans, 1939) as well as Laidler and Shuler (1951; Laidler, 1955). Recently Warshel and Weiss (1980; Warshel, 1981) have used a similar approach to compare enzyme reactions with chemical reactions in solution.

Since both MO and VB approaches have been shown to be equivalent (Shaik, 1981), in principle either may be used. However, the VB approach is simpler to use, more "chemical" in its application and well-suited to questions of transition state structure and charge distribution, intermediate formation, and mechanistic variations within a reaction family. Of particular importance is the fact that the qualitative VB procedure is far more general than the qualitative MO procedure, since it is based on simple Lewis structures. A qualitative MO treatment of elimination reactions, for example, (see Section 3, p. 161) would be too unwieldy and impractical to apply while, in contrast, a VB approach is entirely straightforward in its application.

On the other hand, the MO procedure, while difficult to apply in many cases, provides information which is inaccessible with a simple VB approach. Specifically, stereochemical questions and problems of reactivity which require the use of symmetry arguments depend on an MO analysis since the MO formalism rests heavily on symmetry considerations.

The approach taken in this review, therefore, is to use, wherever possible, the VB approach because of its generality, "chemical" formalism, and simplicity. For those cases where an MO approach is dictated by the nature of the problem, those features of the MO analysis, which are essential to achieve understanding, will be blended into the VB treatment. This seems to us the most useful tack and such a double-barrelled approach will be used for example in pericyclic reactions (Section 3, p. 173) where incorporation of symmetry arguments is crucial to obtain a proper understanding of the reaction process.

Substituent and solvent effects on reactivity

Understanding the nature of the configurations which go into a description of the reaction profile provides us with a tool for analysing both substituent and solvent effects on reactivity. If one assumes that configurations respond to substituent and solvent effects in much the same way as ground states do – that is, electron-releasing groups stabilize positive charge, electron-withdrawing groups stabilize negative charge, high charge density groups are solvated more strongly than neutral or low charge density groups – then we have a tool for predicting substituent and solvent effects on reactivity.

Substituent effects. To a first approximation one need only consider substituent effects on charge centres. These have been shown to be about an order of magnitude larger than those on neutral centres (Pross and Radom, 1981) so such an approximation is reasonable.

For a reaction adequately described by just two configurations, reactant and product, the analysis of substituents effects is straightforward and was first treated by Horiuti and Polanyi (1935) almost 50 years ago. Subsequent contributions by Bell (1936) and Evans and Polanyi (1938) have led to these general ideas being jointly termed the Bell–Evans–Polanyi principle (Dewar, 1969). The treatment of multiconfiguration reactions is analogous and is illustrated in Fig. 12. Let us discuss this in detail.

A reaction described by the three configurations, φ_R, φ_P and φ_I is shown. If a substituent is now introduced into the system which has the effect of stabilizing φ_P (indicated by the two arrows), then φ_P is displaced to lower energy along the entire reaction co-ordinate. The effect of such a perturbation is predicted to be *(a)* to lower the reaction barrier since the intended crossing of φ_P and φ_R is now lower in energy, *(b)* to lead to an earlier transition state since the intended crossing point is now earlier along the reaction co-ordinate, *(c)* to reduce the contribution of the intermediate configuration φ_I in describing the transition state using (58), since φ_I is now relatively higher in energy with respect to the new intended crossing point and therefore contributes less to the ground state wave-function, ψ_G. Since *(a)* provides energetic information, *(b)* provides structural information, and *(c)* provides

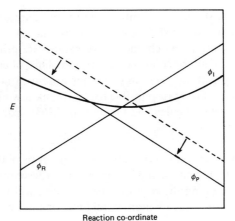

FIG. 12 Energy diagram for a general reaction described by three configurations that shows the effect of stabilizing the product configuration, φ_P (indicated by the two arrows). The perturbation is predicted to (1) lower the reaction barrier, (2) lead to an earlier transition state, and (3) decrease the contribution of φ_I in the ground state profile (not shown)

information on electronic charge distribution with regard to the transition state, albeit at a qualitative level, the CM approach enables a qualitative assessment to be made regarding the effect of a substituent on a particular reaction.

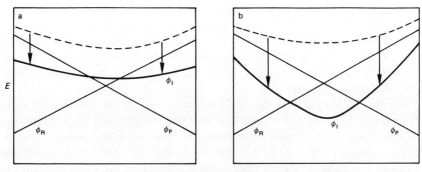

FIG. 13 Energy diagram for a general reaction described by three configurations that shows the effect of stabilizing the intermediate configuration φ_I (a) moderately (indicated by arrows) and (b) strongly (indicated by arrows). For (a) the reaction remains concerted but the transition state takes on "intermediate" character. For (b) the reaction actually becomes a two-step process that proceeds via the intermediate, described by φ_I

If the substituent stabilizes the intermediate configuration, as indicated in Fig. 13a, then the transition state is predicted to take on more of the character of that configuration. By character, we mean here primarily the *charge* characteristics of the intermediate configuration. This, of course, enables the reaction transition state to take on character absent in both reactants and products and is a key factor responsible for the breakdown in the Leffler relation (59) (Leffler, 1953; Leffler and Grunwald, 1963). Equation (59), which

$$\delta G^{\neq} = \alpha \delta G_P + (1-\alpha)\delta G_R \tag{59}$$

treats substituent effects on the transition state in terms of the corresponding effects on reactants, products, and a parameter, α (describing the position of the transition state along the reaction co-ordinate), will be inapplicable for reactions that require three or more configurations in order to characterise them.

If the intermediate configuration is strongly stabilized in a given situation then an actual intermediate is likely to be generated. This is illustrated in Fig. 13b. The reaction becomes a two-step reaction, each step of which becomes a two-configuration profile. Thus the initial barrier comes about through the avoided crossing of reactant, φ_R, and intermediate, φ_I, configurations, while the second step – reaction of the intermediate to generate the final product – comes about through the avoided crossing of φ_R and φ_P (Metiu *et al.*, 1979).

The variable influence of this third or "intermediate" configuration is responsible for generating the mechanistic diversity within a reaction class. In the absence of the third configuration, concerted one-step reactions take place (E2, S_N2). As the third configuration begins to play a significant role, the reaction remains concerted, but the transition state takes on the character of that intermediate configuration (E2–E1cB-like, loose S_N2, etc.). Further stabilization of the intermediate configuration leads to a stepwise reaction through some intermediate (E1, E1cB, S_N1). Relevant applications will be discussed in Section 3 (p. 139).

The reason for the general confusion surrounding the question of substituent effects on transition state structure and energy is not hard to fathom. Whereas most ground states are normally reasonably well-described by a single configuration, e.g. $R\cdot \cdot X$ for R—X, Na^+Cl^- for NaCl, transition states of reactions invariably need to be described by *two or more configurations,* generally near the avoided crossing region. Furthermore, and most importantly, since transition states are maxima along the reaction co-ordinate their nature is highly variable. Another way of phrasing this is to say that a family of ground states is far more limited in its structural form than a family of transition states. To give a specific example, the nature of the C—Cl bond (energy, polarity, bond length) is essentially independent of the alkyl moiety, while the nature of the C—Cl bond in the transition state of an S_N2 exchange process will be highly variable and strongly dependent on the nature of the alkyl moiety. This variability of S_N2 transition states will be discussed in detail in Section 3. When this variability in transition state structure is coupled with the severe experimental difficulties in characterizing transition states, it is not surprising that the overall question of the transition state (energy, geometry, charge distribution) remains relatively poorly understood.

The CM model suggests that a widespread view relating ground and transition state substituent effects may be invalid. It is widely held that a transition state incorporating some fractional charge, α, at a particular site will respond to a substituent at that site in much the same way as a ground state molecule might. In other words, if the energy perturbation of the ground state incorporating a full charge by the substituent is δE, then the corresponding perturbation in the transition state with a fractional charge, α, is expected, on the basis of straightforward charge–substituent dipole interactions, to be αδE. However, as we have just noted, substituents in transition states do not operate on frozen structures. Substituent effects on transition states introduce charge and geometric changes which make energetic predictions less straightforward, and the simple charge–dipole model, assuming constant charge and geometry, unreliable. The meaning of the Brønsted parameter α will be discussed in Section 4 (p. 177).

Solvent effects. Solvation perturbations may be treated in a similar fashion to substituent perturbations. This can be readily illustrated by examining how the gas phase dissociation profile of the R—X bond, discussed earlier, differs in solution. At the simplest level of approximation one might assume that neutral groups will be unaffected by solvation, and that it is only the effects on charged groups that need to be considered (Warshel and Weiss, 1980). If this is the case then $R^+:X^-$ is expected to be strongly stabilized with respect to R· ·X. This is illustrated in Fig. 14. As in Fig. 5, the bold lines represent the configuration curves. Note that at infinity $R^+:X^-$ is now lower in energy than R· ·X. While in the gas phase (Fig. 5), $I_R - A_X$ is commonly *c.* 6–7 eV (i.e. $R^+:X^-$ is higher in energy), the strong ionic solvation (a typical figure is *c.* 100 kcal mol^{-1} or 4 eV per ion, i.e. *c.* 8 eV in total) becomes dominant so that $R^+:X^-$ may be stabilized *below* R· ·X at infinity. The result is that now the configuration curves cross so that once again, a state avoided crossing (as for NaCl) is observed.

The avoided crossing means that, for R—X dissociation in solution, ground and excited configurations at reactant and product ends switch about. Moving along the reaction profile, the ground configuration, R· ·X, becomes the *excited* configuration, while concomitantly, the excited configuration,

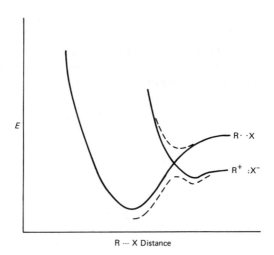

Fig. 14 Energy diagram illustrating the avoided crossing of $R\cdot\ \cdot X$ and $R^+:X^-$ configurations for dissociation of the R—X bond in solution. At infinite R----X distances $R^+:X^-$ is stabilized compared to $R\cdot\ \cdot X$ due to solvation of $R^+:X^-$

$R^+:X^-$, becomes the ground configuration. What this means, of course, is that dissociation of RX in solution is expected to yield ions (due to the $R\cdot\ \cdot X/R^+:X^-$ avoided crossing) while gas-phase dissociation leads to radicals (since at all geometries the dominant configuration is $R\cdot\ \cdot X$).

Inspection of Fig. 14 provides a naive, but nonetheless interesting explanation, as to why ion-pair intermediates may form in S_N1 processes. One can see that after the transition state is overcome, a minimum on the reaction surface is reached. Since this minimum is part of the ionic configuration it would properly be described as an ion-pair intermediate. Thus the reason why R—X may pass through an intermediate called a contact or intimate ion-pair (Winstein et al., 1965) on the pathway toward full dissociation becomes more readily discernible.

An interesting consequence of the R—X avoided crossing in solution is that its very existence in certain systems may actually depend on the magnitude of solvation. Since in the gas phase dissociation tends to be homolytic and in solution heterolytic, it is conceivable that in certain cases the solvent may determine whether homolysis or heterolysis takes place.

An illustration of this effect may be the brominating action of N-bromosuccinimide (NBS). In weakly polar solvents, NBS tends to generate Br atoms while in polar solvents, Br^+ tends to be formed (March, 1968). While it is now considered that NBS itself is not the primary source of Br atoms (Walling et al., 1963; Russell and Desmond, 1963; R. E. Pearson and Martin, 1963; Skell et al., 1963) the basic observation, that the dependence of the nature of the active brominating species on solvent polarity, is nicely rationalized in terms of configuration crossings.

Symmetry considerations

In previous sections we have discussed the energetics of the configuration curves and the manner in which they mix. However, a crucial factor which governs the mixing of configurations is the *symmetry* of the respective configurations. A proper and detailed discussion of the role of symmetry in configuration mixing is outside the scope of this review. Nonetheless, some fundamental aspects of symmetry relations are essential in order to achieve a qualitatively sound appreciation of the key factors governing the nature of reaction profiles. For a detailed theoretical account of the crossings between electronic states, see Teller (1937), Herzberg and Longuet-Higgins (1963), Herzberg (1966), and Carrington (1972). However, much of the simplified analysis presented here is based on the particularly lucid publications of Salem and coworkers (Salem *et al.*, 1975; Salem, 1982). For a general review on the role of symmetry in chemical reactions, see R. G. Pearson (1976).

A question of particular importance is whether the intersection of two configuration curves will result in a *real* crossing, as indicated in [6a], or an avoided crossing as indicated in [6b]. The general rule which governs the

[6a]   [6b]

nature of the crossing is that *real* crossings may always occur between curves of different spin or spatial symmetry. If however, at some point along the reaction profile the symmetry element which differentiated between the two curves is destroyed the crossing becomes *avoided*. Let us explore this idea in greater detail.

A particular reaction profile whose wave function is ψ is described by a linear combination (60) of two configurations, φ_1 and φ_2.

$$\psi = c_1\varphi_1 + c_2\varphi_2 \qquad (60)$$

Applying the variational principle to obtain the optimum energy requires a solution of the secular equations which may be written in determinant form as (61). H_{11} and H_{22} are the energies of φ_1 and φ_2 respectively and are defined by (62) and (63); β is the off-diagonal matrix element (64).

$$\begin{vmatrix} H_{11} - E & \beta \\ \beta & H_{22} - E \end{vmatrix} = 0 \qquad (61)$$

$$H_{11} = <\varphi_1|H|\varphi_1> \tag{62}$$

$$H_{22} = <\varphi_2|H|\varphi_2> \tag{63}$$

$$\beta = <\varphi_1|H|\varphi_2> \tag{64}$$

The solution to the secular equation (61) is given by (65).

$$E = \frac{1}{2}\{H_{11} + H_{22} \pm [(H_{11} - H_{22})^2 + 4\beta^2]^{1/2}\} \tag{65}$$

A real crossing will result if at the crossing point (i.e. $H_{11} = H_{22}$) there is only *one* solution to (65). Inspection of (65) shows this will occur when $\beta = 0$. On the other hand if $\beta \neq 0$ then E is given by (66) and the crossing is *avoided*.

$$E = H_{11} \pm \beta \tag{66}$$

Through interaction of the two configurations (via β) a high- *and* a low-energy combination result.

For the interaction of two configurations which are related by a single electron transfer (e.g. DA and D^+A^-) then, as previously discussed [see (20)], the magnitude of β is essentially proportional to the overlap between the two atomic orbitals which differ in one-electron occupancy (Salem *et al.*, 1975; Epiotis, 1978). The role of symmetry now becomes clear. If these two orbitals are orthogonal then $\beta \approx 0$, and the extent of mixing of the two configurations will be slight. As an example of this effect consider the interaction of two MO configurations, DA and D^+A^-, for the system of two ethylene molecules. The MO configurations are represented in Scheme 2.

Scheme 2

Since the two configurations are related by a single electron transfer ($\pi_D \to \pi_A^*$) the magnitude of β will be given by (67).

$$\beta = <DA|H|D^+A^-> \approx <\pi_D|H|\pi_A^*> \tag{67}$$

Using the approximation of (21) we may write (68).

$$\beta \approx K <\pi_D|\pi_A^*> \tag{68}$$

Since for the face-on approach of two ethylene molecules the π orbital of one ethylene (π_D) and the π^* orbital of the second ethylene (π_A^*) have opposite symmetry their overlap (S_{π,π^*}) equals zero as in [7a]. As a consequence, the value of β will be close to zero, i.e. DA and D^+A^- do not mix appreciably. Of course, for the cycloaddition reaction of ethylene and butadiene, π_D *can* overlap with π_A^* so that DA and D^+A^- may interact effectively [7b].

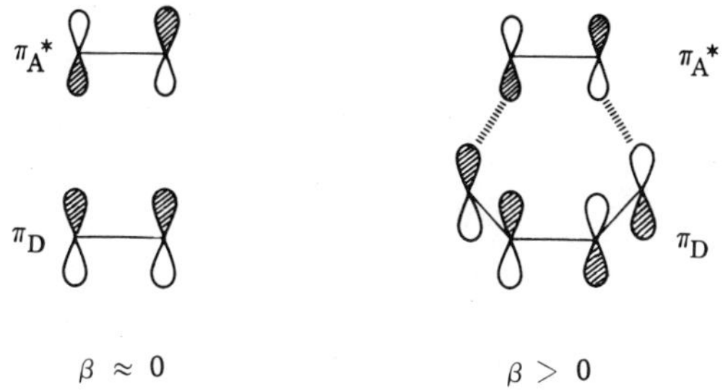

$\beta \approx 0$ $\beta > 0$

[7a]   [7b]

This lack of mixing of DA and D^+A^- for two ethylenes, in contrast to the situation for ethylene and butadiene, where mixing can take place, will be utilized (Section 3) to analyse the relationship of allowed and forbidden pericyclic reactions in configuration terms.

A second example from Salem's work which illustrates the role of symmetry is the photochemical hydrogen abstraction by ketones, termed the Norrish type II reaction (Norrish, 1937). A typical case is illustrated in (69).

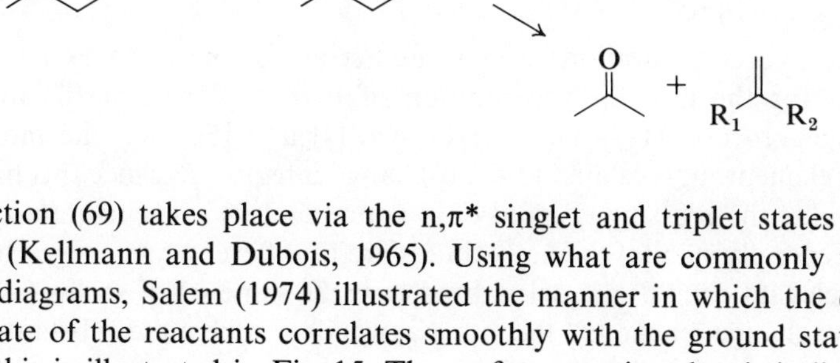

(69)

Reaction (69) takes place via the n,π^* singlet and triplet states of the ketone (Kellmann and Dubois, 1965). Using what are commonly termed Salem diagrams, Salem (1974) illustrated the manner in which the excited n,π^* state of the reactants correlates smoothly with the ground state products: this is illustrated in Fig. 15. The surface crossing that is indicated is

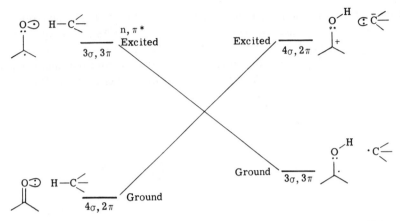

FIG. 15 State correlation diagram for hydrogen abstraction by ketones. (Adapted from Salem, 1974)

allowed because the state symmetries of the ground state (4σ,2π), [8], and the excited state (3σ,3π), [9], are different, viz. A' for [8], and A" for [9]. The difference in symmetry between [8] and [9] stems from the conversion of a σ-electron in [8] to a π-electron in [9].

[8]   [9]

The point here is that the entire symmetry classification of reaction (69) is based on the presumption that a plane of symmetry is conserved thoughout the hydrogen abstraction process (the plane incorporating the C—H bond and the carbonyl group). If one or more atoms are displaced slightly so as to destroy the symmetry plane the crossing becomes *avoided*. A comparison of the in-plane and slightly out-of-plane cases is illustrated in Fig. 16a and b (Salem et al., 1975).

If the two configurations which interact are of the bond exchange type such as for the [2 + 2] cycloaddition of two ethylenes, or the hydrogen exchange process, $H_2 + D_2 \rightarrow 2HD$, e.g. [4] and [5], then the interaction matrix element, β, is related to the exchange integral, K. Since this has some finite value the crossing is *avoided*. In actual fact its range is quite variable, being generally small if the key orbitals are of different symmetry. However, if the relevant orbitals are delocalized over the same set of atoms (as for the above exchange reactions), then its magnitude can be quite substantial (1 or 2 eV) (Salem et al., 1975).

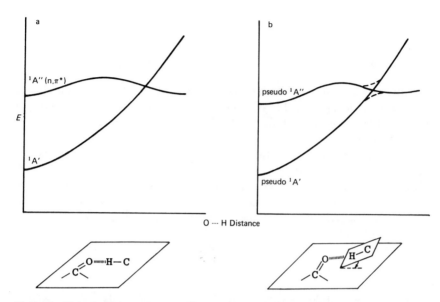

FIG. 16 Energy diagram for the hydrogen abstraction reaction of ketones showing the interaction of the ground state and n,π* excited state curves. (a) In-plane approach. A real crossing occurs when the carbonyl group and the C—H bond lie in a plane since the two state curves are of opposite symmetry. (b) Out-of-plane approach. An avoided crossing (dotted lines) occurs when the symmetry plane is destroyed and the two states can mix. (Adapted from Salem et al., 1975)

In summary, therefore, the extent of configuration mixing of two configurations, which are related by an electron transfer, will be governed primarily by the overlap of the two orbitals associated with the electron transfer. The role of symmetry in this situation is paramount. For cases where the two configurations are not related by a single electron transfer (e.g. DA and D*A*) the matrix element is governed by bielectronic terms in the Hamiltonian and is variable in magnitude. Finally, singlet and triplet states cannot mix and such crossings are real (ignoring spin-orbit coupling effects).

Experimental support of the CM model

Kochi and coworkers have published a number of papers in the last five years that lend considerable support to the key elements of the CM model (Fukuzumi et al., 1979, 1980; Fukuzumi and Kochi, 1980a,b, 1981a,b, 1982; Hilinski et al., 1983).

In one of the earlier studies (Fukuzumi et al., 1979), the thermal and photochemical addition of organometals to carbon–carbon double bonds, e.g. R_4Sn to tetracyanoethylene TCNE, (70), was investigated. An interesting feature of the reaction is that it may be induced both thermally *and*

$$R_4Sn + \underset{NC}{\overset{NC}{>}}C=C\underset{CN}{\overset{CN}{<}} \longrightarrow R_3Sn-\underset{\underset{NC}{|}}{\overset{\overset{NC}{|}}{C}}-\underset{\underset{CN}{|}}{\overset{\overset{CN}{|}}{C}}-R \qquad (70)$$

photochemically, and that both processes appear to proceed from the charge transfer complex, [R_4Sn, TCNE], and via the intermediate formation of the corresponding ion-pair, [$R_4Sn^{\ddot{+}}$ TCNE$^{\ddot{-}}$]. Let us consider these facts in terms of the CM model.

Since this reaction constitutes a bond exchange process, such as the H_2—D_2 exchange reaction (57), the product configuration is $^3D^*\,^3A^*$. However, due to the excellent donor-acceptor ability of the R_4Sn—TCNE pair the charge transfer configuration, D^+A^-, is, by comparison, low in energy. A schematic plot of these two configurations, together with DA, is shown in Fig. 17.

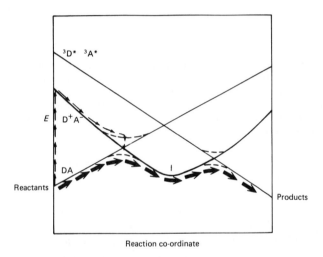

FIG. 17 Energy diagram showing the DA, D^+A^-, and $^3D^*\,^3A^*$ configurations for the addition of R_4Sn to tetracyanoethylene (70). The thermal pathway is indicated by the bold arrows and the photochemical pathway by the light arrows. Both pathways proceed via the intermediate I, described by the [$R_4Sn^{\ddot{+}}$ TCNE$^{\ddot{-}}$] species

The thermal pathway is indicated by the bold arrows. The ground state surface is generated by the avoided crossing of DA with D^+A^- leading to the intermediate, I. Since I is predominantly described by D^+A^-, the intermediate is just the ion-pair species [$R_4Sn^{\ddot{+}}$ TCNE$^{\ddot{-}}$]. This intermediate then reacts to form products on the surface generated from the $D^+A^-/^3D^*\,^3A^*$ avoided crossing. Note that while DA/D^+A^- overlap (related to overlap of

σ_{Sn-R} with $\pi_{C=C}$) may be precluded in a square-type approach, [10a], due to symmetry considerations (in the same way as for two ethylenes), a sideways approach [10b] will improve overlap significantly.

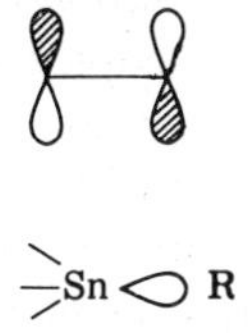

[10a] 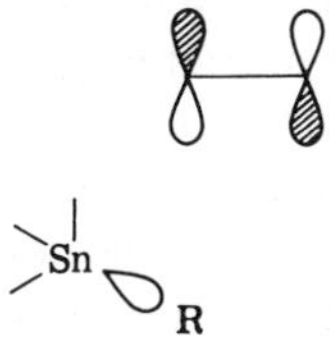 [10b]

The photochemical pathway is indicated by the light arrows (Fig. 17) and involves excitation of the charge transfer complex DA up to the D^+A^- surface. The reaction complex then funnels down onto the ground state surface so as to follow the thermal route. Clearly, both pathways involve formation of the ion-pair intermediate, and are intimately related. We see therefore that the CM model constitutes a useful framework for understanding the relationship between ground state and excited state reactions.

The actual relationship between reactivity (barrier height) and the energy of the excited state has also been vividly illustrated by Kochi and coworkers (Fukuzumi and Kochi, 1980a,b, 1981a, 1982; Fukuzumi et al., 1980). Kochi found that in a number of reaction types, such as electrophilic aromatic substitution (S_E2), cleavage of alkylmetals by halogen, cycloadditions and electron transfer processes of alkylmetals, the reaction rate is related to the charge transfer bands of the reactants. For example, in the electrophilic substitution of a family of arenes by halogen or mercuric trifluoroacetate (71), the

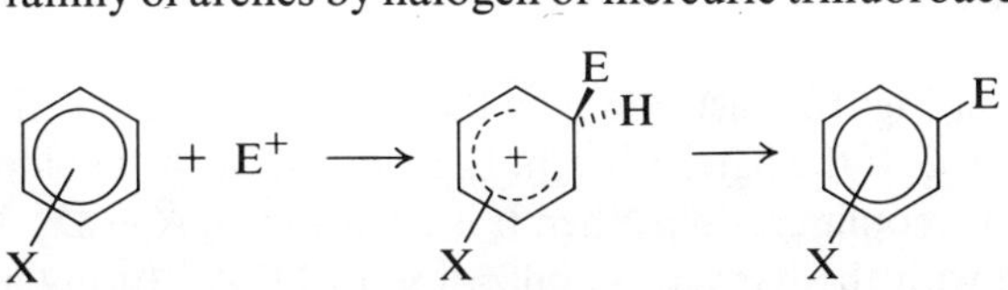

$$\qquad \qquad (71)$$

[11]

relative reactivity of any arene with a particular electrophile is correlated by

$$\Delta h\nu = -RT \ln k_X/k_0 \qquad (72)$$

(Fukuzumi and Kochi, 1981a). In (72) k_X is the second order rate constant of the particular arene with the electrophile E^+, k_0 is the corresponding rate constant for benzene, and $\Delta h\nu$ is given by (73), in which ν_X is the

$$\Delta h\nu = h\nu_X - h\nu_0 \qquad (73)$$

charge-transfer band frequency for the particular arene and electrophile, and v_0 the corresponding frequency for the benzene reference and electrophile. The significance of Kochi's result is that *ground state reactivity correlates directly with the frequency of spectroscopic excitation of the reacting molecules.* In other words barrier heights and spectroscopic excitation energies are in some way related. The above result has two obvious consequences. First, reactivity may be estimated from a compilation of spectral data. This is of technical utility. Secondly, and more importantly, the correlation provides information about the factors governing barrier heights. Let us see how this correlation lends support to the CM model.

The product configurations for the rate determining step of an electrophilic substitution process (71) are those configurations which describe the Wheland intermediate [11]. One such VB configuration is [12], while others may be generated by delocalizing the positive charge around the aromatic ring ([13] and [14]).

[12] [13] [14]

Since [12]–[14] are generated by an electron transfer from the arene to E^+, together they represent the configurational contributors to the *charge-transfer state* of the arene-E^+ system. Thus the first step of the electrophilic substitution reaction is seen to be generated by the avoided crossing of ground and charge transfer states (described by DA and D^+A^- configurations).

Inspection of Fig. 18 illustrates why the barrier height and charge-transfer bands are related. If the barrier height is governed by the avoided crossing of DA and D^+A^- configurations, then it is clear that DA → D^+A^- excitation will correlate with the barrier. Stabilization of D^+A^- (using a better arene donor) will decrease the charge transfer band frequency ($\Delta h v$) and correspondingly decrease the barrier height ($\Delta E^{\neq}$) so that these two parameters are related (Fig. 18). This is just a simple application of a diagram of the Bell–Evans–Polanyi type where the anchor point for the excited configuration is given a physical meaning – in this case the energy of the charge transfer state of the reactants.

Since $\Delta h v$ manifests itself in the thermodynamics of the reaction, and $\Delta E^{\neq}$ is a kinetic term, the relationship between the two represents a Brønsted coefficient, and the ratio, $\Delta E^{\neq}/\Delta h v$, a measure of the Brønsted parameter, α. The fact that the Brønsted parameter for (72) equals 1 is somewhat surprising

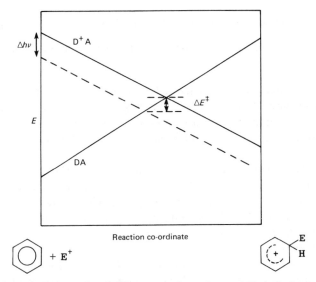

FIG. 18 Energy diagram for the first step of an electrophilic substitution reaction illustrating the crossing of DA and D^+A^- configurations. The effect of a substituent that stabilizes the D^+A^- configuration (e.g. by improving the arene donor ability) is indicated by the dotted line. The diagram illustrates the correlation between $\Delta h\nu$, the difference in excitation energies for the perturbed and unperturbed systems, and $\Delta E^{\neq}$, the difference in activation energy for the two systems. (Avoided crossing deleted for clarity)

(Fukuzumi and Kochi, 1981a). On the basis of a Bell–Evans–Polanyi diagram, an α-value closer to 0.5 would have been expected. However we do not feel that the α-value of 1 necessarily constitutes evidence for the ion-pair-like character of the transition state. On the basis of the CM model the transition state is described by (74). This would imply only c. 50% ion-pair character in

$$TS = (1/\sqrt{2})(DA \longleftrightarrow D^+A^-) \tag{74}$$

the TS. In our view the fact that α = 1 reflects the slope characteristics of DA and D^+A^- rather than the degree of charge development, though this point requires further clarification. The significance of α-values in S_N2 and proton transfer processes is discussed in Section 3 (pp. 149 and 167 respectively). The general significance of α-values is discussed further in Section 4.

In summary then we see that Kochi's work provides a direct experimental link between reactivity and reactant excited state energies. While at the present time the only excited state that has been correlated in this way is the charge transfer state, the CM model predicts that similar correlations for additional excited states are likely to exist. The significance of Kochi's work in relation to cycloaddition reactions is discussed on p. 176.

Rules for determining the reaction profile and the nature of the transition state

At this point it is worth summarizing the major rules necessary to apply the CM model.

1 The reaction profile is generated from a linear combination of VB or MO configurations. For simplicity VB configurations are preferred.

2 Two key configurations are those describing reactants and products. Additional configurations are obtained by seeking out chemically "sensible" intermediates.

3 The change in energy of any configuration along the reaction co-ordinate is governed by the electron distribution within that configuration. Thus, the interaction between A and X in the species AX is attractive for $A \cdot \downarrow \uparrow \cdot X$, but repulsive for $A \cdot \uparrow \uparrow \cdot X$, $A: \cdot X$ and $A: :X$. Interaction between the forms $A: \cdot X$ and $A \cdot :X$ is stabilizing, however, and may lead to a stable three-electron bond.

4 The reaction complex at any point along the reaction co-ordinate, including the transition state, will be described by a mixture of configurations in proportion to their relative stability. Configurations of low energy will mix into the complex more than configurations of high energy.

5 The reaction mechanism and hence the reaction co-ordinate will, themselves, be governed by the nature of the configurations from which the profile is built up. Thus, for example, for the set of nucleophilic substitution reactions a large contribution of the carbocation configuration will endow the entire reaction co-ordinate with S_N1 character.

6 The character of the transition state will reflect the extent to which the configurations mix into its wave-function. Thus, in general, the transition state will be endowed with the characteristics of the configurations of lowest energy in the vicinity of the transition state.

7 An intermediate is likely to be formed in a reaction whenever the configuration describing that intermediate is similar or lower in energy than reactant and product configurations in the region of the transition state.

8 Stabilization of any configuration, with respect to the reactant configuration, through a substituent or solvent effect, will generally lead to an energy lowering of the entire reaction profile. This results in an enhanced reaction rate and to the transition state acquiring more of the character of that configuration.

9 Configurations of different spin or spatial symmetry may not mix. For configurations which are related by a single electron transfer the extent of mixing is related to the overlap of the orbitals which differ in one-electron occupancy. For example, DA and D^+A^- for the face-on dimerization of two ethylenes do not mix significantly since the π_D and π_{A^*} orbitals cannot

overlap, while for the cycloaddition of ethylene and butadiene these configurations do mix.

3 Applications

In this section we apply the CM model to specific reaction types so as to understand these reactions in a more fundamental manner.

PHOTOCHEMICAL PROCESSES

In contrast to the limited application of configuration mixing and correlation diagrams to ground state reactivity, use of these tools in the discussion of photochemical reactivity is comparatively widespread. Indeed, essentially all mechanistic treatments of such processes are couched in these terms (see for example, Longuet-Higgins and Abrahamson, 1965; van der Lugt and Oosterhoff, 1969; Dauben *et al.*, 1975; Turro, 1978; Michl, 1972; Zimmerman, 1982). The purpose of this section, therefore, is to explore the manner in which the leading workers in this area have utilized the correlation diagram concept. The relation of these ideas to ground state reactions and the intimate relationship between thermal and photochemical processes will then become more apparent.

The application of correlation diagrams to photochemical processes goes back to the early work of Laidler and Shuler (1951). More recently, Longuet-Higgins and Abrahamson (1965) showed how orbital-symmetry correlations may be converted into state-symmetry correlations. Here the fundamental consideration is that states of the same symmetry do not cross, i.e. the result is an *avoided* crossing, while states of different symmetry do cross. The principles involved are illustrated in Fig. 19 for the dimerization of

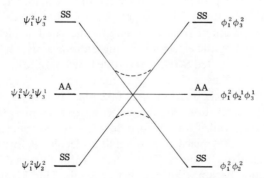

FIG. 19 State correlation diagram for the dimerization reaction of two ethylene molecules

two ethylene molecules. Since both the ground configuration of two ethylenes $\psi_1^2\psi_2^2$, as well as the diexcited configuration, $\psi_1^2\psi_3^2$ are totally symmetric, SS, an avoided crossing results between these two curves (dotted lines of Fig. 19). On the other hand, the monoexcited antisymmetric (AA) configuration, $\psi_1^2\psi_2^1\psi_3^1$, correlates with the antisymmetric configuration of the cyclobutane product, $\varphi_1^2\varphi_2^1\varphi_3^1$. On this basis the thermal reaction is expected to possess a large barrier ($\psi_1^2\psi_2^2 \to \varphi_1^2\varphi_2^2$) while the photochemical process will proceed smoothly and adiabatically, i.e. wholly along the excited state surface ($\psi_1^2\psi_2^1\psi_3^1 \to \varphi_1^2\varphi_2^1\varphi_3^1$).

While this state-correlation diagram makes the correct predictions concerning thermal and photochemical reactivity it does contain certain flaws. Firstly, the treatment seems to suggest that allowed photochemical processes yield products in an excited state. Experimentally, however, relaxation of the excited products to the ground state is not observed. A second problem has been pointed out by Dauben (1967) for the closely related photochemical ring closure of butadiene to cyclobutene (75).

$$\text{butadiene} \xrightarrow{h\nu} \text{cyclobutene} \tag{75}$$

Dauben noted that the first excited singlet state for cyclobutene is some 50–60 kcal mol^{-1} higher in energy than the corresponding singlet state for butadiene. Dauben concluded therefore that despite the implications of both the Woodward–Hoffmann rules (1965) and the Longuet–Higgins and Abrahamson approach, *the formation of the spectroscopic excited singlet state of cyclobutene from the corresponding singlet state of butadiene is unlikely.* What then is the precise pathway for the photochemical transformation?

The solution to this problem was provided in a brilliant paper by van der Lugt and Oosterhoff (1969). Their suggestion was that the driving force for reaction (75) comes about from a symmetric diexcited singlet state, S_2, of butadiene described by the electronic configuration, $\psi_1^2\psi_3^2$, and *not* from the antisymmetric monoexcited $S_1(\psi_1^2\psi_2^1\psi_3^1)$ state which is of lower energy (at the reaction beginning). This is illustrated in the correlation diagram of Fig. 20. So, while photochemical excitation *does* take place into the S_1 state the reaction proceeds by the reaction complex switching over from S_1 to the S_2 state. This switching over, while formally forbidden due to the opposite symmetries of the two states, takes place due to small molecular deformations which destroy the molecular symmetry. Thus what is formally a real crossing of S_1 and S_2 in practice becomes an avoided crossing. This breakdown in the prediction of the symmetry rules is significant. It makes the point that, while symmetry considerations provide an important guide to understanding the

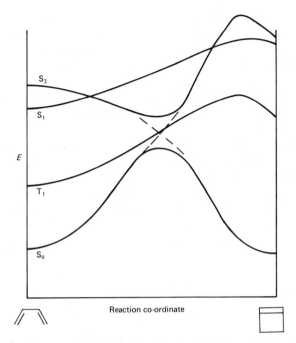

FIG. 20 Energy diagram that illustrates the S_0, T_1, S_1 and S_2 states for the photochemical disrotatory ring closure of butadiene. (Diagram adapted from Salem, 1982)

manner in which configurations mix, its predictive powers are nevertheless limited.

What is it about the S_2 state that makes it play such a crucial role in governing the photochemical process? The answer becomes readily apparent when we consider the simple VB representation for the doubly excited S_2 state. The VB representation, [15] is that of a double-triplet which together

[15]

comprises an overall singlet. This VB configuration is so important because it is the cyclobutene product configuration. It is immediately apparent that the electrons have been promoted in such a way so as to facilitate construction of a C(2)—C(3) double bond (a singlet interaction), and a C(1)—C(4) single bond (again a singlet interaction), precisely those interactions necessary to generate cyclobutene. So while [15] represents an excited configuration for butadiene, it represents a ground configuration for cyclobutene. Clearly then any excited reactant configuration which constitutes the primary descriptor of the product molecules will play a dominant role in describing the reaction pathway, even if this contribution only manifests itself late along the reaction co-ordinate. The very fact that the product configuration is of lowest energy at the product end of the reaction means that it must play a crucial role in governing both thermal and non-adiabatic photochemical processes. At some point the reaction complex will need to be largely described by this product configuration.

The reason for the basic form of Fig. 20 now becomes clearer. If we delete the T_1 curve from Fig. 20 then schematically it closely resembles the diagram of Fig. 11. At the reaction beginning, S_0 is described primarily by the reactant configuration, φ_R, while S_2 is described primarily by the product configuration, φ_P. At the reaction end the reverse is true. It is now S_2 that is composed mainly of φ_R, while S_0 is composed mainly of φ_P. The S_1 state is built up primarily from φ_I. We see therefore that the ground state reaction profile comes about from the avoided crossing of φ_R and φ_P configurations, and for this reason the S_2 surface has a minimum above the ground state surface maximum.

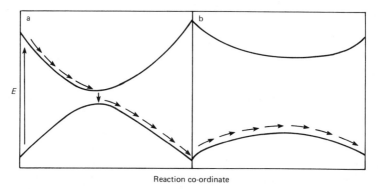

FIG. 21 Energy diagram for ground and excited state curves for (a) a thermally forbidden, photochemically allowed reaction, and (b) a thermally allowed, photochemically forbidden reaction

The reason that thermally allowed reactions are photochemically forbidden and vice versa also becomes more readily apparent (van der Lugt and Oosterhoff, 1969). As we have just noted the existence of a large ground state barrier for a particular reaction (i.e. the reaction is thermally "forbidden") implies that there is another potential surface of the same symmetry that has a well not far above this barrier (as in Fig. 21a). Photoinduced reactions may occur through the funnelling of the reaction complex from the excited surface well on to the ground state surface in a radiationless transition. Since the efficiency of the decay process is very much dependent on the energy separation between the surfaces (Devaquet, 1975), the photochemically allowed process will occur when the thermal barrier is high and close in energy to the excited state well (Fig. 21a). The thermal process will take place when the barrier is low in energy and far from the upper state surface (Fig. 21b). This, in turn, precludes the photochemical process. The way in which this occurs will be discussed in Section 3 (p. 173), where the configurations for thermally allowed and forbidden cycloaddition are presented.

The above reasoning concerning the importance of the second excited singlet state of butadiene during the cycloaddition reaction has far-reaching implications. It suggests that the simplified correlation diagrams used by Salem (1974) and Dauben et al. (1975), while useful for relating photoexcited reactants to products in different cases, do not provide a physical representation of the reaction pathway. Let us illustrate this point with the photochemical hydrogen abstraction by a ketone (69) discussed previously.

On the basis of the Salem diagram for this reaction (Fig. 15) the n,π* excited ketone [9] is considered to correlate smoothly with the product [16]. If

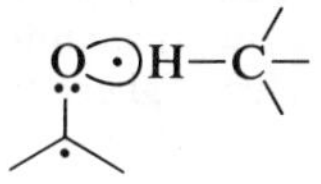

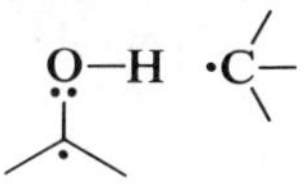

we consider earlier arguments, however, it is clear that the configuration describing the reaction complex composed of the n,π* excited ketone and ground C—H bond, [18], is *not* the true product configuration. While the ketone has been adequately prepared for bonding through excitation, the C—H bond in *its singlet state is unprepared for the hydrogen abstraction that is to follow*. A simple VB representation of the actual product configuration would be [19]. Here we see that in addition to n,π* excitation of the ketone

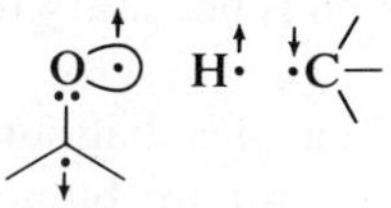

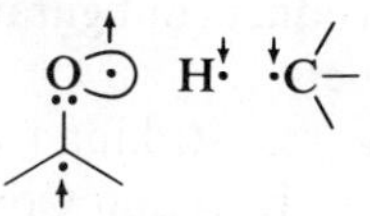

(shown here as the triplet), the C—H bond has also been prepared for bond-breaking by excitation *to the triplet state*. In this configuration the hydrogen atom is free to couple with the oxygen radical, and the methyl radical with the carbon radical centre – these interactions both being singlet interactions. It is apparent therefore why [19] is termed the *product* configuration. The reactants have been electronically excited to precisely the required electronic

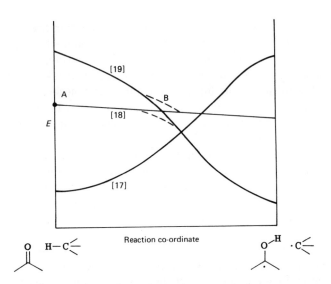

FIG. 22 Energy diagram for ground [17], n,π* excited [18], and product [19] configurations for the photochemical hydrogen abstraction reaction of ketones. Reaction proceeds from reactants up to point A (by photochemical excitation), along the n,π* excited configuration curve, [18], undergoes an avoided crossing with [19] (point B) and down to product

configuration which is the primary descriptor of the products. The above description has certain similarities to the $H_2 + D_2$ exchange process discussed earlier (Section 2, p. 120) in which the ditriplet (overall singlet) state was the appropriate excited state to correlate with products.

A correlation diagram which incorporates this additional configuration, [19], is shown schematically in Fig. 22. For simplicity only, three configurations, [17], [18] and [19] appear. Configurations [17] and [18] are respectively reactant and intermediate configurations used in the Salem diagram, [19] is the product configuration, which is necessary for a proper description of the products.

The first striking feature is that this diagram is remarkably similar to the van der Lugt and Oosterhoff picture for butadiene ring closure without the T_1 surface (Fig. 20). Initially the ketone is photo-excited to the n,π* state (point A). From this point the reaction proceeds along this "intermediate" configuration, [18], until this surface intersects the strongly descending product configuration, [19]. Since both [18] and [19] are 3σ, 3π states, the crossing between them is avoided (point B). From this point the reaction complex descends along the product configuration surface, crosses the reactant configuration (a true crossing since [17] and [19] are of different symmetry) finally to yield products.

It is interesting that *ab initio* calculations conducted by Salem (1974) are consistent with the three-configuration picture in which [19] is the strongly descending configuration, rather than the Salem diagram in which [18] is presumed to descend strongly. Energy surfaces for the 4σ, 2π and the excited 3σ, 3π states were explored. It was found, predictably, that the 4σ, 2π state rises sharply in energy along the reaction co-ordinate. On the other hand, the 3σ, 3π surface is relatively flat, in contrast to the prediction of the Salem diagram. This is due to the dominant influence of the intermediate configuration [18] in the early stages of the reaction. It is only once the reaction pathway switches from [18] to [19] that the energy is expected to drop sharply. The flatness of the 3σ, 3π surface is, therefore, consistent with the three-configuration picture (Fig. 22).

The implication of the above analysis may be summarized as follows: the Oosterhoff pattern, where a doubly excited configuration is the driving force for butadiene ring closure, is considered to be of wide generality. Both thermal and photochemical processes are governed by the nature of the product configuration, and for many organic reactions this configuration begins as an excited configuration of higher energy than other excited configurations which do *not* correlate smoothly with the products.

NUCLEOPHILIC SUBSTITUTION

In this section we apply the CM model to the family of nucleophilic substitution reactions. The aim of the discussion is to illustrate how reactivity trends and mechanistic variations within this class of reactions are readily accounted for in terms of the model (Pross and Shaik, 1981, 1982b,c, 1983; Shaik and Pross, 1982a,b; McLennan and Pross, 1984; Pross *et al.*, 1984; Pross, 1984).

In order to study the reaction profile for a nucleophilic substitution reaction (76), we first choose our basis set of VB configurations. The major configurations are obtained by rearranging the four key electrons which govern the reaction in all possible, energetically sensible forms. These are indicated

$$N:^- + R\text{—}X \rightarrow N\text{—}R + :X^- \qquad (76)$$

in [20]–[23]. Configuration [20] is the reactant configuration because it is the major descriptor of the reactant molecules, while configuration [21] is the product configuration since it is the major descriptor of the product molecules (i.e. N—R and X^-). We see immediately that [20] and [21] are the two key configurations from which the reaction profile is generated and involve

N:⁻ R· ·X   N· ·R :X⁻
[20]   [21]

$$\text{N:}^- \quad \text{R}^+ : \text{X}^- \qquad\qquad \text{N} \cdot \quad \text{R:}^- \cdot \text{X}$$
$$[22] \qquad\qquad\qquad\qquad [23]$$

the switching of an R· ·X bond pair to an N· ·R bond pair. There are, however, two additional configurations, [22] and [23], termed the carbocation and carbanion configurations, respectively, both of which are intermediate configurations, and which may also play a role in certain cases. Using the set of configurations [20]–[23] all nucleophilic substitution reactions, including S_N1, S_N2 and radical anion pathways, may be understood. Let us see how this occurs.

Two-configuration substitution reactions. Methyl derivatives

The S_N2 reaction profile of methyl derivatives may in simplest terms, be considered as resulting from the avoided crossing of reactant [20] and product [21] configurations. This is illustrated in Fig. 23a.

The reactant configuration [20] increases in energy along the reaction co-ordinate because an attractive Heitler–London R· ·X interaction is broken while a repulsive N:⁻ ·R interaction is generated. In similar fashion the product configuration [21] decreases in energy since an attractive N· ·R interaction is formed while the repulsive R· :X⁻ interaction is eliminated. At this point the intermediate configurations are not utilized since we may make the simplifying assumption that for methyl derivatives the carbocationic and carbanionic configurations do not play an important role. Of course this is an assumption which is not strictly valid. A moment's thought tells us that even in the ground state of the reactants, the carbocation configuration does play a role. This was pointed out in the discussion of the ground state description of the R—X bond in (16). However, for $R = CH_3$ we can make the simplifying assumption that the role of the intermediate configurations [22] and [23] is slight due to the relative instability of the CH_3^+ and CH_3^- moieties. Experimentally there is strong evidence to support this supposition. Kevill (1981) has put forward cogent arguments which suggest that little positive charge development takes place in the S_N2 transition state. It has been estimated that the rate of formation of the incipient carbocation in a limiting S_N1 process is increased by a factor of 10^8 by the introduction of a methyl substituent (Fry et al., 1972). Yet in an S_N2 reaction, a methyl substituent (converting a methyl derivative to an ethyl derivative) *slows down* the reaction rate by a factor of c. 20 (Arnett and Reich, 1980). Even taking possible steric effects into account the above data point to little *positive* charge development in S_N2 transition states of methyl derivatives.

A similar analysis suggests the absence of significant *negative* charge development as well. The fact that β-haloethyl derivatives have only a weak deactivating effect, rather than an activating effect on S_N2 reactivity (Hine and Brader, 1953; Hine, 1962; Streitwieser, 1962) implies that there is no significant negative charge development in the transition state. This follows since β-halogen substituents have a powerful stabilizing effect on carbanions (Pross and Radom, 1980). For example, the β-fluoro substituent in $FCH_2CH_2^-$ stabilizes the ethyl anion by $c.$ 22 kcal mol^{-1} based on 4-31G *ab initio* calculations. So, for methyl derivatives at least, our simple two configuration model seems reasonable.

What can we learn from this simple picture of an S_N2 process for methyl derivatives? A number of results immediately follow, some of them quite unexpected and contrary to conventional wisdom.

1 S_N2 reactions are single electron shift processes (Pross and Shaik, 1983).

2 S_N2 charge development is likely to be about 50% advanced regardless of whether the transition state is early or late (Pross and Shaik, 1982c).

3 Normal rate–equilibrium relationships are expected for methyl S_N2 processes but the Brønsted parameter α is *not* a measure of TS charge development in these cases (Pross, 1984).

Let us now discuss these in turn. The first conclusion, a rather controversial one, is that an S_N2 reaction really only involves a *single electron shift*. This possibility has been proposed both in the past and recently (Bank and Noyd, 1973; Chanon and Tobe, 1982), but appears to have been largely dismissed. The CM analysis, however, lends considerable weight to this view, since the only electronic change that is necessary to convert reactant configuration [20] to product configuration [21] is a *single electron shift* (77)

$$N:^- \quad R\uparrow \downarrow X \longrightarrow N\uparrow \quad \downarrow R \quad :X^- \qquad (77)$$

(Pross and Shaik, 1983). The conventional view which describes an S_N2 reaction as a *two-electron* process in contrast to the electron-transfer $S_{RN}1$ pathway (Bunnett, 1978; Kornblum, 1975) is, at best, misleading. The traditional "curly arrow" picture for an S_N2 reaction (78) implies that the nucleophile "attacks" with *two* electrons and that the leaving group leaves with two

$$N:^- \quad R-X \longrightarrow N-R + :X^- \qquad (78)$$

electrons. Such a process is physically unclear. Even on the basis of the conventional picture it should be apparent that two electrons are *not* involved – the formal charges of both nucleophile *and* leaving group change by 1, consistent with a single electron shift process. A configuration picture, however, makes this point immediately apparent and we would accordingly suggest that so-called "two-electron" processes be termed in future *spin-coupled single electron shift processes,* to emphasize that the process involves a *single* electron shift in which electron pairs *remain coupled at all times*. The actual difference between the above process and an electron transfer process in which radicals *are* formed is discussed subsequently.

The second conclusion to come out of the CM analysis, regarding the amount of charge development in S_N2 reactions, is also contrary to established physical organic thinking. For example, in the S_N2 reaction of an anionic nucleophile with a neutral substrate (76), most of the negative charge would be expected to be localized on the nucleophile for an early transition state, as indicated in [24] while for a late transition state most of the charge would be expected to be localized on the leaving group, as indicated in [25].

$$\overset{\delta-}{N}\text{------------------}R\text{-------}X \qquad \overset{\delta\delta\delta-}{N}\text{-------}R\text{------------------}\overset{\delta-}{X}$$
$$\overset{\delta\delta\delta-}{} \qquad\qquad\qquad\qquad\qquad$$

[24] \qquad\qquad\qquad\qquad [25]

We believe that the above supposition is invalid and that *charge development in the transition state, in general, is unrelated to the position of the transition state along the reaction co-ordinate*. Inspection of Fig. 23a suggests that the position of the transition state along the reaction co-ordinate will be in the vicinity of the intersection point of the two VB configuration curves. This means that the wave-function, ψ_{TS}, describing the transition state will incorporate the two configurations in equal amounts (79). On the basis of

$$\psi_{TS} = (1/\sqrt{2}) [(N:^- \quad R \cdot \ \cdot X) + (N \cdot \ \cdot R \ :X^-)] \qquad (79)$$

(79), *the charge on the nucleophile and leaving group will be the same and equal to 0.5*, and this conclusion holds *regardless of the position of the transition state along the reaction co-ordinate*. If the nucleophile or leaving group is modified so as to make the reaction faster, the product configuration is stabilized (dotted line in Fig. 23b), then the transition state becomes earlier in accord with the Bell–Evans–Polanyi principle (Bell, 1936; Evans and Polanyi, 1938; see also Dewar, 1969; Hammond, 1955; Agmon, 1978; Lewis, 1978).

However, the electron shift also takes place earlier (Fig. 23b) so that the transition state remains at the crossing point of [20] and [21] with about 50% charge development. Note that the slight mixing-in of the intermediate configurations, [22] and [23], does not modify the basic conclusion. Mixing in of the carbocationic configuration will tend to increase the net negative charge on both nucleophile and leaving group while mixing in of the carbanion configuration will tend to decrease these corresponding charges. However, if we assume that the mixing in of intermediate configurations does not significantly disturb the equality of the contribution of the major contributors, [20] and [21], then the charge on the leaving group and the nucleophile in the

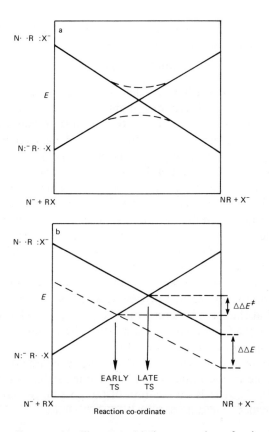

FIG. 23 Energy diagram that illustrates (a) the generation of a simplified S_N2 reaction profile from reactant, $N:^- \ R\cdot \ \cdot X$, and product, $N\cdot \ \cdot R \ :X^-$, configurations. Dotted lines denote avoided crossing (i.e. the reaction profile after configuration mixing). (b) The effect of stabilization of $N\cdot \ \cdot R \ :X^-$ (e.g. by a substituent effect) is indicated by dotted lines. Arrows indicate the positions of the transition states with and without the stabilization

transition state *will* remain the same (though not necessarily 0.5). The relationship between transition state charge development and geometry will be discussed further in Section 4 (p. 186).

The existence of rate-equilibrium relationships in the S_N2 reactions of simple alkyl derivatives is well established (Arnett and Reich, 1980; Lewis and Kukes, 1979; Lewis *et al.*, 1980; Bordwell and Hughes, 1982). A plot of the rate constants for a family of nucleophiles against the pK_a for the nucleophiles generates linear Brønsted plots whose slopes lie in the range 0.3 to 0.5. A typical example, taken from Bordwell's work (Bordwell and Hughes, 1982) is the reaction of a family of aryl thiolates with *n*-butyl chloride in DMSO (80). A plot of the log of the rate constants against pK_a gives a linear correlation with a slope of 0.4.

$$ArS^- + n\text{-BuCl} \rightarrow n\text{-BuSAr} + Cl^- \tag{80}$$

The question that now arises is: What is the molecular significance of the Brønsted parameter? Intuitively one might argue that the magnitude of the Brønsted parameter is a measure of the transition-state charge development. Since the effect of the substituent on the pK_a for the nucleophile reflects the interaction of the substituent with a *full* negative charge, it seems reasonable to assume that the relative effect of the substituent on the transition state may provide a measure of the transition state charge. Thus if the rate constant responds to substituent effects by just half the response of the corresponding equilibrium constant then the transition state charge is deduced to be half of that in the product (or reactant). Analysis of the transition state charge for a particular reaction based on the CM model suggests that the above concept is overly simplistic. For a reaction adequately described by just two configurations, it appears that the concept is invalid. On the basis of the two-configuration description of the S_N2 process (Fig. 23) α is defined by (81). We assume here that changes in energies and free energies are linearly related. It seems that α is not a function of transition state charge but is governed by the *slopes of the configuration curves* (Bell, 1973). If the two curves have identical slopes then the value of α equals 0.5. However, depending on the respective slopes, α may lie in the range 0–1. It is of interest that in Bell's detailed monograph (1973) there is no mention of the possibility that the Brønsted parameter provides a measure of transition state charge.

$$\alpha = \Delta\Delta E^{\neq}/\Delta\Delta E \tag{81}$$

If the intersecting curves are parabolic in form, then the reactivity pattern expected is described by the Marcus equation (Marcus, 1964, 1977). The magnitude of α may then be shown to be that in (82). Here $\Delta G_o^{\neq}$ is the intrinsic barrier for reaction and ΔG° the free energy of reaction.

$$\alpha = \frac{1}{2}(1 + \Delta G^\circ/4\Delta G_o^{\neq}) \tag{82}$$

Equation (82) predicts that for reactions with zero free energy change $\alpha = 0.5$, while for exothermic reactions, $\alpha < 0.5$ and for endothermic reactions, $\alpha > 0.5$. Since according to both Marcus theory and the Bell–Evans–Polanyi model early transition states are related to exothermic reactions and late transition states to endothermic reactions, α may be interpreted as a relative measure of transition state *geometry*. However, in our view even this interpretation should be treated with a measure of healthy scepticism. Even if one accepts Marcus theory without reservation, the α

value may be affected by a number of additional parameters. These include solvation factors (Jencks et al., 1982; Hupe and Jencks, 1977; Hupe and Wu, 1977) which in Marcus terminology may appear as variable work terms (Murdoch, 1972; Agmon, 1980). Alternatively, "perpendicular" effects have recently been shown by Kreevoy and Lee (1984) to influence α-values. In configurational terms this just means the involvement of additional configurations (Pross, 1984) while in Marcus terms this means a variable intrinsic barrier (Marcus, 1969).

In conclusion, for the S_N2 reactions of primary alkyl derivatives, the author is of the view that the Brønsted parameter α does not constitute a measure of transition state charge development, and is unlikely to represent a meaningful measure of transition state geometry. As we shall see in the following section, however, for reactions described by *three* or more configurations, the Brønsted parameter or Hammett ρ parameter may provide a *relative* measure of transition state charge development (Pross, 1984).

Multiconfiguration substitution reactions

Whereas the S_N2 reaction of methyl or primary alkyl derivatives may be simply represented by just two configurations, introduction of an α-substituent into the R group is likely to bring into play one of the possible intermediate configurations [22] or [23]. It is the consideration of these two additional configurations which enables one to analyse and explain in a straightforward fashion: (*a*) mechanistic variations and the possible formation of intermediates within the range of substitution reactions; (*b*) the mechanistic significance of the Brønsted and Hammett parameters α(β) and ρ; (*c*) the basis for non-linear free energy relationships; and (*d*) the significance of anomalous Brønsted parameters ($\alpha > 1$).

Substitution of the central carbon of RX may induce significant mechanistic variations within the range of nucleophilic substitution reactions. For example, introduction of electron-releasing substituents on R will lead to a stabilization of the carbocation configuration [22]. An excellent illustration of this type of process is the reaction of methoxymethyl derivatives (83) taken from Jencks' work (Knier and Jencks, 1980). The carbocation within [22] is stabilized by the resonance interaction (84). Thus the configuration diagram

$$N^- + CH_3OCH_2X \rightarrow CH_3OCH_2N + X^- \tag{83}$$

$$CH_3O-CH_2^+ \longleftrightarrow CH_3\overset{+}{O}=CH_2 \tag{84}$$

for this reaction will be that represented by Fig. 11, and the transition state will be described as the hybrid of all three configurations [20], [21] and [22], as indicated in (85). The reaction remains a single-step S_N2 process but, due to

$$TS \equiv N{:}^-\ R{\cdot}\ {\cdot}X \longleftrightarrow N{\cdot}\ {\cdot}R\ {:}X^- \longleftrightarrow N{:}^-\ R^+\ {:}X^- \tag{85}$$

the influence of the carbocation configuration [22], takes place through an open or "exploded" transition state. For this reaction a β_{nuc} value (obtained from a plot of $\log k_{nuc}$ vs pK_{nuc}) of 0.14 and a β_{lg} (lg ≡ leaving group) value of -0.7 to -0.9 *does* provide a relative measure of transition state charge development and confirms the description provided by (85) (Pross, 1984). Let us elaborate on this point by consideration of the configurational description of the transition state in (85).

If a substituent is introduced into the nucleophile, its effect on reactivity will be *small* since only *one* of the three configurations which describe the transition state contains the nucleophile in its product form (N·). In other words, as far as the nucleophile is concerned the transition state is reactant-like and as a consequence β_{nuc} (or the Hammett ρ parameter) is small in magnitude. For the leaving group, however, the situation is reversed. Since now *two* of the three configurations which make up the transition state, describe the leaving group in its product form (i.e. $X{:}^-$), the effect of substituents located on the leaving group will be *large* in magnitude. Thus for this kind of a reaction, where at least *three* configurations are necessary to describe the transition state adequately, the magnitude of the Brønsted parameter (β_{nuc} or β_{lg}) *does* provide a measure of transition state charge. It should be pointed out however that the magnitude of β should not be interpreted in terms of the precise charge in the transition state. Clearly the actual value of β is governed by curve slopes, as well as the extent of configuration mixing, and therefore should only be utilized as an approximate and relative measure of transition state charge (Pross, 1984).

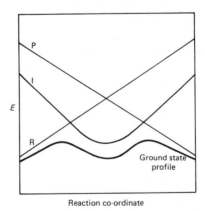

FIG. 24 Energy diagram that illustrates how a two-step reaction profile (bold line) is generated from a reactant R, product P, and low-lying intermediate I, configuration

CONFIGURATION MIXING MODEL

If the intermediate configuration is strongly stabilized, then the situation described schematically in Fig. 24 obtains. An actual intermediate is formed due to the dominant influence of the intermediate configuration. For the case of [22] the intermediate would be described as a nucleophilically solvated ion-pair species. The existence of such intermediates has been proposed by Bentley, Schleyer and coworkers (Bentley et al., 1981). We see therefore that a concerted, one-step process becomes a two-step process as the result of the stabilization of the intermediate configuration. It should be noted, however, that as the intermediate configuration is stabilized, the actual reaction co-ordinate changes. This means that in the limit the reaction profile can no longer be described in terms of a coupled process where the motions of the nucleophile and leaving group are synchronized. For nucleophilic substitution reactions the limit is, of course, the S_N1 pathway, whose first step was described in Fig. 14.

An interesting problem was recently posed by Jencks (1980) concerning the question of concerted vs stepwise reactions. Jencks proposed that concerted pathways are followed when potential intermediates are too high in energy to be generated along the reaction pathway. The present analysis is consistent with this point of view: concerted pathways are enforced when the possible intermediate configurations are too high in energy to mix into the wave function describing the reaction profile. A stepwise reaction, if available, is always preferred to the high energy concerted process.

Let us now consider the effect on an S_N2 reaction of stabilizing the *carbanion configuration*. If the carbanionic configuration [23] is the one that is stabilized – for example, by an α-carbonyl group – then the transition state is expected to take on carbanionic character. This is indeed what is observed for the S_N2 reaction of a family of α-carbonyl derivatives indicated in (86)

$$N^- + R'COCH_2X \rightarrow R'COCH_2N + X^- \tag{86}$$

(McLennan and Pross, 1984). For this reaction the transition state may be described by the resonance forms (87). This is because the carbanion con-

$$TS \equiv N:^- \; R\cdot \;\cdot X \longleftrightarrow N\cdot \;\cdot R \; :X^- \longleftrightarrow N\cdot \; R:^- \;\cdot X \tag{87}$$

figuration, [23], is stabilized through resonance delocalization as shown in (88). If the stability of [23] is, in fact, important in governing the reactivity of

$$\begin{array}{ccc} ^-C-C & \longleftrightarrow & C=C \\ \parallel & & \mid \\ O & & O^- \end{array} \tag{88}$$

α-carbonyl derivatives, then one would expect that the reactivity sequence of a family, RCH_2Cl, with different R substituents, would parallel the relative stability of the corresponding carbanion, RCH_2^-. This pattern is indeed observed and is illustrated by the data in Table 1 (McLennan and Pross, 1984). Both gas-phase and solution acidity data parallel the S_N2 reactivity sequence.

TABLE 1

Relative reactivity of RCH_2Cl with KI in acetone, solution acidities of RCH_3 in DMSO, and relative gas-phase acidities of RCH_3[a]

R	Relative substitution rate with KI in acetone at 50 °	pK_a of RCH_3 (DMSO)	$\delta\Delta G°$acid (kcal mol^{-1})
CH_3CH_2—	1	—	—
Ph—	250	40.9	36.3
EtOCO—	1600	—	44.3
NC—	2800	31.3	44.2
CH_3CO—	33,000	26.5	47.0
PhCO—	97,000	24.7	52.2

[a]Compiled by McLennan and Pross, 1984

Theoretical evidence in support of the above discussion exists. Pross et al. (1984) studied the energies of the α-carbonyl S_N2 transition state in both "on" and "off" conformations, [26] and [27] respectively. In the "on" conformation, interaction between the N---C---X partial bonds and the π_{CO} system is permitted, while in the "off" conformation such an interaction cannot occur. It was found that conformation [26] is substantially more stable than [27], consistent with enolate delocalization in [26]. Furthermore,

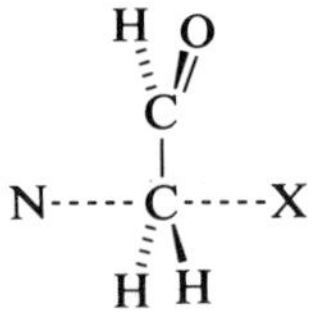

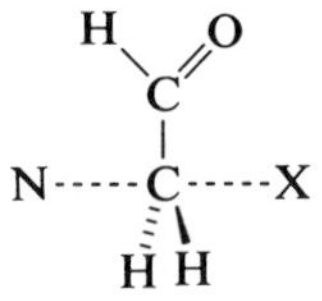

the relatively short C—C bond and relatively long C—O bond in [26] compared to [27] is also consistent with carbanionic delocalization into the carbonyl group. Similar conclusions have been reached by Kost and Aviram (1982) for the BH_2 substituent. The large negative ρ-value (-1.87) for the reaction of substituted pyridines with phenacyl bromide (Forster and Laird, 1982) compared to the corresponding value (-0.85) for dodecyl bromide (Murai et al., 1973) is also consistent with a transition state described by (87), (McLennan and Pross, 1984).

The basis for the curved Hammett plots observed for nucleophilic substitution of ring-substituted benzyl derivatives now becomes apparent. The benzyl system is unique in that both carbocation [22] *and* carbanion [23] configurations may play a role. If an electron-releasing group (e.g. p-OCH$_3$) is incorporated into the ring then [22] will actively participate and the transition state for the reaction will be described by (85). If an electron-withdrawing group (e.g. NO$_2$) is incorporated into the ring then [23] will be the effective "intermediate" configuration and the transition state will be described by (87). Thus for benzyl ring substitution *both* electron-withdrawing *and* -releasing substituents may induce a rate enhancement, since in both cases an intermediate configuration is stabilized. Such an effect will lead to curved Hammett plots. The measured ρ-values for substituents in the nucleophile (e.g. a series of substituted anilines) are entirely consistent with this picture. A large absolute value is observed for the reaction of a p-nitrobenzyl derivative, a small absolute value is observed for a p-methoxybenzyl derivative, (Ballistreri *et al.*, 1976) consistent with transition states described by (87) and (85), respectively (Pross, 1984). Solvent activity coefficients measured by Parker (1969) are also consistent with this picture.

Application of kinetic isotope effect data in the benzyl system is confusing. Chlorine leaving group isotope effects obtained by Hill and Fry (1962) and Grimsrud and Taylor (1970) are constant in certain reaction families but variable in others, making interpretation of the data difficult. The role of solvent in determining leaving-group kinetic isotope effects must also be considered (Thornton, 1964). In any event secondary deuterium isotope effects on C_α obtained by Westaway and Waszczylo (1982) are consistent with a shorter nucleophile–carbon bond with electron-withdrawing groups on the ring, in agreement with the CM analysis. $^{14}C/^{12}C$-Kinetic isotope effect data obtained by Yamataka and Ando (1979) also confirm that ring substituents have a marked influence on the tightness of the transition state.

S_N2 and electron transfer reactions

We may now ask the question: What are the factors which govern whether the reaction between a nucleophile N and a substrate RX will take place by an S_N2 process or by an electron transfer mechanism in which radical species are generated? The first step (89) in the $S_{RN}1$ pathway exemplifies electron transfer.

$$N:^- + A \rightarrow N\cdot + A^{\overline{\cdot}} \tag{89}$$

The fundamental mechanistic difference between the S_N2 and electron transfer processes is whether *after* the single electron shift takes place, an intermediate is formed or not. Any factor capable of *delaying* N---R coupling (77) after the single electron shift may lead to the actual generation of radical intermediates. Let us explore what these factors may be.

Electron transfer from N to RX generates the species $N\cdot(R \therefore X)^-$. The fate of this species depends on the stability of the three-electron bond $R \therefore X$. If the species is stable then $N\text{---}R$ coupling does not readily occur and two radicals are generated, $N\cdot$ and $(R \therefore X)^-$. The reason is twofold. Firstly, if the $(R \therefore X)^-$ bond is stable then the $R\text{---}X$ linkage needs to be broken in order to enable $N\text{---}R$ coupling to take place. Second, if $(R \therefore X)^-$ is stable this means that there is a significant contribution of the resonance form $R:^- \cdot X$ to the VB description (49) of the three-electron bond. Since $N\text{---}R$ coupling requires the R group to be present as $R\cdot$, significant contribution of the form $R:^-$ will inhibit the coupling process. Or expressed differently, in order for the nucleophile $N\cdot$ to couple to $R\cdot$, the *delocalized* three-electron bond $(R \therefore X)^-$ must be *localized* to just the $R\cdot :X^-$ form. This localization requirement inhibits the $N\text{---}R$ coupling reaction and enables free radicals to be detected. Thus species capable of generating stable three-electron bonds will tend to undergo electron transfer reactions, while those that generate weak three-electron bonds (primarily described by $R\cdot :X^-$) will tend to undergo S_N2 reactons.

In Section 2 (p. 114) we noted that the stability of the $(R \therefore X)^-$ three-electron bond is largely governed by the electronegativity difference between R and X. The more electronegative the R group, the stronger the three-electron bond. What this means is that as R becomes more electronegative its electron affinity *increases* but its ability to release the leaving group decreases. The effect then of increasing electron-withdrawing groups on R is to bring about a shift in mechanism from an S_N2 reaction to an electron transfer reaction. On this basis CH_3Cl is expected to undergo an S_N2 process, while CF_3Cl or CCl_4 are more likely to undergo electron transfer reactions. Indeed CX_4 molecules are known to react with amines by electron transfer (Wyrzykowska *et al.*, 1978; Iwasaki *et al.*, 1978).

The most common electron transfer process in organic chemistry, the $S_{RN}1$ reaction, takes place somewhat differently. This particular reaction [e.g. (90)]

$$p\text{-}O_2N\text{-}C_6H_4\text{-}CH_2Cl + (CH_3)_2\bar{C}\text{-}NO_2 \longrightarrow p\text{-}O_2N\text{-}C_6H_4\text{-}CH_2\text{-}C(CH_3)_2NO_2 + Cl^- \quad (90)$$

occurs because electron transfer takes place into a low-lying orbital *other* than σ^*_{R-X}. Nitroaromatic substrates, which commonly undergo electron transfer reactions, have a low-lying π^* orbital which can readily accept an odd electron [first step, (89)]. Since $N\text{---}R$ coupling cannot occur while the electron resides in the π^* orbital, radical species are generated, e.g. [28]. In

$$\underset{[28]}{\underset{NO_2}{\underset{|}{\bigcirc}}\text{–}\overset{CH_2Cl}{|}} \xrightarrow{e^-} \underset{NO_2}{\underset{|}{\bigcirc}}\text{–}\overset{CH_2Cl}{|} \longrightarrow \underset{[29]}{\underset{NO_2}{\underset{|}{\bigcirc}}\text{–}\overset{CH_2\dot{-}Cl}{|}} \longrightarrow \underset{[30]}{\underset{NO_2}{\underset{|}{\bigcirc}}\text{–}\overset{CH_2^{\cdot}}{|}} + Cl^- \quad (91)$$

order for the reaction to proceed the electron must relocate into the σ^*_{R-X} orbital, e.g. [29]. This species then may undergo dissociation to produce the primary radical species, e.g. [30] from which products are formed.

In summary, we see that both S_N2 and $S_{RN}1$ processes involve a single electron shift. In the S_N2 case the electron shift occurs synchronously with electron coupling. In the $S_{RN}1$ case the electron shift leads to the formation of radical intermediates.

State correlation diagrams

An alternative manner of building up the reaction profile for an S_N2 reaction is based on the state correlation diagram (Shaik and Pross, 1982b; Shaik, 1983, 1984). This approach is of theoretical interest since it provides further insight into the factors which govern barrier formation.

The procedure is based on state curves rather than configuration curves. This has the advantage that the energies of the available states of a system are generally accessible from simple physical measurements such as ionization potential, electron affinity, and from spectroscopy. On the other hand it suffers from the disadvantage that it is more complex to apply, thereby limiting its usefulness to some extent. Let us illustrate the principles by building up the state correlation diagram for an S_N2 reaction.

The S_N2 profile may be considered to result from the avoided crossing of two state curves. One curve begins as the ground state of the reactants, $N:^- (R-X)$, and ends up as the charge transfer state of the products, $(N\dot{-}R)^- \cdot X$, while the second curve begins as the charge transfer state of the reactants, $N \cdot (R\dot{-}X)^-$, and ends up as the ground state of the products $(N-R) :X^-$. The state correlation diagram is illustrated in Fig. 25. As described earlier, any state may be described in terms of a linear combination of configurations, for example, in VB terminology (92) and (93). Thus the

$$N:^- (R-X) = N:^- R \cdot \cdot X + \lambda(N:^- R^+ :X) \quad (92)$$

$$N \cdot (R\dot{-}X)^- = N \cdot \cdot R :X + \lambda(N \cdot R:^- \cdot X) \quad (93)$$

ground state is described primarily by the reactant [20] and carbocationic [22] configurations, while the charge transfer state is described primarily by the

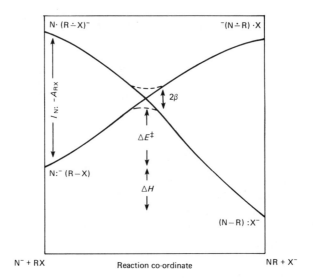

FIG. 25 State correlation diagram for an S_N2 reaction. The lower states are ground states of reactants and products, while the upper states are the corresponding charge transfer states. β is the degree of avoided crossing. $\Delta E^{\neq}$ is the reaction barrier

product [21] and carbanion [23] configurations. Given that the state diagram is composed of a ground state charge transfer state avoided crossing, it re-emphasizes that the S_N2 process involves a single electron shift followed by bond reorganization.

The barrier for the reaction may be seen to be some fraction of the energy gap between the ground and excited states. Quantitatively this may be written as in (94), where f is some fraction, I_N is the ionization potential of the

$$E^{\neq} = f(I_N - A_{RX}) - \beta \qquad (94)$$

nucleophile (N:$^-$), A_{RX} is the electron affinity of the substrate, and β is the avoided crossing constant. I_N and A_{RX} are physical parameters and are thus readily available; f and β are however not easily estimated. The magnitude of f depends on the slopes of the curves. For a small initial slope, f is large, as indicated in [31], while f is small if the slope is large, as shown in [32]. The

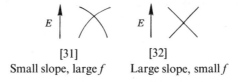

[31]   [32]
Small slope, large f Large slope, small f

slope, and hence f, depend on the degree of delocalization of the species in the charge transfer state. If the charge transfer species are highly delocalized then both N---R coupling and R---X bond-breaking are inhibited, leading to a small energy decrease in the initial stages of the reaction co-ordinate, i.e. a small slope and large barrier.

On the basis of this model, substituent effects on reactivity are assessed by examining the effect of substituents on the energy gap between ground and excited states and on the curve slopes. Large barriers will be observed for substituents which *increase* the gap and decrease the slope and vice versa. For example, introduction of an electron-withdrawing substituent on the leaving group, X, is expected to *decrease* the barrier because both the gap and slope factor f are decreased. The gap decreases because A_{RX} is increased while the factor f decreases because the electron-withdrawing substituent localizes the three-electron bond to the R· :X$^-$ form, thereby weakening it. This is of course, consistent with experiment; electron-withdrawing groups improve the leaving group ability (e.g. $CH_3SO_3^-$ is a poorer leaving group than $CF_3SO_3^-$).

If the substituent is situated on the central carbon the model fails to make a clear-cut prediction. This is because the effect of substitution on the gap and slope factor f are opposed. For example, an electron-withdrawing substituent (e.g. CH_2Cl_2 compared to CH_3Cl) is expected to *decrease* the gap and *increase* the slope factor, so that without more extensive informaton on the effect of the substituent on the above parameters, no definitive prediction is possible. A second limitation of the model is that it does not readily lend itself to treating mechanistic variations. Since its more detailed structure favours its application with just two state curves, in this case ground and charge transfer states, it is less readily extended to situations in which mechanistic variations are explicitly considered (e.g. the S_N2–S_N1 spectrum).

In summary, it appears that the state correlation approach represents a physically more realistic representation of the factors governing the formation of reaction barriers than the corresponding configuration treatment utilized in the bulk of this review. However, its practical utility as a readily applied model is handicapped by its more complex structure and the fact that its two-curve basis does not readily lend itself to the consideration of mechanistic variations.

Reactivity trends in identity exchange processes

We end this section by illustrating that the CM model is useful not only to rationalize existing data but may also make clear predictions concerning questions currently under study.

Lewis (1983) has posed the intriguing question: What is the effect of simultaneous substitution on both nucleophile and leaving group on reactivity for an identity exchange process, such as (95)? In other words, if we introduce a particular substituent into the aromatic ring of both nucleophile

$$ArSO_3^- + CH_3\text{—}O_3SAr \rightarrow ArSO_3\text{—}CH_3 + {}^-O_3SAr \qquad (95)$$

and leaving group, will the reaction rate vary, and, if so, will it be faster or slower? Preliminary results by Lewis suggest that the above reaction has a ρ-value of +0.6. This, of course, means that substitution in the leaving group is more important than in the nucleophile. Thus an electron-withdrawing substituent on the leaving group enhances reactivity *more* than the same substituent on the nucleophile retards the reaction.

Analysis of the S_N2 process in terms of the four configurations [20] to [23] indicates that the effect of substituents on the identity exchange process is not affected by [20] and [21]. This is because for N = X these two configurations are equivalent, and the effect of substitution on reactivity for these two configurations will exactly cancel out. The change in reaction rate will therefore be governed by configurations [22] and [23].

If the carbocationic configuration [22] is dominant over [23] then the reaction will be enhanced by electron-withdrawing substituents, since [23] contains both nucleophile and leaving group in the free anionic form $X:^-$. This will result in the observation of a *positive* ρ-value. Contrariwise, if the carbanion configuration [23] is dominant, then the reaction will be enhanced by electron-releasing substituents and a *negative* ρ-value will result.

The observation of a positive ρ-value (+0.6) for reaction (95) indicates that in the transition state the contribution of [22] is somewhat larger than [23]. Any perturbation on the identity exchange process which enhances the contribution of [22] in comparison to [23] will result in an increased ρ-value. Of course, the reverse applies also; if the perturbation increases the contribution of [23] compared to [22], the ρ-value will decrease and may even become negative if the influence of [23] outweighs that of [22].

Let us now point out some consequences of the above analysis.

1 A carbocationic stabilizing substituent on R which increases the importance of [22] compared to [23] will lead to a larger ρ-value for the identity-exchange family. Thus the ρ-value for the series (96) which incorporates the carbocationic stabilizing substituent CH_3O on R is predicted to be large and positive.

$$X^- + \underset{\underset{OCH_3}{|}}{CH_2X} \rightarrow \qquad (96)$$

2 A carbanionic stabilizing substituent on R is predicted to do the reverse. Thus the reaction (97) is predicted to exhibit a negative ρ-value.

$$X^- + CH_2X \rightarrow \atop | \atop CH_3CO \qquad (97)$$

3 If the family of nucleophiles is greatly strengthened in comparison to the family of arenesulphonates used by Lewis, then the ρ-value is predicted to decrease since the stronger nucleophile is expected to stabilize [23] in comparison to [22]. Thus (98) is predicted to result in a ρ-value less than + 0.6 (obtained for the weak nucleophile $ArSO_3^-$).

$$ArS^- + CH_3-SAr \rightarrow \qquad (98)$$

Before concluding this section we should point out that some recent data by Kurz and El-Nasr (1982) on the substitution of methyl derivatives by substituted pyridines, 4-methylpyridines and 2,6-dimethylpyridines are quite unexpected. Based on $^{15}N/^{14}N$ isotope effects these workers concluded that the N---C bond length appeared to become tighter for the better leaving group. This is contrary to the prediction of the reactivity-selectivity principle (Pross, 1977; Giese, 1977; for an alternative view, see Johnson, 1975) which predicts an early transition state (little N---C bond-making) in reactive systems. In terms of a two-configuration treatment, such as we have used for S_N2 reactions of methyl derivatives, the result is also surprising. The two-configuration treatment, which is equivalent to the Bell–Evans–Polanyi curve crossing model discussed earlier, suggests *more* reactive systems will have less N---C bond-making (in geometric terms) so that Kurz's result is an intriguing one which requires further consideration.

With this exception we can see that the impact of the configuration mixing model on nucleophilic substitution reactions, which constitute the most widely studied organic reaction, is indeed extensive. The model readily rationalizes much available experimental data, relates the entire mechanistic spectrum within a single framework, challenges some fundamental precepts of physical organic chemistry and enables one to make reactivity predictions about reactions yet to be investigated. For such a simple, qualitative theory, this is no mean achievement.

ELIMINATION

Elimination reactions are of special interest with regard to the CM model, because they have been well characterized by the potential energy surface (PES) models. The PES models, developed and applied by Thornton (1967), More O'Ferrall (1970), Harris and Kurz (1970), Jencks (1972), Critchlow (1972), and Jencks and Jencks (1977), have made an enormous contribution to our present understanding of organic reactivity by clarifying the effect of substituents on structures not actually lying along the reaction co-ordinate – the so-called perpendicular effects. In this section, therefore, we will apply the CM model to the range of elimination reactions and compare the conclusions with those obtained by More O'Ferrall (1970) in his classic analysis of the elimination spectrum.

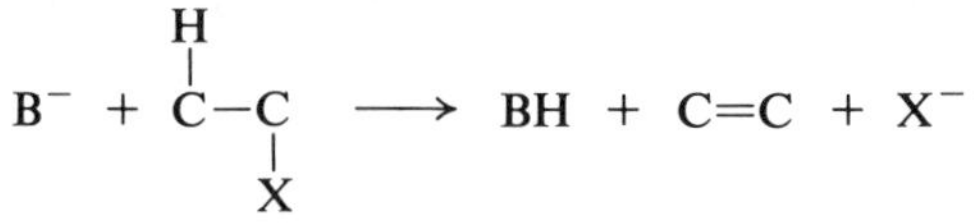

$$B^- + \underset{X}{\overset{H}{C-C}} \longrightarrow BH + C=C + X^- \qquad (99)$$

The reaction profile for an elimination reaction of a substrate with a base, shown in reaction (99), may be built up from a set of five VB configurations (Pross and Shaik, 1982a). These are indicated in [33]–[37] and involve the energetically "sensible" rearrangement of the 6 valence electrons about the five atomic centres.

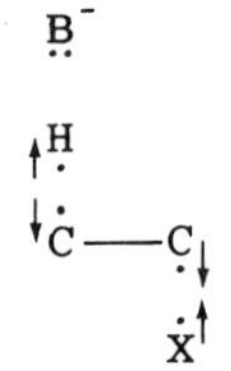

Reactant (R)	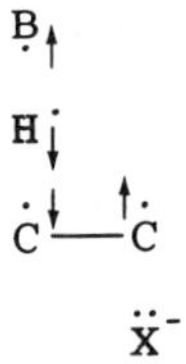 Product (P)	Carbanion (C⁻)	Carbocation (C⁺)
[33]	[34]	[35]	[36]

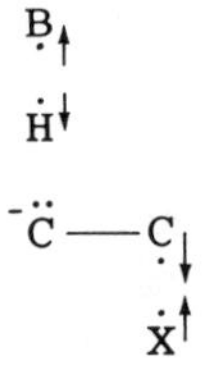

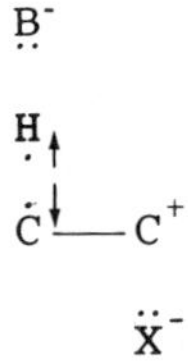

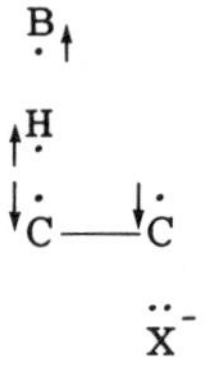

S_N2
[37]

Configuration [33] describes the reactants (spin paired C—H and C—X bonds), while configurations [34]–[37] are generated by the *excitation* of [33]. The product configuration [34] is obtained by *double excitation*. It involves an electron ransfer from the base B to the leaving group X, as well as the decoupling of the C—H bond pair. Together these two excitations prepare the reactants for the bonding that takes place in the products. Thus [34] contains within it the elements for B—H and C=C bond making, as well as C—H and C—X bond breaking. Here again, as for other cases discussed in earlier sections, we see that the product configuration is doubly excited with respect to the reactant configuration. However, despite its initial high energy, it plays an important role because at the product end of the reaction, it is, by definition, the dominant configuration.

The three potential intermediate configurations, any one of which may play an important role in any given elimination reaction, are obtained through *monoexcitation* of [33]. The first of these is [35], the carbanion configuration, since it best describes the potential carbanion intermediate that may be formed in any elimination reaction. In VB terms [35] is generated from [33] by an electron transfer from B^- to C_β. In MO terms it may be thought of as an $n_B \rightarrow \sigma^*_{C-H}$ electron transfer. The second intermediate configuration [36] is the carbocation configuration and is generated by excitation of the C—X bond. In VB terms this occurs via an electron transfer from C to X while in MO terms it is a $\sigma_{C-X} \rightarrow \sigma^*_{C-X}$ excitation. (Recall that the first excited singlet state of the C—X bond, $\sigma^1\sigma^{*1}$ is represented in VB terms by $C^+ X:^-$.) The third intermediate configuration, [37], is the S_N2 product configuration. This is generated from [33] by an electron transfer from B^- to X or in MO terms it is an $n_B \rightarrow \sigma^*_{C-X}$ electron transfer. This configuration is termed S_N2 because it describes the products of an S_N2 reaction, if it were to take place. However, along the elimination pathway it constitutes an intermediate configuration since along the *elimination* reaction co-ordinate, [37] does *not* lead to a ground configuration (the role fulfilled by [34]).

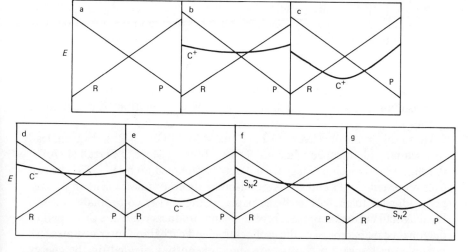

FIG. 26 Energy diagrams showing the mechanistic spectrum of elimination reactions; (a) concerted E2, (b) concerted E2 with E1 character, (c) stepwise E1, (d) concerted E2 with E1cB character, (e) stepwise E1cB. (f) concerted E2C and (g) stepwise E2C–I

Let us now build up the entire mechanistic range of elimination reactions from this set of five configurational building blocks. The various possibilities are illustrated in Fig. 26. The simplest pathway is the concerted E2 pathway. This profile may be generated by the avoided crossing of just reactant (R) and product (P) configurations (Fig. 26a). For a "pure" E2 pathway to be followed, all three intermediate configurations need to be sufficiently high in energy in the transition state region so that they contribute only slightly to the transition state wave-function. The transition state may simply be described by (100).

$$TS(E2) = R \longleftrightarrow P \qquad (100)$$

If the carbocation configuration is stabilized, the configuration picture is that of Fig. 26b. Under these circumstances the transition state takes on some of the character of the carbocation configuration [36], as indicated in (101).

$$TS(E2-E1) = R \longleftrightarrow P \longleftrightarrow C^+ \qquad (101)$$

The mechanism is now described as a concerted E2 pathway with E1 character. If the carbocation configuration is strongly stabilized then the pattern illustrated in Fig. 26c will occur. An actual carbocationic intermediate is generated so that the reaction has become a two-step process. This is the E1 pathway.

If the intermediate configuration that is stabilized is the carbanion configuration [35], then the pattern observed is that illustrated in Fig. 26d. A concerted E2 process with E1cB character takes place and the transition state may be described by (102). If the carbanion configuration is strongly

$$TS(E2-E1cB) = R \longleftrightarrow P \longleftrightarrow C^- \qquad (102)$$

stabilized then a two-step E1cB pathway results. This is indicated in Fig. 26e. The reaction proceeds through the intermediate formation of a carbanion species.

Up to this point the CM model and the More O'Ferrall PES diagram lead to essentially identical conclusions. The analysis by More O'Ferrall (1970) of the elimination mechanism spectrum may be summed up by Fig. 27. The two axes represent C—H and C—X bond-breaking co-ordinates. The third axis, perpendicular to the plane of the paper, is the energy co-ordinate. Two of the diagonal corners represent reactants and products while the other pair of diagonal corners represent the carbocation and carbanion intermediates. All possible mechanistic pathways are simultaneously indicated on the energy surface.

The conceptual significance of the More O'Ferrall diagram is that it divides substituent effects into two classes. Firstly substituent effects which stabilize a corner lying *on* the reaction co-ordinate (i.e. reactants or products) and move the transition state *away* from that corner; these are termed "Hammond" effects. Secondly, substituents which stabilize a corner *perpendicular* to the reaction co-ordinate (i.e. those describing potential intermediates) and move the transition state *toward* that corner. Because this response is opposite to that taken by reactants or product corner, such effects are termed "anti-Hammond" effects. For example in an E2 elimination, stabilization of the product is predicted to make the transition state *less* product-like, while stabilization of the potential carbocation intermediate is predicted to make the transition state *more* carbocationic in character.

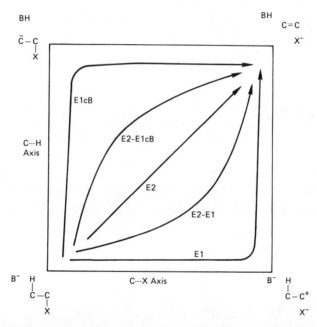

FIG. 27 More O'Ferrall potential energy surface which shows the entire mechanistic range of elimination reaction pathways (from E1 through to E1cB) superimposed on one surface

The predictions of the CM model are exactly the same. In line with a simple Bell–Evans–Polanyi diagram (e.g. Fig. 18), stabilization of the product configuration leads to an earlier transition state, while stabilization of an intermediate configuration leads it increasingly to mix into the transition-state wave-function. For example, stabilization of the carbocationic configuration [36] results in the transition state acquiring *more* of that character so that an E2 process becomes more E1-like (Fig. 26*b*).

The energetic consequences of the CM and PES models are also the same. Stabilization of both parallel and perpendicular structures leads to stabilization of the transition state, in line with predictions of the CM model. We see then that perturbations on reactant and product configurations are equivalent to the PES parallel (Hammond) effects, while perturbations on intermediate configurations are equivalent to the PES perpendicular (anti-Hammond) effects. Thus the CM and PES approaches are, at least for this kind of application, analogous.

As we have mentioned in the context of Marcus theory, the fact that the CM model blends in naturally with other well-established models is reassuring. Since the CM model is derived in a totally different fashion to the PES models, the fact that they both make the same predictions strengthens both. For the CM model, which is still in its infancy, such reinforcement is particularly encouraging.

If we return now to the CM model for elimination reactions we see that there is an additional configuration, the S_N2 configuration [37], that has not as yet been considered. This configuration may be utilized to explain the E2C–E2H mechanistic spectrum (Beltrame *et al.*, 1972; Parker *et al.*, 1972). The experimental observation is that elimination reactions may be effectively catalysed by *weak* bases that are *strong* nucleophiles. Thus Parker and Winstein proposed that eliminations that proceed with a strong base pass through transition state [38], termed an E2H transition state, while those that proceed with a weak base but strong nuclcophile, proceed through transition state [39], termed an E2C transition state. These facts are readily accounted

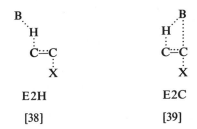

for by the S_N2 configuration, [37]. The CM model suggests that factors that stabilize the S_N2 configuration [37] will accelerate the elimination reaction, i.e. for cases that possess both a good nucleophile and leaving group (Pross and Shaik, 1982a). The corresponding configuration diagram is in Fig. 26*f*. Furthermore, the transition state will take on some of the character of this configuration. This consists of a partial B- - - -C bond as indicated in [39]. Thus the CM model lends weight to the Parker–Winstein proposal.

A key difference between the CM and the PES models now emerges. Whereas both treat substituent effects in terms of parallel and perpendicular perturbations, the CM model has more flexibility with regard to the range of the perpendicular configurations. The S_N2 configuration has no PES equivalent and as such the E2H/E2C spectrum is not within the scope of the PES models. The CM model may explicitly consider any energetically reasonable configuration and is not restricted to the two intermediates at the perpendicular corners that are generated by the stepwise process along either edge of the PES surface. A further example of a configuration which has no PES equivalent is the charge transfer configuration of cycloaddition reactions discussed in Section 3 (p. 174).

Extension of this line of reasoning suggests that there should be a stepwise mechanism based on the S_N2 configuration. Thus, just as a concerted E2/E1 pathway finally turns into a stepwise E1 pathway, the E2H/E2C pathway is predicted, in the limit, to lead to a *radical anion intermediate* which then undergoes elimination. We have termed this, as yet undiscovered pathway, the E2C–I (I = intermediate) pathway (Pross and Shaik, 1982a). The E2C–I pathway constitutes the missing link in the chain of potential elimination processes based on three intermediate configurations (Fig. 26g).

A final comment on the question of the substitution–elimination ratio is desirable. On the basis of the above discussion the S_N2 pathway is expected initially to be preferred. This is because the S_N2 configuration is a monoexcited configuration while the E2 product configuration is doubly excited. However, substitution reactions are highly sensitive to steric effects, so that the slopes of these respective configurations are markedly affected by steric congestion along the reaction co-ordinate. In the absence of such steric effects substitution is indeed favoured. In the presence of significant steric congestion however, elimination dominates (Pross and Shaik, 1982a). This constitutes a further example of the interplay between gap and slope factors.

PROTON TRANSFER

In CM terms the proton transfer reaction is very similar to the S_N2 reaction of methyl derivatives. The two key configurations which are involved are the reactant and product configurations, [40] and [41]. The arguments that were applicable to methyl transfer reactions now may be utilized for proton transfer reactions. These may be summarized as follows:

(a) The reaction may be simply described by just [40] and [41] since configurations incorporating the hydrogen as H^+ or $H:^-$ (i.e. $B_1:^-$ H^+ $:B_2^-$ and $B_1\cdot$ $H:^-$ $\cdot B_2$) may be ignored, due to their high energy.

$$B_1:^- \quad H\cdot \cdot B_2 \qquad B_1\cdot \cdot H \quad :B_2^-$$
$$[40] \qquad\qquad\qquad [41]$$

(b) The proton transfer may be viewed in physical terms as an electron

shift process coupled with a hydrogen atom transfer. In order to effect proton transfer, the *only* configurational change which is required is a single electron shift (103)

$$B_1:^- \quad H\cdot \quad \cdot B_2 \longrightarrow B_1\cdot \quad \cdot H \quad :B_2^- \qquad (103)$$

(c) Since for "normal" proton transfer reactions only two configurations are necessary to describe the reaction profile, the Brønsted parameter is expected to lie in the range 0 to 1.

(d) The Brønsted parameter α is *not* expected to provide a measure of charge development (or destruction) on the two bases since, in analogy with the S_N2 reaction of methyl derivatives, the charge development is likely to be approximately 50% while the magnitude of the Brønsted parameter will be governed by curve slopes.

(e) If curve slopes are patterned according to the general requirements of Marcus theory, then the Brønsted parameter may constitute a relative *geometric* measure of transition state structure.

We will devote the remaining part of this section to a discussion of the proton transfer reaction of nitroalkanes.

Nitroalkane anomaly

Without a doubt, the proton transfer reaction which has attracted most attention over the past 15 years is the reaction of a series of nitroalkanes with a base, as indicated in (104) (Bordwell *et al.*, 1969, 1970, 1972). The reason

$$B^- + ArCH_2NO_2 \rightarrow BH + ArCH=NO_2^- \qquad (104)$$

for the inordinate interest stems from the anomalous Brønsted parameters observed for this reaction (α ∼ 1.5). An α-value greater than 1 signifies that the effect of substituents on rates exceeds that on equilibria, and this is contrary to what is widely believed to be the norm, namely values of α in the range 0–1.

Proton transfer from nitroalkanes is characterized by a second anomaly. While nitroalkanes are unusually strong C—H acids, kinetically they are exceedingly slow (Pearson and Dillon, 1953; Eigen, 1964; Bell, 1973). For example, the thermodynamically favourable proton transfer between acetic acid and the carbanion of nitroethane is 9 orders of magnitude slower than the generally diffusion-controlled proton transfer from oxygen or nitrogen acids (Cox and Gibson, 1974).

A variety of explanations of the nitroalkane anomaly have been proposed (Bordwell *et al.*, 1969, 1970; Bordwell and Boyle, 1972; Bell, 1973; Kresge, 1970, 1973, 1974, 1975; Agmon, 1980; Keeffe *et al.*, 1979; Marcus, 1969;

More O'Ferrall, 1975). The ones that seem to be most widely accepted are those based on solvation (Keefe *et al.,* 1979; Agmon, 1980) as well as the need to rehybridize and delocalize excess negative charge from the carbanionic carbon (Bordwell *et al.,* 1969, 1970, 1972; Bell, 1973; Fuchs and Lewis, 1974; Kresge, 1975; More O'Ferrall, 1975). For the explanations based on solvation and charge delocalization, the anomalous Brønsted parameter is rationalized by assuming there are stabilizing features in the product that are absent in the transition state. Or stated in reverse, there are some *destabilizing* features in the transition state which are absent in reactants and products. Let us clarify this point a little further.

If in the transition state [42] there is significant localization of negative charge on the carbanionic carbon, then anomalous Brønsted parameters are to be expected. This is because a substituent, X, on carbon will have a large effect on the transition-state energy due to its proximity to the negative charge, and a small effect on product [43], where the charge is largely on

$$B^{\delta-}\text{---}H \qquad\qquad B\text{---}H$$

[42]   [43]

oxygen and remote from the substituent. Under these circumstances, α is expected to exceed 1. A similar argument holds for solvation. The nitronate anion [43] is solvated more effectively than the carbanionic transition state [42] so that the substituent effects on [43] are expected to be smaller than on [42].

In a sense, however, the above explanation merely replaces the original problem by a new one. If in the product ion the negative charge prefers to reside on the oxygen atoms, then *why does the delocalization of charge into the nitro groups lag in the transition state*? Presumably, it could be argued that what is energetically attractive for the product ion should be equally attractive for the transition state. Bernasconi and Kanavarioti (1985), following on from Hine (1977), have brought this question into sharper focus by their Principle of Imperfect Synchronization. These workers have pointed out that in certain reactions – typically carbanion-forming reactions – there seems to be a lack of synchronization between bond formation (or cleavage) and charge delocalization. Thus if one indeed attributes the anomalous Brønsted parameters to a build-up of negative charge on carbon then the question which requires a satisfactory answer is: why *does* the negative charge build up on carbon in the transition state?

A solution to this problem is forthcoming within the CM framework. The nitroalkane deprotonation reaction cannot be simply built up by just two interacting configurations, but needs a third "intermediate" configuration in order to provide a meaningful description of the reaction profile. It is this "intermediate" configuration that provides the transition state with character either absent or at least reduced in reactants and products (Pross and Shaik, 1982b). Let us now demonstrate this.

The electrons which need to be rearranged in order to generate the set of VB configurations include the lone pair on the base, B, the C—H bonding pair, and the N—O π-bond. The reactant configuration is illustrated in [44].

$$
\begin{array}{ccc}
\mathrm{B}^- & \mathrm{B}\!\cdot\!\downarrow & \mathrm{B}\!\cdot\!\downarrow \\[4pt]
\mathrm{H}\!\cdot\!\uparrow & \mathrm{H}\!\cdot\!\uparrow & \overset{\cdot\downarrow}{\mathrm{H}} \\[4pt]
\overset{\cdot\uparrow}{\mathrm{C}}-\overset{\cdot\uparrow\ \uparrow\cdot}{\mathrm{N}}-\mathrm{O} & \overset{\cdot\uparrow}{\mathrm{C}}-\overset{\uparrow\cdot}{\mathrm{N}}-\overset{\cdot\cdot}{\mathrm{O}}{}^- & {}^-\overset{\cdot\cdot}{\mathrm{C}}-\overset{\cdot\uparrow\ \uparrow\cdot}{\mathrm{N}}-\mathrm{O} \\
\downarrow & \downarrow & \downarrow \\
\mathrm{O} & \mathrm{O} & \mathrm{O} \\[4pt]
[44] & [45] & [46]
\end{array}
$$

The three electron pairs are distributed so as to describe the reactants, and the arrows that appear adjacent to electrons denote spin. Thus the C—H and N—O electrons each constitute a bonding pair.

The product nitronate anion cannot be described by just a single configuration. Its structure needs to be described by *two* configurations, one in which the negative charge is localized on carbon and the other with the charge on oxygen. These are just the two resonance forms of (105) and in

$$\bar{\mathrm{C}}-\overset{+}{\mathrm{N}}{\diagup\!\!\!\!\diagdown}{}^{\mathrm{O}}_{\mathrm{O}^-} \longleftrightarrow \mathrm{C}=\overset{+}{\mathrm{N}}{\diagup\!\!\!\!\diagdown}{}^{\mathrm{O}^-}_{\mathrm{O}^-} \qquad (105)$$

configuration terms are described by [45] and [46]. Note that the electrons are spin-paired in such a way as to describe the particular resonance form required. Thus [45] describes the oxyanion, while [46] describes the carbanion. The energy of these three configuration is plotted as a function of the reaction co-ordinate in Fig. 28.

The reactant configuration [44] increases in energy along the reaction co-ordinate since a repulsive B: ·H three electron interaction is generated while an attractive C—H bonding pair is broken. On the other hand the two configurations describing the product, [45] and [46], decrease in energy along the reaction co-ordinate. In [45] a repulsive C—H triplet interaction is released while a stabilizing B—H bond pair is generated, while for [46] a

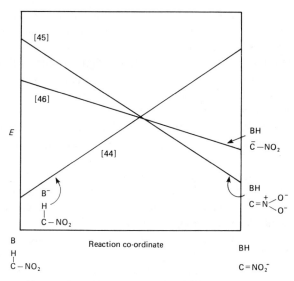

FIG. 28 Energy diagram illustrating the configuration curves that contribute to the deprotonation reaction of a nitroalkane. The major configurations are the reactant configuration [44], the oxyanion configuration [45], and the carbanion configuration [46]

repulsive three-electron C—H interaction is released and a B—H bond pair generated. The relative importance of [45] and [46] at any point along the reaction profile may be qualitatively estimated by assessing their relative energies at reactant and product ends. At the reaction beginning [46] is more stable than [45]; [45] is a high energy configuration since conversion of [44] to [45] involves *double excitation,* namely, electron transfer from B to O and conversion of the singlet C—H bond to the repulsive triplet state. A very approximate figure for each of these processes is $c.$ 10 eV. Formation of [46] from [44], however, involves just a single excitation, an electron transfer from B to C ($c.$ 10 eV). At the reaction end the situation is reversed; the comparison is between the two resonance forms of the nitronate anion (104), and now [45] is *more* stable than [46] because the negative charge prefers to reside on the more electronegative oxygen atoms.

This reversal in the relative stability of [45] and [46] is significant. It means that the relative importance of the carbanionic configuration [46] is greater early along the reaction co-ordinate and decreases as the reaction proceeds, i.e. [46] behaves as an intermediate configuration. In other words, on the basis of the CM model there is likely to be *more negative charge development on carbon in the transition state than in the products.* Thus the CM model provides theoretical support for the proposal, originally put forward by Bordwell, Kresge, and More O'Ferrall, that anomalous Brønsted parameters are due to a build-up of negative charge on carbon. Of course, if the carbanion configuration is strongly stabilized, the possibility arises that a carbanion intermediate will actually be formed. As was discussed in Section 2 (p. 152), substantial stabilization of an intermediate configuration may result in the generation of that intermediate, and the conversion of a one-step reaction with intermediate character in the transition state, to a two-step process via that intermediate.

This is in line with the explanation of Bordwell and Boyle (1975) who proposed the formation of a localized carbanionic intermediate as the cause of the anomalous α-values. However, as we have noted, formation of such an intermediate is not a necessary condition for observing anomalous α-values, although it is clearly a sufficient condition. Thus Bordwell's explanation cannot, *a priori,* be ruled out in all cases, and may indeed apply if the potential carbanionic intermediate is strongly stabilized.

Finally it should be pointed out that the CM picture of nitroalkane deprotonation provides a satisfying explanation for the *slowness* of proton transfer from carbon acids. Throughout this review we have met examples of reactions whose product configuration is a diexcited configuration, e.g. ethylene dimerization, $H_2 - D_2$ bond exchange, and ring closure of butadiene. Such reactions are generally slow since the large energy gap between reactant and product configurations at the reaction beginning carries over to the transition state (this is just the gap factor discussed on p. 159 in the context of S_N2 reactions). In fact, many of these reactions occur only *because of the kinetic assistance of an intermediate configuration.* Proton-transfer reactions from carbon acids fall into this category. For these reactions the predominant product configuration is a *doubly excited* one due to electronic reorganization which is required once the proton is removed. Here again, the reaction is kinetically facilitated by the monoexcited carbanion configuration [46]. On the other hand, the product configuration for *normal* proton transfer is a *monoexcited* configuration (in this case just charge transfer; see configurations [40] and [41]) so the reactions are rapid by comparison. The substitution–elimination product ratio (Section 3) may also be understood in similar terms.

In conclusion, we see that nitroalkane deprotonation may be understood in terms of a three-configuration picture. It is this third configuration which *(a)* enables the reaction to take place kinetically, and *(b)* is responsible for anomalous Brønsted parameters. The above discussion does not presume to exclude other factors, primarily solvation factors, which may contribute toward the generation of anomalous Brønsted parameters (Agmon, 1980). The purpose has been to show that anomalous α-values are an intrinsic chemical phenomenon since for a one-step process there is no necessity for the character of the transition state to be intermediate between that of reactants and products. A detailed discussion of anomalous α-values appears in Section 4 (p. 181).

PERICYCLIC REACTIONS

In previous sections we have seen how the CM model may be utilized to generate reaction profiles for ionic reactions, and it is now of interest to observe whether the same general principles may be applied to the class of pericyclic reactions, the group of reactions that is governed by the Woodward–Hoffmann (1970) rules. In other words, the question we ask is whether the concept of "allowed" and "forbidden" reactions may be understood within the CM framework.

A second point that needs to be clarified is why allowed reactions actually have a barrier at all. In fact, some so-called allowed reactions possess particularly high barriers so that they hardly proceed. One such case, discussed by Houk et al. (1979), concerns the allowed trimerization of acetylene to yield benzene. Despite being extremely exothermic ($\Delta H^\circ = -143$ kcal mol^{-1}) this reaction appears to have an activation barrier in excess of 36 kcal mol^{-1}. Since the orbitals of the reactants correlate smoothly with those of the products, in accord with the Woodward–Hoffmann rules, the origin of barrier formation in allowed reactions generally, needs to be clarified.

An additional point of interest concerns the behaviour of HOMO and LUMO orbitals of reactants in allowed reactions. Fukui (1970, 1975) has pointed out that the frontier-orbital gap actually narrows as the reaction proceeds. This has been confirmed computationally for the cycloaddition of ethylene and butadiene (Townshend et al., 1976), and contrasts with what one might expect based on a static HOMO–LUMO interaction. Such an interaction causes the energy gap between resultant orbitals to *widen,* as indicated in Fig. 29.

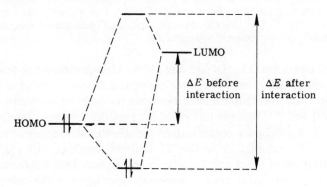

FIG. 29 Static HOMO–LUMO interaction

It is of interest, therefore, to explore to what extent the CM model can accommodate the above facts.

Application of CM theory to explain pericyclic reactions was first attempted by Epiotis and coworkers (Epiotis, 1972, 1973, 1974; Epiotis and Shaik, 1978b; Epiotis *et al.*, 1980). The following analysis is a much-simplified treatment of that approach. Let us compare, therefore, the CM analysis for the [4 + 2] allowed cycloaddition of ethylene to butadiene to give cyclohexene with the [2 + 2] forbidden dimerization of two ethylenes to give cyclobutane. For simplicity only the suprafacial–suprafacial approach is considered, although this simplification in no way weakens the argument.

The forbidden dimerization reaction of two ethylenes may be understood in terms of the CM diagram of Fig. 30. For this process it is helpful to consider the problem in terms of both MO and VB configurations.

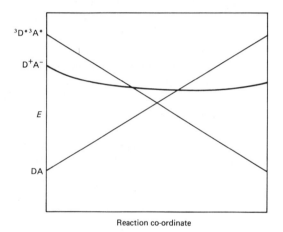

FIG. 30 Energy diagram showing the major configurations that contribute to any cycloaddition reaction, DA, D^+A^-, and $^3D^{*3}A^*$. For allowed reactions, DA and D^+A^- mix, thereby lowering the activation barrier. For forbidden reactions, mixing of DA and D^+A^- is precluded by symmetry

In MO terms the DA, $^3D^{*3}A^*$ and D^+A^- configurations are described by [47] to [49]. In VB terms these three configurations are illustrated by [50] to [53]. Thus [47] and [50] are each representations of the reactant configuration, [48] and [51] representations of the product configuration (recall that $^3D^{*3}A^*$ is the product configuration for any bond exchange process; see the $H_2 - D_2$ exchange in Section 2), and [49] together with [52] and [53] representations of the charge transfer configuration. One VB form is simply inadequate to provide even an approximate description of the charge transfer configuration; both [52] and [53] are required to reflect its intrinsic symmetry.

CONFIGURATION MIXING MODEL

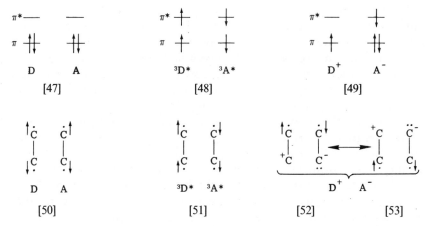

The essential point that distinguishes between allowed and forbidden reactions is the role of the D^+A^- configuration. If the D^+A^- configuration is allowed by symmetry to mix into the transition state wave-function then the transition state will be stabilized and will take on character associated with that configuration. For the ethylene dimerization, D^+A^- is *precluded* from mixing with DA due to their opposite symmetries. As was discussed in detail in Section 2 (p. 130), DA cannot mix with D^+A^- since π and π^* orbitals are orthogonal (106). Thus for ethylene dimerization the concerted process

$$<DA|H|D^+A^-> \approx <\pi|H|\pi^*> \approx 0 \qquad (106)$$

would primarily involve a $DA/^3D^{*\,3}A^*$ avoided crossing with a prohibitively high barrier. Accordingly the reaction does not proceed thermally.

For the allowed reaction between ethylene and butadiene, D^+A^- *can* mix into the transition state wave-function, since now π and π^* orbitals *do* overlap (p. 131). As a consequence the barrier is lowered and the transition state acquires some charge transfer character (107). *Thus the role of symmetry in*

$$TS = C_1(DA) + C_2(D^+A^-) + C_3(^3D^{*\,3}A^*) \qquad 107)$$

cycloaddition reactions manifests itself in the CM model in essentially the same way as in the Woodward–Hoffmann approach. The question of overlap of configurations merely reduces to the more familiar question of orbital overlap.

The effect of substituents on reactivity may be understood in terms of Fig. 31. The ground state profile (unbroken bold line) is the result of the mixing of DA, D^+A^- and $^3D^{*3}A^*$ configurations (107). Substituent effects on reactivity appear to be dominated by their effect on D^+A^-. Substituents that stabilize D^+A^- lower its energy along the entire reaction co-ordinate (broken line) and lead to a lower energy reaction profile (broken bold line).

Since the energy of D^+A^- is governed by the magnitude of $I_D - A_A$, D^+A^- will be stabilized when the diene is substituted by electron-releasing substituents (I_D is lowered) and the ene by electron-withdrawing ones (A_A is increased). Indeed, cycloadditions proceed more rapidly in such cases. For example, 1,2-dicyanoethylene undergoes cycloaddition with cyclopentadiene c. 80 times faster than cyanoethylene (Sauer, 1967). In the former case $I_D - A_A$ is 7.8 eV while in the latter case it is 8.6 eV.

The importance of the D^+A^- configuration in governing reactivity is clearly illustrated in Kochi's work (Fukuzumi and Kochi, 1982). Kochi found that, in analogy with a number of other reactions, the rate of cycloaddition of substituted anthracenes with tetracyanoethylene (TCNE) is linearly related to the charge transfer band of the anthracene–TCNE complex. As discussed previously (Section 2, p. 133) such a correlation emphasizes the relationship between ground and excited surfaces, and may be understood on the basis of Fig. 31. Stabilization of D^+A^- will lead to a *lowering* of the charge transfer frequency band since the frequency, ν, is governed by the DA/D^+A^- energy gap at the reaction beginning. This drop in D^+A^- causes a concomitant drop in the transition state energy, $\Delta E^{\neq}$. It is not surprising, therefore, that a linear relationship (72) exists between $\Delta h\nu$ and $\ln k$.

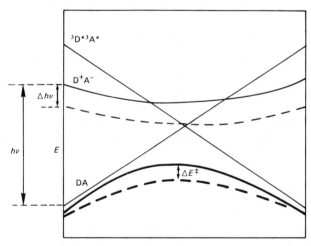

FIG. 31 Energy diagram showing the relationship between $\Delta h\nu$, the change in the charge transfer complex absorption energy for a series of substituted anthracene – TCNE complexes, and the change in activation barrier, $\Delta E^{\neq}$, for the corresponding cycloaddition reactions

An interesting consequence of this analysis has been the suggestion of Epiotis and Shaik (1978b) that intermediates may result in allowed cycloadditions. The way in which such an intermediate is generated is illustrated in Fig. 24. If D^+A^- is stabilized to the extent that the ground-state profile (bold line) proceeds via an intermediate then this intermediate is dominated by the D^+A^- configuration. This means that the intermediate is largely described as the charge transfer ion-pair species, D^+A^-, which *normally only resides on the excited state surface*. This intermediate is, of course, distinct from the regular charge transfer complex, described primarily by DA, that has been detected in cycloadditions (Kiselev and Miller, 1975). A D^+A^- intermediate species appears to have been generated by Kochi and coworkers (Hilinski *et al.*, 1983) in the photochemical excitation of the indene–TCNE charge transfer complex; however, at the present time we are unaware of any firm evidence pointing to the formation of such an intermediate in the course of a thermal cycloaddition reaction.

The reason for frontier orbital narrowing now also becomes clearer. In the early stages of the reaction the reactants climb in energy up the DA surface toward the D^+A^- surface, so that the energy gap between these surfaces narrows. Since the DA–D^+A^- energy gap is just the energy required to transfer an electron from D to A, the energy gap between the configurations and the HOMO–LUMO gap are related. Thus the DA–D^+A^- configuration narrowing is equivalent to HOMO–LUMO narrowing, and serves to enhance the stabilizing interactions DA–D^+A^- (in configuration terms) and HOMO–LUMO (in orbital terms) in the transition state region.

To summarize, we note two key points. Firstly, that the favourable interaction between DA and D^+A^- in allowed reactions does not preclude barrier formation. The CM model, by its very nature, emphasizes barrier formation through the avoided crossing of reactant and product configurations. Second, the CM model gives rise to the Woodward–Hoffmann rules through consideration of the symmetry properties of the DA and D^+A^- configurations. This is as it should be. As we have noted previously, a key test of the CM model is whether it blends in naturally with existing theories that focus on specific areas of reactivity.

4 General consequences

THE SIGNIFICANCE OF THE BRØNSTED PARAMETER

The relationship between rates and equilibria was first proposed by Brønsted and Pedersen (1924) and subsequently treated by Bell (1936) and Evans and Polanyi (1938); it is shown in (108). Some 30 years ago Leffler (1953) provided (108) with a molecular basis by suggesting that the parameter α reflects

$$\log k = \alpha \log K + c \qquad (108)$$

the structure of the transition state. Values of α close to zero were considered to denote transition states structurally similar to reactants, while values of α approaching 1 were considered to denote product-like transition states. Bordwell's work on nitroalkane deprotonation (Section 3) challenged this simple interpretation, since values of α of c. 1.5 were observed. On the basis of Leffler's hypothesis, α-values are expected to be in the range 0–1, so that anomalous α-values (α > 1) require some explanation.

A general analysis of α-values for organic reactions has recently appeared (Pross, 1984). A key point that emerges is that even normal α-values are *not* necessarily reliable measures of transition state structure. We believe that only in well-defined situations is the α-value mechanistically meaningful. Let us now consider the significance of anomalous α-values in greater detail.

For the family of identity exchange reactions such as (109), a plot of $\log k$ against $\log K$ gives a straight line with infinite slope, i.e. the Brønsted

$$\text{Cl}^- + \text{CH}_2\text{Cl}\underset{}{\overset{}{\rightleftarrows}} \text{ClCH}_2 + \text{Cl}^- \qquad (109)$$

(with X-substituted phenyl groups)

parameter α = ∞ (Pross, 1983). This trivial result suggests that the physical organic community need not become unduly excited by anomalous Brønsted parameters of c. 1.5–1.8 (Bordwell *et al.*, 1969, 1970, 1972; Murdoch *et al.*, 1982). An α-value greater than 1 means nothing more than the fact that substituent effects on rates are larger than on equilibria. True, for reactions described by just two configurations this cannot occur. For this reason the Leffler postulate (Leffler, 1953; Leffler and Grunwald, 1963) is applicable only for two-configuration reactions. However, this situation breaks down for three-configuration reactions. Examination of the configurational make-up of any three-configuration reaction makes it apparent that if the position of substitution within the reacting system primarily affects the energy of the intermediate configuration, then anomalous α-values are to be expected (Pross, 1984). This is because changes in the energy of this configuration affect rates with little or no change in equilibria. The bottom line is simply that anomalous Brønsted parameters may be generated in almost any family of reactions and the so-called normality of the α-value is governed more by the site of substitution than the nature of the reaction itself. The anomalous α-value observed in the proton transfer reaction of nitroalkanes is discussed in Section 3 (p. 169).

An important point however, is that even normal α-values should not always be interpreted in terms of the Leffler model. The problem is that the very same factors which generate anomalous α-values may be operative in cases where normal α-values are observed but to a smaller extent. For example the "intrinsic" α-value of a particular system may be 0.5 but the anomaly-generating factors may increase it to 0.9. In such a case there is no warning sign that something is amiss.

Let us now try and pinpoint the factor responsible for generating anomalous α-values, so that we can check for it in systems exhibiting normal α-values.

Misuse of α-values

In this section we generated an α-value of ∞ by taking the family of identity exchange reactions of benzyl derivatives (109) (Pross, 1983). That particular α-value is sufficiently anomalous to deter any possible application to questions of transition state structure. Yet there are cases that are, in principle, quite similar to the identity exchange reaction in which α-values *are* utilized as a measure of transition state structure. In such cases we believe the conclusions may not be valid. A typical example (Bell and Sorenson, 1976) is the addition of hydroxide ion to substituted benzaldehydes (110). For this system the sensitivity of rates and equilibria to substituent effects was similar.

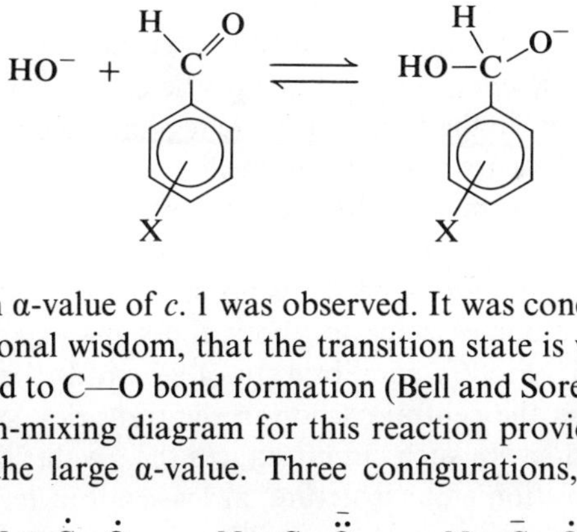

$$HO^- + \underset{X}{\underset{}{C_6H_4}}-C(H)=O \;\rightleftharpoons\; HO-\underset{X}{\underset{}{C_6H_4}}-C(H)(O^-) \tag{110}$$

In other words an α-value of *c.* 1 was observed. It was concluded, consistent with the conventional wisdom, that the transition state is very product-like, at least with regard to C—O bond formation (Bell and Sorenson, 1976).

A configuration-mixing diagram for this reaction provides an alternative explanation for the large α-value. Three configurations, [54] to [56], are

$$\underset{[54]}{N{:}^-\ \dot C{-}\ddot O} \qquad \underset{[55]}{N{\cdot}\ {\cdot}C{-}\bar O} \qquad \underset{[56]}{N{\cdot}\ {:}\bar C{-}\dot O}$$

required in order properly to describe the addition of a nucleophile to a carbonyl group. A plot of these three configurations as a function of the reaction co-ordinate is schematically illustrated in Fig. 32. Configuration [54] is the reactant configuration, [55] is the product configuration and [56] is a carbanion-like intermediate configuration. Together [55] and [56] constitute the VB components of the charge transfer configuration, D^+A^-.

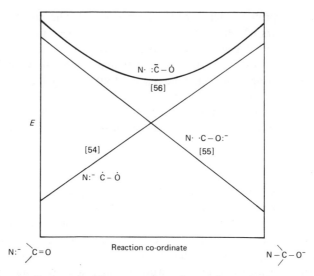

FIG. 32 Energy diagram that illustrates the major configurations, reactant [54], product [55] and carbanion [56], that contribute to the addition of a nucleophile to a carbonyl group (109). In the transition state region there is a build-up of negative charge on carbon due to the carbanion configuration [56]

It is clear that substitution on the central carbon is likely to have a significant effect on the energy of [56] due to the charge on carbon. Since [56] contributes primarily in the transition state region, rates are likely to be significantly affected. Of course, substitution will also affect the energy of the product configuration but to a smaller extent due to the greater distance to the negatively charged oxygen. As a result equilibria are also modified. However, the Brønsted parameter, $\alpha \sim 1$ tells us no more than that the effect of substituents on [56] modifies rates to about the same extent as the effect of substituents on [55] modify equilibria. In other words there is a build-up of negative charge on the central carbon in the transition state that is largely absent in the product. In such circumstances the α-value does not constitute a measure of transition state structure, at least not in terms of earliness or lateness. Had the effect on the transition state been slightly larger, an anomalous α-value would have been observed, in direct analogy with the nitroalkane deprotonation reaction. Note, however, that the magnitude of α *does* provide an indication as to the extent of charge build-up on the site of substitution in the transition state, and in this sense, at least, provides some mechanistic information.

The analogy with the identity exchange reaction (109) is now also evident. "Central" substitution, i.e. at a site in which the formal charge is the same in reactants and products, will lead to large, if not anomalous, α-values.

Reactions (109) and (110) as well as nitroalkane deprotonation are all examples of "central" substitution so that for these reactions α does *not* measure the earliness or lateness of the transition state. "End" substitution, such as in the nucleophile or leaving group of $S_N 2$ reactions, in which the formal charge *does* vary between reactants and products generates α-values which *may* be utilized as a relative measure of transition state charge; this, provided that the reaction is composed of three configurations. Such applications were discussed in Section 3, p. 152.

The prevalence of anomalous α-values
How common are anomalous α-values? The general view appears to be that the overwhelming proportion of α-values are normal and only an insignificant number are anomalous. This view appears to be unjustified. Anomalous α-values are, in principle, as prevalent as normal α-values. Thus in the same way that certain reactions yield Hammett plots that exhibit positive ρ-values and others negative ρ-values, with neither considered to be anomalous, substitution at certain sites of a given system tends to generate normal α-values while substitution at other sites in the same system will generate anomalous α-values. Thus essentially any reaction may be designed to exhibit either anomalous *or* normal α-values. Let us illustrate this point for a family of elimination reactions (111). McLennan and Wong (1974) found that the rates of dehydrochlorination of $Ar_2CH—CCl_3$ derivatives (111) are dependent on the ring substituents while the equilibria are hardly affected.

$$B^- + Ar_2CH—CCl_3 \rightarrow BH + Ar_2C=CCl_2 + Cl^- \qquad (111)$$

This result is, of course, equivalent to an anomalous α-value. As for reaction (109), the substituent is adjacent to a position in the molecule (the β-carbon) whose formal charge remains unchanged. This case represents a further example of "central" substitution. The large substituent effect on rates merely signifies the importance of the carbanionic configuration ([35], Section 3) in describing this reaction. In similar fashion substituents on C_α are also expected to generate an anomalous α-value while substituents on the base or leaving group ("end" substitution) will be expected to generate normal α-values. Thus the general rule would be that "end" substitution tends to generate normal α-values, while "central" substitution will often (but not always) lead to anomalous α-values.

In summary, we see that the interpretation of the Brønsted parameter α is far from straightforward. In addition to the intrinsic factors discussed here, there may also be extraneous factors which affect the magnitude of the Brønsted parameter. These include solvation factors (Jencks *et al.*, 1982; Hupe and Jencks, 1977; Hupe and Wu, 1977), and multistep pathways (Murdoch, 1972). So while in certain cases the Brønsted parameter does appear to provide a measure of transition state charge development, there are many other cases, where such applications are suspect (see also, Scandola *et al.*, 1981). We would therefore caution against routine application of the Brønsted parameter as a tool for estimating transition state properties.

RELATION TO MARCUS THEORY

In the section on nucleophilic substitution processes we saw that the S_N2 process is simply a single electron shift coupled with bond reorganization. The question we wish to consider in this section is how the CM model treats standard electron transfer reactions such as (89) and how the model relates to the Marcus theory, now universally accepted for treating such processes (Marcus, 1964, 1977).

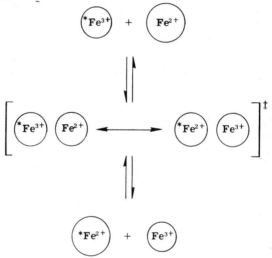

FIG. 33 Schematic representation of the Fe^{2+}/Fe^{3+} electron-transfer reaction. (Adapted from Eberson, 1982)

The qualitative elements of Marcus theory are readily demonstrated. For example, the process of transferring an electron between two metal ions, Fe^{2+} and Fe^{3+}, may be described schematically by Fig. 33 (Eberson, 1982; Albery and Kreevoy, 1978). The reaction may be separated into three discrete stages. In the first stage the solvation shell of both ions distorts so that the energy of the reacting species before electron transfer will be identical to that after electron transfer. For the self-exchange process this of course means that the solvation shell about Fe^{2+} and Fe^{3+} in the transition state must be the same if electron transfer is not to affect the energy of the system. In the second phase, at the transition state, the electron is transferred without change in the energy of the system. In the third phase the solvation shells about the two ions relax to the equilibrium geometry. An energy diagram for such a reaction is shown in Fig. 34. Curve R (reactants) shows the effect of modifying the solvation sphere in the reactants to that existing in the products *without* electron transfer. Thus curve R increases in energy along the reaction (solvation change) co-ordinate. The starting point of curve P (point B) is obtained by transferring the electron from Fe^{2+} to Fe^{3+} *without* changing the solvation sphere (excitation from A to B). Curve P then results by allowing the perturbed solvation shell to relax to that existing in the ground state of the products. The actual reaction pathway involves climbing up the curve R and then switching over to curve P. The electron transfer itself occurs in the crossing region.

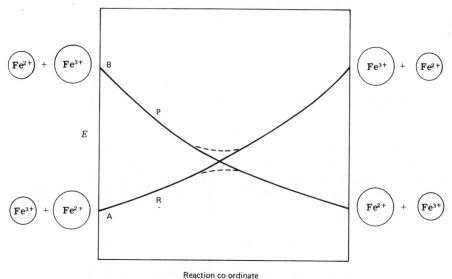

FIG. 34 Energy diagram for the Fe^{2+}/Fe^{3+} electron-transfer reaction. The reaction barrier is generated by the avoided crossing of reactant, R, and product, P, configurations. The region of avoided crossing is described by dotted lines

From this brief description it is quite apparent that the qualitative elements of the Marcus treatment for an electron transfer process are identical to the CM model. In CM terms the reaction involves the avoided crossing of reactant ($Fe^{2+} + Fe^{3+}$) with product ($Fe^{3+} + Fe^{2+}$) configurations, with the reaction co-ordinate just being the distortion-relaxation motion of the solvation sphere. Thus in CM terms *any* electron transfer reaction involves the avoided crossing of the DA (donor-acceptor) and D^+A^- configurations, and for such reactions at least, based on the equivalence with Marcus theory, the CM model has a solid foundation.

Of course, Marcus theory does not stop at this juncture but attempts to provide a *quantitative* relationship between reaction kinetics and thermodynamics. On the basis of Marcus theory the barrier to a particular reaction $\Delta G^{\neq}$, may be described as a function of a parameter called the intrinsic barrier, $\Delta G_o^{\neq}$, and the free energy of the reaction ΔG°. The particular relationship is presented in (112). The basic idea here is that the barrier height is composed of two contributions, a kinetic component termed the intrinsic

$$\Delta G^{\neq} = \Delta G_o^{\neq} + 1/2\Delta G^\circ + [(\Delta G^\circ)^2/16\Delta G_o^{\neq}] \tag{112}$$

barrier, which is the barrier height in the absence of any thermodynamic contribution, and a thermodynamic component which depends on the free energy of reaction. The thermodynamic contribution reduces the barrier for the case of exergonic reactions and increases the barrier for endergonic reactions. Thus Marcus theory places on a quantitative basis what the Bell–Evans–Polanyi (and of course the CM) model approach qualitatively. One way in which Marcus theory may be shown to provide this quantitative relationship (112) is by assuming that the intersecting curves are part of two parabolas of equal curvature (McLennan, 1976). However, due to its general nature it may be derived in a number of other ways as well (Murdoch, 1972, 1983a,b; Murdoch and Magnoli, 1982).

One difficulty that arises within the Marcus formulation is that the intrinsic barrier term is ideally treated as a constant. Earlier applications of Marcus theory were based on this assumption (Cohen and Marcus, 1968; Kreevoy and Konasewich, 1970; Kreevoy and Oh, 1973; Albery *et al.*, 1972). However, as regards methyl transfer, this assumption is clearly invalid.

Solution studies (Albery and Kreevoy, 1978), gas phase data (Pellerite and Brauman, 1980, 1983) and theoretical calculations (Wolfe *et al.*, 1981) all indicate an enormous variation in the barriers for identity exchange of CH_3X (113). For example, in (113) the experimental gas-phase barrier for

$$X^- + CH_3X \rightarrow XCH_3X^- \tag{113}$$

$X = HC \equiv C^-$ is 41.3 kcal mol^{-1} while the corresponding value for $X = Cl$ is only 10.2 kcal mol^{-1} (Pellerite and Brauman, 1983).

Application of the CM model to S_N2 reactions (Shaik and Pross, 1982a,b) suggests that variations in the curve slopes with a change in nucleophile and leaving group are indeed expected to modify intrinsic barriers. For other organic reactions a similar conclusion is to be expected. This is because introduction of a third or "intermediate" configuration is likely to generate variations in rate which do not manifest themselves in the equilibria (Section 4). Such variations will express themselves as a variable intrinsic barrier. In this sense the CM explanation for anomalous Brønsted parameters in nitroalkane deprotonation contains within it the explanation derived from Marcus theory. Marcus (1969) suggested that the anomalous Brønsted parameters come about as a result of a variation in the intrinsic barrier. The CM description adds to this suggestion by explaining *why* the intrinsic barrier varies. The variation in the intrinsic barriers comes about through the variable influence of the carbanion configuration (p. 153). Application of the CM model suggests therefore that the phenomenon of variable intrinsic barriers within even limited families of organic reactions is likely to be quite general. While this precludes application of the Marcus relationship in its most general form (constant intrinsic barriers for a given family), it does allow it to be used in a more restricted manner. Thus in order to calculate the intrinsic barrier for a general substitution reaction (114), one may obtain an "operational" figure for the intrinsic barrier for that reaction by taking an average of the identity barriers for reactions (115) and (116). Such a procedure has indeed been followed by those groups studying methyl transfer.

$$A + BC \rightarrow AB + C \tag{114}$$

$$A + BA \rightarrow AB + A \tag{115}$$

$$C + BC \rightarrow CB + C \tag{116}$$

In summary, then, it appears that the Marcus relationship and the CM model are closely related. The former provides a quantitative framework for tackling reactivity problems while the latter is qualitative in nature. The CM model however makes up for this deficiency by providing added insight into the factors governing reactivity through a physical description of the reaction barriers in any given case. In this sense the CM model is more "chemical" in its format than the functional form of the Marcus equation.

Even with the above modification of averaging identity barriers, Marcus theory suffers a troubling deficiency; its inability to adequately treat "anti-Hammond" effects. As has been discussed by Kurz (1983), Marcus theory does not deal with anti-Hammond effects in a direct fashion. Conventional application of Marcus theory suggests that free energy perturbations will almost always result in just Hammond effects. The only way that anti-Hammond effects may be accommodated routinely within the theory is by variation in the intrinsic barrier. Since Marcus theory, at least as it is applied within organic chemistry, does not provide a means of predicting variations in intrinsic barrier, relating anti-Hammond effects to variations in intrinsic barrier is of no predictive value and hence of questionable theoretical utility. (For recent attempts to resolve this problem, see Agmon, 1984; Murdoch, 1983a.)

Kurz (1983) has attempted to correct this deficiency of Marcus theory by introducing a slight modification into the intersecting parabola model. Anti-

Hammond behaviour may be observed if the intersecting parabolae are of different curvature. The physical interpretation of this difference in curvature is that bond-making and bond-breaking processes are not necessarily synchronous as is assumed in conventional Marcus theory. While such a modification in the theory may overcome the inherent problem of treating anti-Hammond effects it does make application of the Marcus theory more difficult by the introduction of additional unknowns into the free energy relationship of (112).

THE RELATIONSHIP BETWEEN GEOMETRIC AND CHARGE PROGRESSION ALONG THE REACTION CO-ORDINATE

In Section 3 (p. 148) we saw that both early and late transition states for S_N2 reactions of methyl derivatives are likely to have progressed about 50% in terms of their charge development. This proposal is quite contrary to the generally accepted view and in this section we wish to discuss how general this disparity between charge and geometric progression is likely to be.

The concept of a reaction co-ordinate in organic chemistry is one that is not simply defined, since for a polyatomic system of n atoms it may depend on as many as $3n - 6$ parameters for complete characterization. The reaction co-ordinate used by organic chemists in their representation of a reaction profile has only a qualitative significance. For example, that for an S_N2 process on a methyl derivative is generally equated with the change in either the N- - -R or R- - -X distance while the changes in C—H bond lengths and bond angles are usually ignored. This imprecision in the normal usage of the reaction-co-ordinate concept must affect any analysis of the relationship between charge and geometric progression so that much of the discussion that follows is necessarily qualitative.

We begin by considering the relationship between charge and geometry for a process whose reaction co-ordinate is well characterized, namely, the formation of Na^+Cl^- from Na and Cl atoms. For this reaction the Na- - -Cl distance fully characterizes the reaction co-ordinate; there are no additional parameters to be considered.

In Section 2 (p. 116) the reaction of Na and Cl was described in terms of an avoided crossing of ionic and covalent configurations (Fig. 7). When the Na and Cl atoms approach to within ~ 11 Å an electron transfer takes place and two ions, Na^+ and Cl^-, result. A plot of geometric progression *vs* charge development for this reaction is illustrated in Fig. 35*a*. For this reaction there is clearly no relationship between charge development and geometric progression. The entire charge formation takes place within a limited portion of the reaction co-ordinate. Thus, if this reaction is in any way typical, there may be no real basis for the simplistic view that charge and geometric progression are linearly related. A linear relationship would give rise to the graph shown in Fig. 35b.

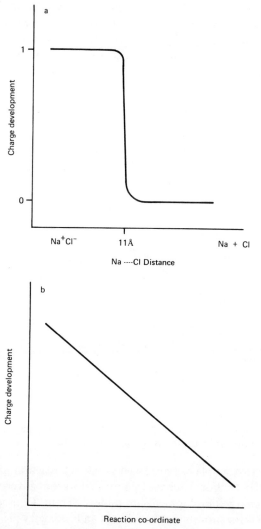

FIG. 35 *(a)* Plot of geometrical progression *vs* charge development for the reaction of Na and Cl to give Na^+Cl^-. *(b)* Hypothetical plot for a reaction in which geometrical progression and charge development are linearly related

A second example which points out the fallacy in relating charge and geometric progression may be seen in the nucleophilic substitution of benzyl derivatives (p. 155). Since such reactions require at least three configurations for a proper description of the reaction profile, the transition state may well take on charge character that is *absent* in reactants and products. For the case of benzyl substitution there may well be positive charge development on the central carbon atom, as indicated in (85). The corresponding graph of charge development on carbon *vs* reaction co-ordinate is illustrated in Fig. 36. This correlation is obviously also a far cry from the proportionality plot of Fig. 35*b*.

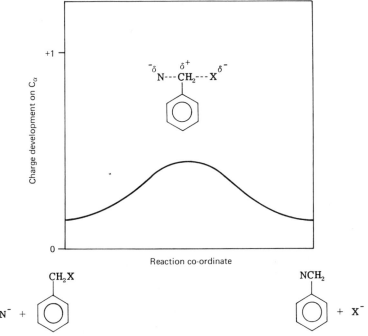

FIG. 36 Plot of geometrical progression *vs* charge development on the benzyl carbon for a nucleophilic substitution reaction of a benzyl derivative

From these two simple cases, as well as the S_N2 reaction of methyl derivatives discussed in Section 3, we see that the factors that govern geometric progression and charge development are quite different. Consequently there is no general basis for presuming that charge and geometry progress in a linear fashion. The factor that governs the degree of charge development at any point along the reaction co-ordinate, including the transition state, is just the charge distributions of the configurations that make up the reaction profile at any given point. This means that any attempt to characterize the degree of charge development in the transition state of a particular reaction must be based on specifying the important configurations from which the transition state wave-function is built up, as well as their relative contributions.

The above separation of charge and geometric progression of the transition state has at least one disturbing consequence. Reaction transition states are commonly characterized by various parameters. These include kinetic isotope effects, the Brønsted parameter, and solvent activity coefficients. The question immediately arises: do these measures of transition state structure measure charge or geometric progression? On the basis of the previous discussion, they can, at best, measure one but not both of these parameters. Let us first consider the Brønsted parameter α.

For outer-sphere electron transfer reactions which are well-described by just two configurations, DA and D^+A^-, the slope of rate-equilibrium plots, α, may reflect the geometry. This is a consequence of Marcus theory. The value of α cannot be a measure of charge since, by definition almost, electron transfer takes place at the transition state, suggesting a charge development of 0.5. So for these reactions the transition states may always be described as half-way along the reaction co-ordinate according to the charge criteria, but at variable positions (measured by α) according to the geometric criteria.

For most organic reactions, which require at least three configurations for a proper description, $\alpha(\beta)$ may constitute a measure of *charge* development. Thus β, β_{nuc} and β_{lg} values in nucleophilic substitution and elimination reactions appear to be useful relative measures of charge development in these systems (see, for example, Jencks, 1980; Hupe and Jencks, 1977; Gandler and Jencks, 1982). Similarly Abraham Z-values (Abraham, 1974; Abraham and Grellier, 1976) and solvent activity coefficients (Parker, 1969) are also likely to reflect the development of charge in the transition state since both are based on charge-solvent interactions.

Use of kinetic isotope effects (KIE) as a measure of transition state structure appears to be rather more complex. The physical basis of the KIE rests on the effect of isotopes on the vibrational frequency of key bonds in the ground and transition states, which in turn depend on the force constants of those bonds. Thus the KIE depends on the extent to which these force constants are different in the transition state from what they were in the ground state Since the magnitude of the force constants is likely to be governed by both charge *and* geometric factors, both of these are expected to influence the KIE. For example, at any given R- -X distance, the force constant of the R—X bond is expected to be significantly influenced by the relative mix of Heitler–London (R· ·X) and zwitterionic (R^+X^-) forms in the R—X wavefunction. As a result the measure of transition state structure provided by kinetic isotope effects is likely to represent a complex hybrid of both the charge and geometric influences.

Clearly, work is required to clarify further the nature of the various experimental techniques for measuring transition-state structure. However, we now believe that the common view that transition state structure may generally be described in terms of just *one* parameter, and all experimental measures provide an estimate of this parameter, is unsatisfactory, and should be modified. Even as a simple working hypothesis there appears to be no real basis for such an assumption.

5 Conclusion

This review has attempted to illustrate the relevance and the widespread utility of the CM model. Indeed, the author believes it is difficult to specify any area of structural or mechanistic chemistry where the CM approach is not applicable. The reason is not hard to find: the CM model has its roots in the Schrödinger equation and as such its relevance to chemistry cannot be easily overstated. Even the fundamental chemical concept of a covalent bond derives from the CM approach. The covalent bond (e.g. in H_2) owes its energy to the configuration mix: $H\uparrow\downarrow H \longleftrightarrow H\downarrow\uparrow H$. A wave-function for the hydrogen molecule based on just *one* spin-paired form does not lead to a stable bond. Both spin forms are necessary. Addition of ionic configurations improves the bond further and in the case of heteroatomic bonds generates polar covalent bonds.

For problems of structure the CM model is not the first choice. The development of qualitative MO models over recent years has made an enormous impact on these questions so that, as a whole, this subject is relatively well understood. With regard to questions of reactivity, however, the situation is quite different. By its very design, qualitative MO theory is most readily applied to static problems where the interacting groups are stationary with respect to one another. There is no simple way in which motion along a reaction co-ordinate can be readily accommodated within such a framework.

The CM model is well-suited to fill this role. It can be effectively applied to problems of reactivity and it is on this particular aspect that this review has focused. By providing a simple, qualitative description of any reaction profile, the CM model may be utilized to consider the energetic, structural and charge characteristics of transition states. Since a clearer understanding of the transition state is the key to a greater appreciation of reactivity as a whole, we believe that the CM model provides a useful conceptual framework for considering reactivity problems.

In order to be applied most effectively, knowledge of both VB and MO concepts and the way they interrelate is essential. The two approaches complement each other naturally so that the insight obtained through a double-barrelled approach is maximized. The VB approach, by virtue of its more chemical form, provides the basic features, while incorporation of MO concepts enables stereochemical aspects and the role of symmetry to be assessed. Of particular significance is the fact that the CM model represents a unified approach to organic reactivity. Whereas Marcus theory, frontier molecular orbital theory, resonance theory, and potential energy surface models represent discrete and in most cases, mutually unrelated models that cater specifically to a certain type of problem, the CM model embraces all of these separate approaches. Indeed, all of the above models may be rediscovered within the general CM framework. In the final analysis, however, we believe that the fact that a single three-curve diagram such as Fig. 11 may be utilized to accommodate such apparently divergent phenomena as the elimination mechanistic spectrum, the reason for anomalous rate–equilibrium relationships, the Woodward–Hoffmann rules, the factors governing concerted *vs* stepwise reactions, the relationship between ground and excited state reactivity, curved Hammett plots, and rotational barriers, to mention just a few, is in itself convincing testimony to its fundamental nature and general utility. The author believes, therefore, that application of the CM model to other aspects of chemical reactivity is also likely to prove of value in the continuing pursuit of greater chemical understanding.

Acknowledgements

The author is grateful to Professors N. Agmon, S. Efrima, J. Michl and S. Shaik for helpful discussions; to Professor J. Kurz for comments; and to Mrs Yerushah Israel for typing the manuscript.

References

Abraham, M. H. (1974). *Prog. Phys. Org. Chem.* **11**, 1
Abraham, M. H. and Grellier, P. L. (1976). *J. Chem. Soc. Perkin 2* 1735
Agmon, N. (1978). *J. Chem. Soc. Faraday 2* **74**, 338
Agmon, N. (1980). *J. Am. Chem. Soc.* **102**, 2164
Agmon, N. (1984). *J. Am. Chem. Soc.* **106**, 6960
Albery, W. J., Campbell-Crawford, A. N. and Curran, J. S. (1972). *J. Chem. Soc. Perkin 2* 2206
Albery, W. J. and Kreevoy, M. M. (1978). *Adv. Phys. Org. Chem.* **16**, 87
Arnett, E. M. and Reich, R. (1980). *J. Am. Chem. Soc.* **102**, 5892
Ballistreri, F. P., Maccarone, E. and Mamo, A. (1976). *J. Org. Chem.* **41**, 3364
Bank, S. and Noyd, D. A. (1973). *J. Am. Chem. Soc.* **95**, 8203
Baughan, E. C. and Polanyi, M. (1941). *Trans. Faraday Soc.* **37**, 648
Bell, R. P. (1936). *Proc. Roy. Soc. A* **154**, 414
Bell, R. P. (1973). "The Proton in Chemistry", 2nd edn, p. 207. Cornell University Press, USA
Bell, R. P. and Sorenson, P. E. (1976). *J. Chem. Soc. Perkin 2* 1594
Beltrame, P., Biale, G., Loyd, D. J., Parker, A. J., Ruane, M. and Winstein, S. (1972). *J. Am. Chem. Soc.* **94**, 2240

Bentley, T. W., Bowen, C. T., Morten, D. H. and Schleyer, P. v. R. (1981). *J. Am. Chem. Soc.* **103**, 5466
Bernasconi, C. F. and Kanavarioti, A. (1985). Submitted.
Bordwell, F. G., Boyle, W. J. Jr., Hautala, J. A. and Yee, K. C. (1969). *J. Am. Chem. Soc.* **91**, 4002
Bordwell, F. G., Boyle, W. J. Jr. and Yee, K. C. (1970). *J. Am. Chem. Soc.* **92**, 5926
Bordwell, F. G. and Boyle, W. J. Jr. (1972). *J. Am. Chem. Soc.* **94**, 3907
Bordwell, F. G. and Boyle, W. J. (1975). *J. Am. Chem. Soc.* **97**, 3447
Bordwell, F. G. and Hughes, D. L. (1982). *J. Org. Chem.* **47**, 3224
Brønsted, J. N. and Pedersen, K. J. (1924). *Z. Phys. Chem. (Leipzig)* **108**, 185
Bunnett, J. F. (1978). *Acc. Chem. Res.* **11**, 413
Carrington, T. (1972). *Disc. Faraday Soc.* **53**, 27
Chanon, M. and Tobe, M. L. (1982). *Angew. Chem. Int. Ed. Engl.* **21**, 1
Cohen, A. O. and Marcus, R. A. (1968). *J. Phys. Chem.* **72**, 4249
Cox, B. G. and Gibson, A. (1974). *J. Chem. Soc. Chem. Comm.* 638
Critchlow, J. E. (1972). *J. Chem. Soc. Faraday Trans.* **68**, 1774
Dauben, W. G. (1967). 13th Chemistry Conference of the Solvay Institute, "Reactivity of the Photoexcited Organic Molecule", p. 171. Interscience, New York
Dauben, W. G., Salem, L. and Turro, N. J. (1975). *Acc. Chem. Res.* **8**, 41
Devaquet, A. (1975). *Pure Appl. Chem.* **41**, 455
Dewar, M. J. S. (1969). "The Molecular Orbital Theory of Organic Chemistry", p. 284. McGraw-Hill, New York
Eberson, L. (1982). *Adv. Phys. Org. Chem.* **18**, 79
Eigen, M. (1964). *Angew. Chem. Int. Ed. Engl.* **3**, 1
Epiotis, N. D. (1972). *J. Am. Chem. Soc.* **94**, 1924, 1935
Epiotis, N. D. (1973). *J. Am. Chem. Soc.* **95**, 1191, 1214
Epiotis, N. D. (1974). *Angew. Chem. Int. Ed. Engl.* **13**, 751
Epiotis, N. D. (1978). "Theory of Organic Reactions." Springer-Verlag, Heidelberg
Epiotis, N. D. and Shaik, S. S. (1977a). *J. Am. Chem. Soc.* **99**, 4936
Epiotis, N. D. and Shaik, S. S. (1977b). In "Progress in Theoretical Organic Chemistry" (Csizmadia, I. G., ed.) Vol. 2. Elsevier, Amsterdam
Epiotis, N. D. and Shaik, S. S. (1978a). *J. Am. Chem. Soc.* **100**, 1
Epiotis, N. D. and Shaik, S. S. (1978b). *J. Am. Chem. Soc.* **100**, 9
Epiotis, N. D. and Shaik, S. S. (1978c). *J. Am. Chem. Soc.* **100**, 29
Epiotis, N. D., Shaik, S. S. and Zander, W. (1980). In "Rearrangements in Ground and Excited States" (De Mayo, P., ed.) Vol. 2. Academic Press, New York
Evans, M. G. (1939). *Trans. Faraday Soc.* **35**, 824
Evans, M. G. and Polanyi, M. (1938). *Trans. Faraday Soc.* **34**, 11
Evans, M. G. and Warhurst, E. (1938) *Trans. Faraday Soc.* **34**, 614
Fleming, I. (1976). "Frontier Orbitals and Organic Chemical Reactions." Wiley-Interscience, New York
Forster, W. and Laird, R. M. (1982). *J. Chem. Soc. Perkin 2* 135
Fry, J. L., Engler, E. M. and Schleyer, P. v. R. (1972). *J. Am. Chem. Soc.* **94**, 4628
Fuchs, R. and Lewis, E. S. (1974). In "Investigation of Rates and Mechanisms of Reactions", 3rd edn (Lewis, E. S., ed.) Part 1. Wiley, New York
Fukui, K. (1970). *Fortschr. Chem. Forsch.* **15**, 1
Fukui, K. (1975). "Theory of Orientation and Stereoselection." Springer-Verlag, Heidelberg
Fukuzumi, S. and Kochi, J. K. (1980a). *J. Phys. Chem.* **84**, 2246
Fukuzumi, S. and Kochi, J. K. (1980b). *J. Am. Chem. Soc.* **102**, 2141
Fukuzumi, S. and Kochi, J. K. (1981a). *J. Am. Chem. Soc.* **103**, 7240

Fukuzumi, S. and Kochi, J. K. (1981b). *J. Am. Chem. Soc.* **103**, 2783
Fukuzumi, S. and Kochi, J. K. (1982). *Tetrahedron* **38**, 1035
Fukuzumi, S., Mochida, K. and Kochi, J. K. (1979). *J. Am. Chem. Soc.* **101**, 5961
Fukuzumi, S., Wong, C. L. and Kochi, J. K. (1980). *J. Am. Chem. Soc.* **102**, 2928
Gandler, J. R. and Jencks, W. P. (1982). *J. Am. Chem. Soc.* **104**, 1937
Gerhartz, W., Poshusta, R. D. and Michl, J. (1976). *J. Am. Chem. Soc.* **98**, 6427
Giese, B. (1977). *Angew. Chem. Int. Ed. Engl.* **16**, 125
Goddard, W. A., III, Dunning, T. H., Jr., Hunt, W. J. and Hay, J. P. (1973). *Accounts Chem. Res.* **6**, 368
Grimsrud, E. P. and Taylor, J. W. (1970). *J. Am. Chem. Soc.* **92**, 739
Hammond, G. S. (1955). *J. Am. Chem. Soc.* **77**, 334
Harris, J. C. and Kurz, J. L. (1970). *J. Am. Chem. Soc.* **92**, 349
Heitler, W. and London, F. (1927). *Z. Phys.* **44**, 455
Herzberg, G. (1950). "Diatomic Molecules", 2nd edn, p. 372. Van Nostrand, Princeton, New Jersey
Herzberg, G. (1966). "The Electronic Spectra of Polyatomic Molecules", p. 442. Van Nostrand, Princeton, New Jersey
Herzberg, G. and Longuet-Higgins, H. C. (1963). *Trans. Faraday Soc.* **35**, 77
Hilinski, E. F., Masnovici, J. M., Amatore, C., Kochi, J. K. and Rentzepis, P. M. (1983). *J. Am. Chem. Soc.* **105**, 6167
Hill, J. W. and Fry, A. (1962). *J. Am. Chem. Soc.* **84**, 2763
Hine, J. (1962). "Physical Organic Chemistry", 2nd edn, pp. 175–180. McGraw-Hill, New York
Hine, J. and Brader, W. H. (1953). *J. Am. Chem. Soc.* **75**, 3964
Hine. J. (1977). *J. Am. Chem. Soc.* **15**, 1
Horiuti, J. and Polanyi, M. (1935). *Acta Physicochim. U.R.S.S.* **2**, 505
Houk, K. N., Gandour, R. W., Strozier, R. W., Rondan, N. G. and Paquette, L. A. (1979). *J. Am. Chem. Soc.* **101**, 6797
Hupe, D. J. and Jencks, W. P. (1977). *J. Am. Chem. Soc.* **99**, 451
Hupe, D. J. and Wu, D. (1977). *J. Am. Chem. Soc.* **99**, 7653
Iwasaki, T., Sawada, T., Okuyama, M. and Kamada, H. (1978). *J. Phys. Chem.* **82**, 371
Jencks, W. P. (1972). *Chem. Rev.* **72**, 705
Jencks, W. P. (1980). *Acc. Chem. Res.* **13**, 161
Jencks, D. A. and Jencks, W. P. (1977). *J. Am. Chem. Soc.* **99**, 7948
Jencks, W. P., Brant, S. R., Gandler, J. R., Fendrich, G. and Nakamura, C. (1982). *J. Am. Chem. Soc.* **104**, 7045
Johnson, C. D. (1975). *Chem. Rev.* **75**, 755
Keeffe, J. R., Morey, J., Palmer, C. A. and Lee, J. C. (1979). *J. Am. Chem. Soc.* **101**, 1295
Kellman, A. and Dubois, J. T. (1965). *J. Chem. Phys.* **42**, 2518
Kevill, D. N. (1981). *J.C.S. Chem. Comm.* 421
Kiselev, V. D. and Miller, J. G. (1975). *J. Am. Chem. Soc.* **97**, 4036
Knier, B. L. and Jencks, W. P. (1980). *J. Am. Chem. Soc.* **102**, 6789
Kornblum, N. (1975). *Angew. Chem. Int. Ed. Engl.* **14**, 734
Kost, D. and Aviram, K. (1982). *Tetrahedron Letters* 4157
Kreevoy, M. M. and Konasewich, D. E. (1970). *J. Phys. Chem.* **74**, 4464
Kreevoy, M. M. and Oh, S. (1973). *J. Am. Chem. Soc.* **95**, 4805
Kreevoy, M. M. and Lee, I-S. H. (1984). *J. Am. Chem. Soc.* **106**, 2550
Kresge, A. J. (1970). *J. Am. Chem. Soc.* **92**, 3210
Kresge, A. J. (1973). *Chem. Soc. Rev.* **2**, 475

Kresge, A. J. (1974). *Canad. J. Chem.* **52**, 1897
Kresge, A. J. (1975). *In* "Proton Transfer Reactions" (Caldin, E. and Gold, V. eds). Chapman and Hall, London
Kurz, J. L. (1983). *J. Org. Chem.* **48**, 5117
Kurz, J. L. and El-Nasr, M. M. (1982). *J. Am. Chem. Soc.* **104**, 5823
Laidler, K. J. (1955). "The Chemical Kinetics of Excited States". Clarendon Press, Oxford
Laidler, K. J. and Shuler, K.E. (1951). *Chem. Rev.* **48**, 153
Leffler, J. E. (1953). *Science* **117**, 340
Leffler, J. E. and Grunwald, E. (1963). "Rates and Equilibria of Organic Reactions", pp. 156–7. Wiley, New York.
Levine, R. D. and Bernstein, R. B. (1974). "Molecular Reaction Mechanics." Oxford University Press, New York.
Lewis, E. S. (1978). *Top. Curr. Chem.* **74**, 31
Lewis, E. S. (1983). Abstracts of the Tenth Anniversary Meeting of the Royal Society of Chemistry Organic Reaction Mechanisms Group, Maynooth, July 12–15
Lewis, E. S. and Kukes, S. (1979). *J. Am. Chem. Soc.* **101**, 417
Lewis, E. S., Kukes, S. and Slater, C. D. (1980). *J. Am. Chem. Soc.* **102**, 1619
Libit, L. and Hoffmann, R. (1974). *J. Am. Chem. Soc.* **96**, 1370
Longuet-Higgins, H. C. and Abrahamson, E. W. (1965). *J. Am. Chem. Soc.* **87**, 2045
March, J. (1968). "Advanced Organic Chemistry: Reactions, Mechanism and Structure", p. 537. McGraw-Hill, New York
Marcus, R. A. (1964). *Annu. Rev. Phys. Chem.* **15**, 155
Marcus, R. A. (1969). *J. Am. Chem. Soc.* **91**, 7224
Marcus, R. A. (1977). "Special Topics in Electrochemistry" (Rock, P. A., ed.). Elsevier, Amsterdam
McLennan, D. J. (1976). *J. Chem. Ed.* **53**, 348
McLennan, D. J. and Pross, A. (1984). *J. Chem. Soc. Perkin 2* 981
McLennan, D. J. and Wong, R. J. (1974). *J. Chem. Soc. Perkin 2* 1818
Metiu, H., Ross, J. and Whitesides, G. M. (1979). *Angew. Chem. Int. Ed. Engl.* **18**, 377
Michl, J. (1972). *Mol. Photochem.* **4**, 243, 257, 287
Michl, J. (1974). *Topics Curr. Chem.* **46**, 1
Michl, J. (1975). *Pure Appl. Chem.* **41**, 507
Michl, J. (1977). *Photochem Photobiol.* **25**, 141
More O'Ferrall, R. A. (1970). *J. Chem. Soc. B* 274
More O'Ferrall, R. A. (1975). *In* "Proton Transfer Reactions" (Caldin, E. and Gold, V. eds). Chapman and Hall, London
Mulliken, R. S. (1952a). *J. Am. Chem. Soc.* **74**, 811
Mulliken, R. S. (1952b). *J. Phys. Chem.* **56**, 801
Mulliken, R. S. and Person, W. B. (1969). "Molecular Complexes." Wiley-Interscience, New York
Murai, K. Takeuchi, S. and Kimura, C. (1973). *Nippon Kagaku Zasshi* 95
Murdoch, J. R. (1972). *J. Am. Chem. Soc.* **94**, 4410
Murdoch, J. R. (1983a). *J. Am. Chem. Soc.* **105**, 2159
Murdoch, J. R. (1983b). *J. Am. Chem. Soc.* **105**, 2667
Murdoch, J. R. and Magnoli, D. E. (1982). *J. Am. Chem. Soc.* **104**, 3792
Murdoch, J. R., Bryson, J. A., McMillen, D. F. and Brauman, J. I. (1982). *J. Am. Chem. Soc.* **104**, 600
Nagakura, S. (1963). *Tetrahedron, Suppl. 2* **19**, 361
Norrish, R. G. W. (1937). *Trans. Faraday Soc.* **33**, 1521

Ogg, R. A. Jr. and Polanyi, M. (1935). *Trans. Faraday Soc.* **31**, 604
Parker, A. J. (1969). *Chem. Revs.* **69**, 1
Parker, A. J., Ruane, M., Palmer, D. A. and Winstein, S. (1972). *J. Am. Chem. Soc.* **94**, 2228
Pauling, L. and Wilson, E. B. Jr. (1935). "Introduction to Quantum Mechanics." McGraw-Hill, New York
Pearson, R. E. and Martin, J. C. (1963). *J. Am. Chem. Soc.* **85**, 3142
Pearson, R. G. (1976). "Symmetry Rules for Chemical Reactions." Wiley-Interscience, New York
Pearson, R. G. and Dillon, R. L. (1953). *J. Am. Chem. Soc.* **75**, 2439
Pellerite, M. J. and Brauman, J. I. (1980). *J. Am. Chem. Soc.* **102**, 5993
Pellerite, M. J. and Brauman, J. I. (1983). *J. Am. Chem. Soc.* **105**, 2672
Pross, A. (1977). *Adv. Phys. Org. Chem.* **14**, 69
Pross, A. (1983). *Tetrahedron Letters* 835
Pross, A. (1984). *J. Org. Chem.* **49**, 1811
Pross, A. and Radom, L. (1980). *Aust. J. Chem.* **33**, 241
Pross, A. and Radom, L. (1981). *Prog. Phys. Org. Chem.* **13**, 1
Pross, A. and Shaik, S. S. (1981). *J. Am. Chem. Soc.* **103**, 3702
Pross, A. and Shaik, S. S. (1982a). *J. Am. Chem. Soc.* **104**, 187
Pross, A. and Shaik, S. S. (1982b). *J. Am. Chem. Soc.* **104**, 1129
Pross, A. and Shaik, S. S. (1982c). *Tetrahedron Letters* 5467
Pross, A. and Shaik, S. S. (1983). *Accounts Chem. Res.* **16**, 363
Pross, A., Aviram, K., Klix, R. C., Kost, D. and Bach, R. D. (1984). *Nouv. J. Chim.* **8**, 711
Russell, G. A. and Desmond, K. M. (1963). *J. Am. Chem. Soc.* **85**, 3139
Salem, L. (1973). *Pure Appl. Chem.* **33**, 317
Salem, L. (1974). *J. Am. Chem. Soc.* **96**, 3486
Salem, L. (1982). "Electrons in Chemical Reactions: First principles." Wiley-Interscience, New York
Salem, L. and Rowland, C. (1971). *Angew. Chem. Int. Ed. Engl.* **11**, 92
Salem, L., Leforestier, C., Segal, G. and Wetmore, R. (1975). *J. Am. Chem. Soc.* **97**, 479
Sauer, J. (1967). *Angew Chem. Int. Ed. Engl.* **6**, 16
Scandola, F., Balzani, V. and Schuster, G. B. (1981). *J. Am. Chem. Soc.* **103**, 2519
Shaik, S. S. (1981). *J. Am. Chem. Soc.* **103**, 3692
Shaik, S. S. (1983). *J. Am. Chem. Soc.* **105**, 4359
Shaik, S. S. (1984). *J. Am. Chem. Soc.* **106**, 1227
Shaik, S. S. and Pross, A. (1982a). *Bull. Soc. Chim. Belg.* **91**, 355
Shaik, S. S. and Pross, A. (1982b). *J. Am. Chem. Soc.* **104**, 2708
Skell, P. S., Tuleen, D. L. and Readio, P. D. (1963). *J. Am. Chem. Soc.* **85**, 2850
Streitwieser, A. Jr. (1962). "Solvolytic Displacement Reactions", pp. 15-17. McGraw-Hill, New York
Teller, E. (1937). *J. Phys. Chem.* **41**, 109
Thornton, E. R. (1964). "Solvolysis Mechanisms", pp. 197-206. Ronald Press, New York
Thornton, E. R. (1967). *J. Am. Chem. Soc.* **89**, 2915
Townshend, R. E., Ramunni, G., Segal, G., Hehre, W. J. and Salem, L. (1976). *J. Am. Chem. Soc.* **98**, 2190
Turro, N. J. (1978). "Modern Molecular Photochemistry" Benjamin/Cummings, California

Turro, N. J., McVey, J., Ramamurthy, V. and Lechtken, P. (1979). *Angew. Chem. Int. Ed. Engl.* **18**, 572
van der Lugt, W. Th. A. M. and Oosterhoff, L. J. (1969). *J. Am. Chem. Soc.* **91**, 6042
Walling, C., Rieger, A. L. and Tanner, D. D. (1963). *J. Am. Chem. Soc.* **85**, 3129
Wang, J. T. and Williams, F. (1980). *J. Am. Chem. Soc.* **102**, 2860
Warshel, A. (1981). *Acc. Chem. Res.* **14**, 284
Warshel, A. and Weiss, R. M. (1980). *J. Am. Chem. Soc.* **102**, 6218
Westaway, K. C. and Waszczylo, Z. (1982). *Canad. J. Chem.* **60**, 2500
Winstein, S., Appel, B., Baker, R. and Diaz, A. (1965). *In* "Organic Reaction Mechanisms". *Chem. Soc. Spec. Publ.* **19**, 109
Wolfe, S., Mitchell, D. J. and Schlegel, H. B. (1981). *J. Am. Chem. Soc.* **103**, 7694
Woodward, R. B. and Hoffmann, R. (1965). *J. Am. Chem. Soc.* **87**, 395, 2046, 4388
Woodward, R. B. and Hoffmann, R. (1970). "The Conservation of Orbital Symmetry." Verlag Chemie-Academic Press, Weinheim
Wyrzykowska, K., Grodowski, M., Weiss, K. and Latowski, T. (1978). *Photochem. Photobiol.* **28**, 311
Yamataka, H. and Ando, T. (1979). *J. Am. Chem. Soc.* **101**, 266
Zimmerman, H. E. (1982). *Acc. Chem. Res.* **15**, 312

Gas-phase Nucleophilic Displacement Reactions

JOSÉ MANUEL RIVEROS,[1] SONIA MARIA JOSÉ[1] and KEIKO TAKASHIMA[2]

[1] *Institute of Chemistry, University of São Paulo, São Paulo, Brazil*
[2] *State University of Londrina, Londrina, Pr., Brazil*

1 Introduction 198
2 Experimental techniques 200
    Ion cyclotron resonance spectrometry 201
    Flowing afterglow 203
    High pressure mass spectrometry 204
3 General features of gas-phase ion-molecule reactions 204
4 Gas-phase S_N2 reactions involving negative ions 206
    Thermochemical considerations 206
    General aspects of gas-phase S_N2 reactions 207
    Stereochemistry 209
    Nucleophilicity and leaving group ability 211
    Effect of solvation on the gas-phase reaction 212
    Mechanism of the gas-phase S_N2 reaction 213
    Potential energy surfaces for gas-phase S_N2 reactions 214
    Recent theoretical developments 218
5 Some examples of gas-phase S_N2 reactions involving positive ions 220
6 Nucleophilic displacement reactions by negative ions in carbonyl systems 222
    General features 222
    Model for nucleophilic displacement reactions at carbonyl centres 227
7 Gas-phase nucleophilic reactions of carbonyl systems involving positive ions 229
    General features 229
    Mechanism 231
8 Nucleophilic displacement reactions in aromatic systems 234
9 Conclusions 237
References 237

1 Introduction

The search for meaningful trends in chemical reactivity and their correlation with molecular parameters is one of the fundamental goals of physical organic chemistry. A wealth of data on rates of organic reactions has been gathered over the years to provide experimental support for the proposed basic mechanisms. On the theoretical side, qualitative and empirical description has given way to sophisticated methods for calculations of chemical reactivity which allow for a dynamic interplay between theory and experiment.

Despite the impressive knowledge accumulated on some of the more common organic reactions, the question of true intrinsic reactivity is still at large in most cases. It is known that rates and mechanisms can be influenced by solvent effects, and that such changes can seldom be accommodated by rigorous theoretical treatments which are only applicable to isolated species. Thus, the concept of intrinsic reactivity should be, in principle, derived from chemical behaviour in a solvent-free environment. This statement is particularly relevant for reactions involving ionic species which are subject to strong electrostatic interactions with the solvent.

The development of a number of mass spectrometric techniques in the late 1960s paved the way for the study of bimolecular chemical reactions of ions with neutral molecules in the gas phase (see, for example, Harrison *et al.,* 1966; McDaniel *et al.,* 1970; Franklin, 1972). While the systems initially investigated were concerned primarily with positive ions and with relatively simple reactions, the underlying potential of such techniques was soon to be explored in a dramatic way.

The early work on gas-phase acidities and basicities (Brauman and Blair, 1970; Brauman *et al.,* 1971), obtained by probing the preferred direction of proton transfer reactions in the gas phase, established a benchmark in physical organic chemistry. The results pointed out, for example, that polarizability arguments could satisfactorily account for the smooth trends observed within a homologous series, unlike the pK-values in condensed media where solvation effects may reverse the relative orders of acidities and basicities.

The differences between gas-phase and solution reactivity can be viewed as the result of two distinct contributions due to solvation:

(i) Energetics of solvation. The effect of solvation on the thermodynamics of a reaction is best illustrated by the simple heterolytic bond dissociation of H_2O in (1) and (2). Large variations in the energetic parameters will

ΔG°_{298} (kcal mol^{-1})

$H_2O(g) \rightarrow H^+(g) + OH^-(g)$ 384.2   (1)

$H_2O(aq) \rightarrow H^+(aq) + OH^-(aq)$ 21.4   (2)

obviously be responsible for significant changes in the driving force of a reaction in both media. Further examples are shown in (3) and (4).

$$NH_3 + H_3O^+ \rightarrow NH_4^+ + H_2O \qquad \Delta H°_{300}(g) = -32 \text{ kcal mol}^{-1} \qquad (3)$$
$$\Delta H°_{298}(aq) = -12 \text{ kcal mol}^{-1}$$

$$OH^- + HCOOCH_3 \rightarrow HCOO^- + CH_3OH \quad \Delta H°_{298}(g) = -43 \text{ kcal mol}^{-1} \qquad (4)$$
$$\Delta H°_{298}(aq) = -11 \text{ kcal mol}^{-1}$$

(ii) Dynamic and mechanistic effects. Reactions in solution proceed by the initial diffusion of the reagents through the solvent cage followed by chemical reaction. The activation energy of the reaction depends in general on the solvent, and it is dictated by the relative stability of the reagents and the transition state. Proton transfer reactions of strong acids and bases are usually diffusion-limited in solution. These same reactions proceed essentially at the collision frequency in the gas phase. Thus, rate constants in these cases are comparable in the gas phase and in solution. For reaction (3) $k_{aq}(298 \text{ K}) = 4.3 \times 10^{10} \text{ M}^{-1}\text{s}^{-1}$ (Emerson et al., 1960) and $k_{gas}(300 \text{ K}) = 1.4 \times 10^{12} \text{ M}^{-1}\text{s}^{-1}$ (Hemsworth et al., 1974). More complex reactions display spectacular differences, with gas-phase reactions still exhibiting rate constants near the collision limit. A typical example is the S_N2 reaction (5) (Bowie, 1980).

$$Cl^- + CH_3Br \rightarrow Br^- + CH_3Cl \qquad \Delta H°(gas) = -6 \text{ kcal mol}^{-1} \qquad (5)$$
$$\Delta H°(aq) \sim 0 \text{ kcal mol}^{-1}$$

Solvent	H_2O	CH_3OH	DMF	gas
$k_{298}/M^{-1}s^{-1}$	4.9×10^{-6}	6.6×10^{-6}	4.0×10^{-1}	1.3×10^{10}

The field of gas-phase organic reactions of ionic species has blossomed in the last 15 years to the point where a single review article becomes a formidable task. The scope of the present report covers gas-phase reactions initiated by nucleophilic attack with special emphasis on displacement reactions. This choice was dictated by the analogy with well-known reactions in solution, which may be considered as a classical chapter of physical organic chemistry.

Because the techniques used to study ionic reactions in the gas phase are familiar primarily to chemists in the field, an initial discussion of the experimental methods is necessary in order to gain a proper perspective of this subject.

2. Experimental techniques

Three basic techniques, and variations thereof, have been used in recent years to study aspects of gas-phase ion-molecule reactions pertinent to organic systems; they are ion cyclotron resonance spectrometry, flowing afterglow, and high pressure mass spectrometry. The essential feature of these techniques is that ions produced under vacuum are allowed to undergo from few to many collisions with neutrals before they are neutralized at the walls of the instrument.

Positive ions are usually generated in these techniques, as in mass spectrometry, by electron bombardment. Photo-ionization, laser desorption and chemical ionization have also been used for specific applications. When the ionizing energy is above the ionization potential of the parent compound, this procedure may result in the formation of several fragment ions. Collisions of parent and fragment ions with neutral molecules may then give rise to a sequence of chemical reactions.

Negative ions are also produced by electron bombardment through a variety of electron attachment or dissociative capture processes (Dillard, 1973). They may also be generated indirectly as the result of an ion-molecule reaction. The cross-sections for negative ion formation are at least an order of magnitude less than those for positive ions, a fact which accounts for the much weaker ion currents that are produced. Some of the common precursors for well-known negative ions are shown in (6)–(13). For several of these

$$NF_3 + e^- \rightarrow F^- + \cdot NF_2 \qquad (6)$$

$$RONO + e^- \rightarrow RO^- + NO \qquad (7)$$

$$NH_3 + e^- \rightarrow NH_2^- + H\cdot \qquad (8)$$

$$\begin{cases} N_2O + e^- \rightarrow O^{\bar{\cdot}} + N_2 & (9) \\ O^{\bar{\cdot}} + RH \rightarrow OH^- + R\cdot & (10) \end{cases}$$

$$\begin{cases} H_2O + e^- \rightarrow H^- + \cdot OH & (11) \\ H^- + H_2O \rightarrow OH^- + H_2 & (12) \end{cases}$$

$$CCl_4 + e^- \rightarrow Cl^- + \cdot CCl_3 \qquad (13)$$

cases, the appearance of negative ions occurs only at well-defined electron energies in a true resonant process. Since few negative ions possess reasonable stability (i.e. few molecular species or radicals have positive electron affinities), the mass spectra of negative ion systems are usually very simple.

Another common feature of the various techniques is that, due to the short lifetime of the ions before neutralization, they are only applicable to reactions that, by solution standards, are very fast. Thus, the techniques are limited to the observation of bimolecular reactions which proceed with rate constants larger than $10^8 \, M^{-1}s^{-1}$. Furthermore, the techniques can only measure the abundance of reagent and product ions. The neutrals participating in the reaction must be inferred from mass balance, except for a few cases where special techniques have been developed to establish the identity of neutrals.

ION CYCLOTRON RESONANCE SPECTROMETRY

Ion cyclotron resonance (icr) (Baldeschwieler, 1968; Lehman and Bursey, 1976) is a mass spectrometric method in which ions are detected according to their m/e value by the characteristic orbiting frequency of a charged species in a uniform magnetic field (14). Ions of a given charge polarity are trapped in the icr cell (Fig. 1) by the application of suitable small potentials (less than 1 V) on the electrodes. The trapping time can vary from periods of milliseconds to seconds. Even though this technique operates at very low gas pressure (10^{-8} to 10^{-5} torr), the long residence time of the ions allows for collisions between ions and neutral molecules resulting in chemical reactions.

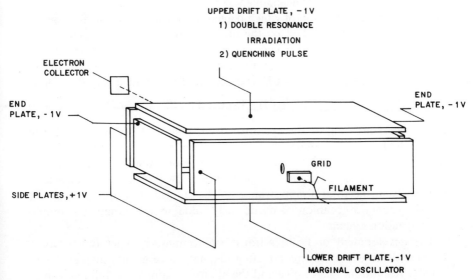

FIG. 1 Typical design of an icr cell used in kinetic experiments. The cell is placed inside a magnetic field oriented perpendicular to the side plates

A mass spectrum (positive or negative) is obtained by maintaining a fixed radio frequency (RF) field and varying the magnetic field. Ions are brought into resonance with the RF field when condition (14) is satisfied, where

$$B = 2\pi v \, (m/e) \quad (14)$$

B = magnetic field intensity and v = frequency of the RF field. At a frequency of 153.5 kHz, ions of $m/e = 100$ come into resonance at $B = 1$ T. This mass spectrum exhibits (primary) ions formed directly by electron bombardment, as well as those formed by chemical reactions. The appearance and abundance of secondary, or tertiary ions, depend on the pressure and the storage time of the ions.

A variety of pulsed techniques are particularly useful for kinetic experiments (McIver and Dunbar, 1971; McMahon and Beauchamp, 1972; McIver, 1978). In these experiments, ions are initially produced by pulsing the electron beam for a few milliseconds. A suitable combination of magnetic and electric fields is then used to store the ions for a variable period of time, after which the detection system is switched on to resonance to measure the abundance of a given ionic species. These techniques allow the monitoring of ion concentration as a function of reaction time. Since the neutrals are in large excess with respect to the ions, a pseudo first-order rate constant can be obtained in a straightforward fashion from these data. The calculation of the rate constant must nevertheless make proper allowance for the fact that ion losses in the icr cell are not negligible.

The actual identification of which reagent ions are responsible for a given product ion is accomplished by the ion cyclotron double resonance technique. For example, the occurrence of the gas-phase reaction (15) can be verified by this technique. At a magnetic field of $B = 0.7$ T, and with the detection set at 307 kHz, the ion $^{35}Cl^-$ is detected. If a second strong RF field is

$$F^- + CH_3Cl \rightarrow CH_3F + Cl^- \tag{15}$$

introduced operating at the cyclotron frequency of F^- ($v = 565.5$ kHz at 0.7 T), the F^- species can be selectively removed from the cell by the increase of its orbiting radius. If a chemical reaction is responsible for the appearance of Cl^- in the mass spectrum, the removal of F^- will result in a decrease, or eventually in the total disappearance, of the chloride ion signal. Thus, the double resonance technique is particularly valuable in untangling a complicated reaction scheme.

The development of Fourier transform techniques in icr (Comisarow, 1978) and the use of superconducting magnets have greatly increased sensitivity, resolution, mass range and the ability to sample the full mass spectrum in one pulse.

Rate constants of ion-molecule reactions measured by icr suffer from two drawbacks: *(a)* accurate absolute pressure measurements below 10^{-5} torr are difficult to make, and in practice these measurements are probably no better than 20%; *(b)* at the working pressures of icr, and in the presence of even small trapping fields, the degree of thermalization of the ions before chemical reaction is a point of contention.

FLOWING AFTERGLOW

Flowing afterglow (FA) was developed in the early 1960s primarily to collect data on atmospheric ion chemistry (Ferguson *et al.*, 1969). The instrumentation (Fig. 2) consists basically of a plasma created in a long tube (usually 1 m long) which is carried by a fast flowing gas like helium (usually around 0.1 to 1 torr).

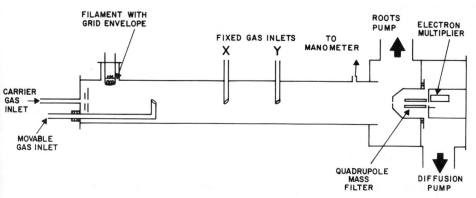

FIG. 2 Schematic diagram of a flowing afterglow instrument

Ions (positive and negative) generated by electron impact are allowed to thermalize with the carrier gas before the neutral reagent gas is introduced further along the tube. Ions (positive or negative) are sampled at the end of the tube through a small sampling orifice by a quadrupole mass-filter and an electron multiplier housed in a differentially pumped high vacuum region. Application of this technique to a wide variety of organic reactions has been reviewed by DePuy and Bierbaum (1981).

Kinetic experiments are usually carried out by recording the ion concentration as a function of neutral reagent-gas flow. The pressure of the reagent gas is usually maintained in the range of 10^{-3} to 10^{-4} torr. This method allows for the direct determination of rate constants without the need for a solution to the complicated hydrodynamic equations which result from the ion motion in the plasma tube (Bohme *et al.*, 1973). An alternative method for rate-constant determination makes use of a movable injector for the reagent gas along the flow tube. Kinetic data are then extracted from experiments in which ion concentrations are measured as a function of the position of the injector, and thus of reaction time. The flowing afterglow method is believed to yield reliable rate constants because of its higher pressure of operation. This implies that the system is certainly under a Maxwellian distribution.

The flowing afterglow requires considerably larger amounts of reagent gases than icr because of the high flowing rates. On the other hand, this feature has been heralded as a potential means of preparative chemistry on a micro-scale (Smith et al., 1980).

A new development, the selected-ion-drift tube or SIFT technique (Smith and Adams, 1979), enables the isolation of reactions due to a given set of ions, of a selected m/e value, with a neutral. Ions are produced outside the flow cell under high vacuum, and injected into the flow tube through a mass filter. This filter eliminates all ions except the ones selected (according to a specific m/e value), and removes the neutral precursor of the ion in question by differential pumping.

HIGH PRESSURE MASS SPECTROMETRY

Several types of instruments can be grouped under this heading, all of which essentially consist of modified mass spectrometers operating at pressures which range from 10^{-3} to 10 torr (see, for example Solomon et al., 1974; Cunningham et al., 1972). This technique (HPMS) is particularly useful in the determination of equilibrium constants of acid–base systems in the gas phase (McMahon and Kebarle, 1977), and of equilibrium constants for successive degrees of ion solvation in the gas phase (Arshadi et al., 1970). Due to the high pressure, it is generally difficult to avoid three-body clustering reactions which result in very stable associated ions.

Rate constants are usually determined in the pulsed mode by switching on the electron gun for several microseconds followed by reaction of the ions in a field-free region. Magnetic mass analysis coupled with an electron multiplier in which the pulses are collected in a multiscaler provide a time development of the particular ions under analysis. There have been few reactions relevant to this review which have been studied by HPMS.

This technique has been used successfully in the study of the temperature dependence of ion-molecule reactions over a wide range (Meot-Ner and Field, 1975; Meot-Ner, 1979a). This particular aspect makes it a potentially useful technique in the determination of activation energies in slower ion-molecule reactions.

3 General features of gas-phase ion-molecule reactions

Exothermic gas-phase ion-molecule reactions are generally characterized by very large rate constants. The two examples quoted in the introduction can be considered as typical cases suitable for observation by present day experimental techniques. The fact that these reactions exhibit such rate constants has been taken as an indication that activation energies are either nil or amount at best to less than 5 kcal mol^{-1}. Admittedly, few experiments have been dedicated to the temperature dependence of ion-molecule reactions although it represents a field of some theoretical interest (Meot-Ner, 1979b).

Early attempts to describe bimolecular gas phase ion-molecule reactions were based on the classical collision dynamics of a point charge and a structureless polarizable neutral molecule. The collision process is dominated by the long range attractive ion-induced dipole potential $V(r)$ given by (16),

$$V(r) = -e^2 \alpha/2r^4, \qquad (16)$$

where e is the charge of the ion, and α the polarizability of the molecule, and r the distance between the two particles. The resulting trajectories, first calculated by Langevin (1905), show that for limiting values of relative initial velocity and angular momentum, the colliding pair spirals towards the scattering centre. These limiting conditions define the concept of a capture or orbiting collision, and a corresponding collision cross-section. The reaction cross-section was initially assumed to be equal to the Langevin capture cross-section in the absence of an activation energy, an approximation which may be regarded as valid for processes which require no extensive reorganization of the electron clouds. The rate constant calculated in this fashion, for a Maxwell–Boltzmann distribution of velocities, yields the famous Giomousis and Stevenson (1958) expression (17), in which μ is the reduced mass of the

$$k = 2\pi(e^2 \alpha/\mu)^{1/2} \qquad (17)$$

collision partners. The success of this straightforward theory was indicated by the reasonable agreement (within an order of magnitude) between the calculated values and the experimental rate constants of simple reactions.

A more refined treatment of the classical problem takes into consideration the presence of a dipole moment in the neutral molecule. The strong ion-permanent dipole attraction depends on the orientation of the molecule with respect to the ion and the corresponding alignment of the dipole moment. The alignment of the dipole moment is nevertheless perturbed by the random rotational motion of the molecule. A procedure developed by Su and Bowers (1973a,b) allows for averaging the dipole orientation during the collision process. The introduction of this concept, coupled with the definition of reaction cross-section as being that of an orbiting collision, leads to the ADO (average dipole orientation) collision rate constant, given by (18), where μ_D

$$k_{ADO} = (2\pi e/\mu^{1/2})[\alpha^{1/2} + C\mu_D(2/\pi kT)^{1/2}] \qquad (18)$$

represents the permanent dipole of the molecule, k is the Boltzmann constant, and C the dipole locking factor (or degree of orientation) which varies from 0 to 1. The value of C can be determined from the polarizability and dipole moment of the molecule (Su and Bowers, 1973b).

The success of the ADO theory for proton transfer reactions has been demonstrated for a number of cases (Hemsworth *et al.*, 1974; Mackay *et al.*, 1976). Theoretical rate constants are found to agree within 20% of the experimental value. Further refinements of this theory (Su *et al.*, 1978), including conservation of angular momentum, bring the calculated rate constants within experimental error (Su and Bowers, 1979).

The rate constant calculated by the ADO theory has been frequently considered as an upper limit for complex ion-molecule reactions in the gas phase. Consequently, the ratio k_{exp}/k_{ADO} is usually referred to as the efficiency of the reaction. The fact that so many reactions display efficiencies in the range 0.1–1 suggests that the attractive ion-dipole (induced and permanent) potential is the dominant feature in gas-phase ion-molecule reactions.

4 Gas-phase $S_N 2$ reactions involving negative ions

Gas-phase reactions which result in nucleophilic displacement at a saturated, or an unsaturated, carbon centre have been observed in positive and negative ion chemistry. By far, the most widely occurring case is the formal analog of the $S_N 2$ reaction initially reported by Bohme and Young (1970). The experimental determination of rate constants for $S_N 2$ reactions has received a great deal of attention as has the mechanistic point of view including the interpretation of the potential energy surface for the gas-phase reaction.

Nucleophilic displacement reactions are often competitive with other processes promoted by a nucleophile, such as addition-elimination, or proton abstraction and base-induced elimination in which the nucleophile acts as a strong base. This particular situation is especially true in reactions that also involve attack at an unsaturated carbon centre. The delicate interplay between these different mechanisms is in itself a matter of great interest, and as yet it has defied attempts to rationalize it on a quantitative basis.

THERMOCHEMICAL CONSIDERATIONS

Displacement reactions observed in the gas phase are generally exothermic or thermoneutral as in the case of simple isotope exchange. This requirement is consistent with the limited dynamic range of the experimental techniques which precludes the observation of reactions with sizable activation energies. The relevant thermochemical data for negative ions have become available in recent years through the determination of electron affinities (Janousek and Brauman, 1979), and indirectly from gas-phase acidity scales (Bartmess and McIver, 1979). Relative gas-phase acidities available at present (Bartmess *et al.*, 1979; Cumming and Kebarle, 1978) are an important consideration in these systems since proton abstraction is generally faster than displacement at a carbon centre.

The exothermicities of S_N2 reactions in the gas phase tend to be in excess of 20 kcal mol^{-1}. For the prototype reactions (19) the enthalpy change can be

$$X^- + CH_3Y \rightarrow Y^- + CH_3X \qquad (19)$$

calculated simply from differences in electron affinities (*EA*) and bond dissociation energies (*D°*) as in (20). Table 1 lists a convenient set of values of

$$\Delta H° = EA(X) - D°(CH_3-X) - EA(Y) + D°(CH_3-Y) \qquad (20)$$

$EA(X) - D°(CH_3-X)$ for a number of common nucleophiles. Heats of reaction can then be readily calculated from these data for methyl derivatives.

TABLE 1

Thermochemical information relevant to reaction (19)[a]

X$^-$	$EA(X)-D°(CH_3X)$/kcal mol^{-1}	X$^-$	$EA(X)-D°(CH_3\text{-}X)$/kcal mol^{-1}
		CN$^-$	−29
O$^-$	−60	SH$^-$	−20
F$^-$	−30	NH$_2^-$	−62
Cl$^-$	1	NO$_2^-$	−5
Br$^-$	8	CH$_3$O$^-$	−41
OH$^-$	−49	CH$_3$S$^-$	−31

[a]The present data have been collected from several sources, and their accuracy ranges from ± 0.5 to ± 3 kcal mol^{-1}, depending on X

GENERAL ASPECTS OF GAS-PHASE S_N2 REACTIONS

Reactions which can be classified as typical S_N2 processes have been studied extensively in the gas phase, both by the icr and flowing afterglow techniques. The early study of Bohme and Young (1970) by flowing afterglow clearly established that the gas-phase reactions of O$^-$, OH$^-$, CH$_3$O$^-$, CH$_3$CH$_2$O$^-$, (CH$_3$)$_2$CHO$^-$ and (CH$_3$)$_3$CO$^-$ with CH$_3$Cl proceed essentially at the collision limit at 295 K, indicating no activation energy. While later experiments resulted in a revision of some of the values for the rate constants, the data, quoted in Table 2, show two important trends. *(i)* The rate constant becomes smaller as the alkyl group becomes larger in reactions of type (21). This trend

$$RO^- + CH_3Cl \rightarrow CH_3OR + Cl^- \qquad (21)$$

has been interpreted as due to the increasing charge delocalization in the nucleophile with larger alkyl groups (Brauman and Blair, 1968). *(ii)* The reaction of associated ions or monosolvated anions, formed by the three-body clustering reactions of RO⁻ with ROH in the flowing afterglow system, suffers a drastic reduction in rate constant.

TABLE 2

Rate constants for the reactions of unsolvated and solvated alkoxide ions with methyl chloride at 295.5 K^a

	$k_{exp}(RO^-)/10^{11}M^{-1}s^{-1}$	$k_{exp}(RO^-\cdot ROH)/10^{11}M^{-1}s^{-1}$
CH_3O^-	9.6	3
$C_2H_5O^-$	7.8	2.4
$(CH_3)_2CHO^-$	6.6	≤1.8
$(CH_3)_3CO^-$	4.8	≤1.2

^{a}Bohme and Young, 1970

These examples, as well as those listed in Table 3, illustrate the fact that the reaction of simple anions with methyl chloride and methyl bromide are extremely fast in the gas phase. The discrepancies in the value of the absolute rate constants according to the experimental procedure used is disturbing, and even the trends are not entirely similar. The fact that the rate constants obtained by FA techniques are consistently higher suggests that the systems under icr conditions may not be fully thermalized. It is known that highly exothermic ion-molecule reactions are characterized by rate constants which display a sensitive negative dependence with increasing translational energy

TABLE 3

Rate constant of gas-phase S_N2 reactions of methyl halides at room temperature

	icr $k_{exp}/10^{11}M^{-1}s^{-1}$	FA^d $k_{exp}/10^{11}M^{-1}s^{-1}$
$CH_3Cl + OH^-$	3.51 ± 0.17^a	9.0 ± 0.6
$+ F^-$	4.8 ± 0.5^b	11
$+ CH_3O^-$	3.6 ± 0.3^a	7.8 ± 0.6
$+ CH_3S^-$	0.46 ± 0.07^b	0.66 ± 0.06
$CH_3Br + OH^-$	11.4^c	6.0 ± 1.2
$+ F^-$	3.6 ± 0.3^a	7.2 ± 0.6
$+ CH_3O^-$	4.38 ± 0.11^a	6.6 ± 0.6
$+ CH_3S^-$	0.84 ± 0.12^a	3.1 ± 0.3
$+ Cl^-$	0.48 ± 0.06^a	0.13 ± 0.01

For the above reactions, the efficiencies (k_{exp}/k_{ADO}) range from 0.01 to 0.6
^{a}Brauman et al., 1974
^{b}Pellerite and Brauman, 1980
^{c}Olmstead and Brauman, 1977
^{d}Tanaka et al., 1976

GAS-PHASE DISPLACEMENTS   209

resulting in smaller rate constants. The differences in measured rate constants are yet to be resolved, and revisions of a given constant by the same author have been known to appear in the literature over the years.

STEREOCHEMISTRY

An important question regarding S_N2 reactions in the gas phase concerns the stereochemistry and the extent to which a Walden inversion occurs at the reaction site. Since the experimental techniques monitor exclusively ion concentration, the actual nature of the neutrals produced in the reaction is subject to some doubt. An indirect method to ascertain the nature of the products is to assess the thermochemistry of other possible reaction channels. In the case of methyl derivatives, the alternatives are few and result in highly endothermic reactions, as exemplified in (22) and (23). For more complicated systems, this argument may not be satisfactory or may not yield an unequivocal answer.

$$F^- + CH_3Cl \rightarrow CH_3F + Cl^- \qquad \Delta H° = -32 \text{ kcal mol}^{-1} \qquad (22)$$

$$F^- + CH_3Cl \rightarrow HF + CH_2 + Cl^- \qquad \Delta H° = 64 \text{ kcal mol}^{-1} \qquad (23)$$

Some attempts have been made in recent years to establish the identity of the neutral products of ion-molecule reactions. Smith *et al.* (1980) carried out a flowing afterglow experiment in which neutral products were collected at the end of the flow tube during periods ranging from 15 minutes to 1 hour. Analysis by glc of the collected fractions shows neopentyl fluoride as the only neutral product in reaction (24).

$$(CH_3)_3CCH_2Cl + F^- \rightarrow (CH_3)_3CCH_2F + Cl^- \qquad (24)$$

An earlier set of experiments using icr techniques (Lieder and Brauman, 1974, 1975) also made use of the principle of collecting the neutrals during a certain amount of time followed by *in situ* mass spectroscopic analysis. Several examples were studied, among them cases which could give rise to more than one set of reaction products. Ethyl derivatives in particular can react according to two possible mechanisms, (25) and (26). The experiment

$$F^- + C_2H_5Cl \rightarrow Cl^- + C_2H_5F \qquad \Delta H° = -29 \text{ kcal mol}^{-1} \qquad (25)$$

$$F^- + C_2H_5Cl \rightarrow Cl^- + C_2H_4 + HF \qquad \Delta H° = -19 \text{ kcal mol}^{-1} \qquad (26)$$

with analysis of the neutrals shows that the first reaction is clearly more important than the elimination channel. On the other hand, a similar set of experiments, carried out to account for the unusual acceleration of a presumed displacement reaction in going from the methyl to the ethyl derivative of carboxylic acids, showed the preferential participation of the elimination channel (Faigle et al., 1976; José and Riveros, 1977).

An elegant and clever experiment along these lines (Lieder and Brauman, 1974) confirms that the gas phase S_N2 reaction proceeds by a predominant *backside* attack. Analysis of neutral products in the icr experiment shows that for reaction (27), backside attack, and consequently inversion at the reaction centre, occurs to the extent of $92 \pm 6\%$. For reaction (28) the same type of experiment indicates that inversion of configuration amounts to $87 \pm 7\%$ of the reaction.

$$Cl^- + HO\text{-cyclohexyl-Br} \longrightarrow HO\text{-cyclohexyl-Cl} + Br^- \quad (27)$$

$$Cl^- + HO\text{-cyclohexyl-Br} \longrightarrow HO\text{-cyclohexyl-Cl} + Br^- \quad (28)$$

The above experiments are generally difficult to perform and the interpretation of the results may not necessarily be straightforward. The low abundance of the neutral products collected and the likelihood of mass spectral interference between reagents and products make these techniques applicable only to special cases. An independent approach to this problem has been proposed by Marinelli and Morton (1978) who have used an electron-bombardment flow reactor allowing in principle for larger collection of neutral products followed by glc and mass spectral analysis.

A well-known feature of S_N2 reactions in solutions is the high sensitivity to steric effects. For example, the relative rates in acetone solution for reaction (29) suffer dramatic changes in going from $R = CH_3$ ($k_{rel} = 1$) through C_2H_5

$$Cl^- + RI \rightarrow RCl + I^- \quad (29)$$

($k_{rel} = 0.089$) and $(CH_3)_2CH$ ($k_{rel} = 0.0028$) to $(CH_3)_3CCH_2$ ($k_{rel} = 1.2 \times 10^{-6}$) (Ingold, 1957). Neopentyl halides are known in fact to be extremely stable towards nucleophilic displacement. There are few examples in the gas phase that allow for a full comparison since alkyl halides with β-hydrogens may be prone to undergo elimination reactions simultaneously with displacement. For example, the reaction of F^- with C_2H_5Cl proceeds with a rate constant $k_{rel} = 0.86$, compared to that of CH_3Cl (José, 1976), whereas $(CH_3)_3CCH_2Cl$ reacts with $k_{rel} = 0.61$ (Olmstead and Brauman, 1977). A larger variation is observed with CH_3S^- as the nucleophile where the ratio $k_{(CH_3)_3CCH_2Cl}/k_{CH_3Cl} = 0.10$. Thus, while steric effects are noticeable in

GAS-PHASE DISPLACEMENTS 211

these reactions, the magnitude of the changes do not correspond to the size of the effect in solution. A rationale for such a behaviour has been proposed based on the theory of these reactions to be discussed in a later section.

NUCLEOPHILICITY AND LEAVING GROUP ABILITY

An early attempt to establish the gas-phase nucleophilicities of H^-, F^-, OH^- and NH_2^- towards methyl chloride was carried out by Young et al. (1973) using the flowing afterglow technique. Since the rate constants for the different reactions were approximately the same within experimental uncertainty, no clear-cut conclusion could be obtained from these experiments. The concept of nucleophilicity in reactions which proceed with large rate constants is based on the efficiency of the S_N2 reactions for the different nucleophiles. This efficiency has been estimated more recently by assuming that the ADO theory adequately describes the collision rate constant. With this assumption, the experiments reported by Bohme et al. (1974) establish the following nucleophilic trends as a function of the substrate:

For CH_3Cl;
$$NH_2^- > OH^- \simeq F^- > C_2H^- > H^- \gg CN^-.$$
For CH_3F;
$$OH^- > NH_2^- > H^- \gg C_2H^-, CN^-.$$

An icr study carried out at about the same time (Brauman et al., 1974) established the sequence for CH_3Br;
$$F^- > CH_3S^- > Cl^-.$$

A more complete picture of nucleophilicity and leaving group ability can be obtained from the data of Olmstead and Brauman (1977) and Tanaka et al. (1976). These results are compiled in Tables 4 and 5.

TABLE 4

Efficiencies of gas-phase S_N2 reactions by icr[a]

	Substrate			
Nucleophile	$CH_3OC_6H_5$	CH_3Cl	CH_3Br	CH_3OCOCF_3
OH^-	0.03	0.68	0.84	0.47
F^-	0.03	0.35	0.28	0.39
CH_3O^-	0.012	0.25	0.40	0.43
CH_3S^-		0.045	0.091	0.25
Cl^-		0.003	0.0070	0.021
CN^-	~0.003	0.0005	0.01	0.01

[a]From Olmstead and Brauman, 1977

TABLE 5

Efficiencies of gas-phase S_N2 reactions by flowing afterglow[a]

Nucleophile	Substrate		
	CH_3F	CH_3Cl	CH_3Br
CH_3NH^-	0.03	0.85 ± 0.17	0.72 ± 0.014
F^-		0.83 ± 0.16	0.59 ± 0.09
O^-	0.048 ± 0.09	0.71 ± 0.14	0.48 ± 0.09
CH_3O^-	0.0074 ± 0.0016	0.68 ± 0.14	0.61 ± 0.11
NH_2^-	0.0076 ± 0.0015	0.63 ± 0.13	0.48 ± 0.09
OH^-	0.11 ± 0.003	0.62 ± 0.13	0.43 ± 0.09
H^-	0.0020 ± 0.0004	0.35 ± 0.07	0.43 ± 0.09
CH_3S^-	0.0006	0.65 ± 0.012	0.35 ± 0.06
SH^-		0.0089 ± 0.0018	0.18 ± 0.03
NO_2^-		0.000059	0.0006
CN^-		0.0002	0.016 ± 0.003

[a]From Tanaka et al., 1976

Despite some important differences in the values obtained by the two techniques, which cannot be explained satisfactorily, there are a number of important conclusions: *(a)* The efficiencies, and thus nucleophilicities, for a given substrate are not solely dependent on the exothermicity of the process as can be seen by comparison with the data in Table 1. *(b)* The negative ions with a localized charge are better gas-phase nucleophiles, leading to an intrinsic order of nucleophilicity of $F^- > Cl^- > Br^-$; this is the reverse of the usually accepted order in solution but in agreement with that observed in aprotic solvents (Winstein et al., 1960). *(c)* As in the case of nucleophilicities, where some reversals occur, the data available at present do not allow for an absolute scale of leaving group ability. While the efficiency of the reaction corresponds with exothermicity for CH_3O^-, reversals are observed with OH^-

It is clear from the raw data that gas-phase S_N2 reactions are not amenable to treatment by linear-free energy relationships of the type observed in solution (Pearson et al., 1968).

EFFECT OF SOLVATION ON THE GAS-PHASE REACTION

The data shown in Table 2 indicate that association of the ion in the gas phase lowers significantly the rate constant of the S_N2 reaction. An even better example of this behaviour can be seen in the recent experiments of Bohme and Mackay (1981). The use of flowing afterglow techniques, in a sample rich in water vapour, allowed the measurement of the rate constant for reaction (30) as a function of successive degrees of hydration. These

$$OH^-(H_2O)_n + CH_3Br \rightarrow Br^-(H_2O)_n + CH_3OH \quad (30)$$

$$n = 0 \quad k = (6.0 \pm 0.12) \times 10^{11} M^{-1} s^{-1}$$
$$n = 1 \quad k = (3.8 \pm 1.5) \times 10^{11} M^{-1} s^{-1}$$
$$n = 2 \quad k = (1.2 \pm 0.6) \times 10^{9} M^{-1} s^{-1}$$
$$n = 3 \quad k < 1.2 \times 10^{8} M^{-1} s^{-1}$$
$$n = \infty \quad k_{aq} = 1.4 \times 10^{-4} M^{-1} s^{-1}$$

experiments help to bridge the gap between the gas-phase and solution reactivity. A proper interpretation will be reserved for the discussion of the model for gas-phase $S_N 2$ reactions.

The fact that these reactions are indeed a plausible approach to interconnect gas-phase and solution reactivity has been contested by Henchman et al. (1983). The basis for their argument is that for $n = 1$, reaction (30) yields $Br^- + H_2O + CH_3OH$ as products, and not the solvated anion. Thus, they conclude that bulk solvent effects cannot be properly extrapolated from the data of reaction (30). While the rigorous argument is correct, the rate constant trend is very useful to show that successive solvation of the reagent anion will slow down the reaction even on a thermochemical basis.

MECHANISM OF THE GAS-PHASE $S_N 2$ REACTION

The existing evidence indicates that the gas-phase reaction, in analogy with the solution process, proceeds by a mechanism which can be represented as in (31).

$$X^- + CH_3Y \longrightarrow [\overset{\delta^-}{X} \cdots \underset{HH}{\overset{H}{C}} \cdots Y^{\delta^-}] \longrightarrow XCH_3 + Y^- \quad (31)$$

The observation that chloride ions can attach to alkyl halides to yield chloride cluster ions in high-pressure mass spectrometry generated considerable interest regarding the structure of the transition state of gas-phase $S_N 2$ reactions (Dougherty et al., 1974, Dougherty and Roberts, 1974, Dougherty, 1974). Ionic species like $CH_3Cl_2^-$, $CH_3Br_2^-$, $CHCl_4^-$, CCl_5^-, $CH_3I_2^-$ and CH_3BrCl^-, were produced from reactions such as (32). The fact that bridge-

$$Cl^- + CH_3Cl \rightarrow CH_3Cl_2^- \quad (32)$$

head halides (e.g. 1-bromoadamantane) failed to attach halide ions provided strong evidence that the $CH_3Cl_2^-$ species represents a backside association. Equilibrium constants for reactions of type (32) provided some valuable data on the relative stability of these species (Table 6).

TABLE 6

Gas-phase enthalpies and entropies of association of halide ions[a]

Reaction		$-\Delta H$/kcal mol^{-1}	ΔS/e.u.
$CH_3Cl + Cl^-$	$\rightarrow CH_3Cl_2^-$	8.6 ± 0.2	15.3 ± 1.0
$CH_2Cl_2 + Cl^-$	$\rightarrow CH_2Cl_3^-$	15.5 ± 0.3	22.0 ± 1.1
$CCl_4 + Cl^-$	$\rightarrow CCl_5^-$	14.2 ± 0.7	27.8 ± 0.7
$CH_3Br + Br^-$	$\rightarrow CH_3Br_2^-$	9.2 ± 0.5	14.0 ± 1.1
$CH_3I + I^-$	$\rightarrow CH_3I_2^-$	9.0 ± 0.2	16.4 ± 0.2
$CH_3Br + Cl^-$	$\rightarrow CH_3BrCl^-$	10.9 ± 0.5	12.8 ± 2.2
$CH_3I + Cl^-$	$\rightarrow CH_3ICl^-$	9.8 ± 0.2	7.3 ± 3.5
$C_2H_5Br + Br^-$	$\rightarrow C_2H_5Br_2^-$	11.6 ± 0.5	19.2 ± 1
$(CH_3)_2CHBr + Br^-$	$\rightarrow (CH_3)_2CHBr_2^-$	12.1 ± 0.5	19.9 ± 1
$(CH_3)_3CBr + Br^-$	$\rightarrow (CH_3)_3CBr_2^-$	12.4 ± 0.5	19.3 ± 1
$(CH_3)_3CCH_2Br + Br^-$	$\rightarrow (CH_3)_3CCH_2Br_2^-$	14.4 ± 05.	25.2 ± 1

[a]Dougherty et al., 1974; Dougherty and Roberts, 1974; Dougherty, 1974

The conclusion from these experiments is that the associated ions [RXY]$^-$ are stereochemically related to the S_N2 transition state. Furthermore, little variation occurs in the stability of the associated ions for CH_3Cl, CH_3Br and CH_3I, and steric hindrance is apparently of small consequence in the enthalpic and entropic contributions of these equilibria. These data can be compared directly to the solution values for enthalpy and entropy of activation to show that solvation effects on both parameters are indeed responsible for the large variations in rates observed in solution.

Despite this information, the question remains as to the exact structure of the associated ions and their possible electronic resemblance to an S_N2 transition state. An icr experiment (Riveros et al., 1973) generated a $CH_3Cl_2^-$ species of mixed isotopic composition from the ion-molecule reaction (33).

$$CO^{35}Cl^- + CH_3{}^{37}Cl \rightarrow [CH_3{}^{37}Cl^{35}Cl]^- + CO \qquad (33)$$

Tertiary ion-molecule reactions of $[CH_3{}^{37}Cl^{35}Cl]^-$ with species of higher chloride affinity like CH_3CF_3 show that $[CH_3{}^{37}Cl^{35}Cl]^-$ transfers only $^{35}Cl^-$, *the chlorine originally in the precursor ion $COCl^-$*, to yield $(CH_3CF_3)^{35}Cl^-$. This experiment clearly demonstrates that the two chlorines are not equivalent in the $CH_3Cl_2^-$ species. Given the fact that bridgehead halides do not attach a halide ion, and discounting a structure with a halogen—halogen bond, the evidence provided by this experiment supports the idea of a loose ion-neutral complex of the type $[Cl^- \cdots CH_3Cl]$.

POTENTIAL ENERGY SURFACES FOR GAS-PHASE S_N2 REACTIONS

Gas-phase S_N2 reactions have been shown to span a range of efficiencies of at least three orders of magnitude. This range may be considerably larger but is at present inaccessible to the experimental techniques available. This variation in efficiencies has been used on several occasions to derive a pseudo-activation energy for the reactions using (34), by assuming the pre-exponential factor to be equal to k_{ADO} (Bohme *et al.*, 1974; Tanaka *et al.*, 1976). This approach is likely to be meaningless, and activation energies obtained in this fashion must be disregarded in view of the present interpretation of S_N2 reactions.

$$k = k_{ADO}\exp(-E_a/RT) \tag{34}$$

The results concerning the stability and likely structure of the $CH_3X_2^-$ adducts, and the range of observed efficiencies (1 to 10^{-3}) led Olmstead and Brauman (1977) to propose that simple S_N2 processes can be represented by a double potential well separated by a central energy barrier. This central energy barrier is located *below* the energy of the reactants for the very fast reactions. This model has gained wide acceptance, and it is illustrated for a thermoneutral exchange process in Fig. 3. The energy profile displays an initial decrease in energy due to the ion-dipole attraction, followed by the formation of a loose ionic aggregate, $CH_3X_2^-$, with thermochemical stability represented by the quantity E_o. This quantity corresponds to the values quoted in Table 6, with typical values ranging between 10 and 15 kcal mol^{-1} for alkyl halides. Reaction efficiencies are identified in this model with differences in the secondary energy barrier, E_o'.

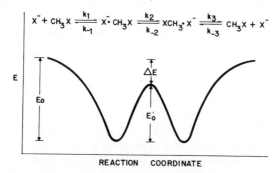

FIG. 3 Energy diagram for a gas-phase thermoneutral S_N2 reaction

In the scheme of Fig. 3, the rate constant k_1 is identified with the collision rate constant, k_{ADO}, $k_{-1} = k_3$ and $k_2 = k_{-2}$. Thus the experimental rate constant is given by (35). For exothermic reactions, it is reasonable to assume that $k_{-2} \ll k_3$, and the observed rate constant will take the form of (36).

$$k_{exp} = k_1 k_2/(k_{-1} + 2k_2) \tag{35}$$

$$k_{exp} = k_1 k_2/(k_{-1} + k_2) \tag{36}$$

These expressions show that derivation of an activation energy from (34) would be inconsistent with the reaction model.

The proposed energy diagram finds theoretical support in calculations carried out at the Hartree–Fock level by Duke and Bader (1971) and by Dedieu and Veillard (1972). In the latter case, the calculation for the $F^- + CH_3F$ system shows, for example, *(a)* an initial decrease in total energy; *(b)* the presence of a stable intermediate with F^- at about 3 Å from the carbon centre, and *(c)* an activation energy of 7.9 kcal mol^{-1} at the symmetric transition state. A more recent calculation by Morokuma (1982) for the $Cl^- + CH_3Cl$ reaction confirms the qualitative energy diagram proposed in Fig. 3. Furthermore, the calculation draws attention to the fact that successive molecules of hydration result in important relative changes in the diagram.

The relationship between the potential energy diagram of Fig. 3 and that advocated for the solution reaction is easily explained on the basis of the large stabilization (especially in polar solvents) of the reagent and product ions. The transition state, with a highly delocalized charge, is expected to suffer a much smaller stabilization by solvation. Furthermore, the initial decrease in potential energy as ion and neutral approach is washed away in solution by the energy required to desolvate the reactant ion. These changes lead to the familiar bell shaped energy diagram for reaction in solution.

In the model advanced by Olmstead and Brauman (1977), it is argued that the prototype thermoneutral halide exchange occurs at efficiencies below 0.5 because of entropic reasons, even when no formal energy barrier exists for the overall process. The theoretical model in this case assumes that the complex $[X^- \cdots CH_3X]$ is formed at a rate given by the ADO collision rate constant. The decomposition of $[X^- \cdots CH_3X]$ is then treated by unimolecular statistical rate theories (RRKM) where competition between return to reactants and passage over the central barrier is dictated by the density of states associated with each path. The forward reaction is expected to have a smaller pre-exponential factor than the back reaction because the former requires a tight transition state. This condition can offset differences between E_o and E'_o in determining the branching ratio. By appropriate modelling of the two transition states, and in particular by a suitable choice of the vibrational frequencies, it is possible to estimate the efficiency of the reaction, given by

$$k_{exp}/k_{ADO} = 1/[1 + (k_{-1}/k_2)] \tag{37}$$

(37), as a function of the relative barrier heights, $\Delta E = E'_o - E_o$. While the quantitative aspects of this model are subject to refinement, the calculations show that even barriers substantially below the potential energy of the reagents can lead to a sizable decrease in reaction efficiency. This model can also be used to obtain ΔE from the experimental reaction efficiencies. As an example, the first calculations show that for the $Cl^- + CH_3Br$ reaction, the relative barrier height, ΔE, ranges from -0.3 to -0.7 kcal mol^{-1}, whereas ΔE for $CH_3O^- + CH_3Br$ is estimated to be around -9 kcal mol^{-1}.

The above model has been further explored to account for reaction efficiencies in terms of a scheme where nucleophilicities and leaving group abilities can be rationalized by a structure–reactivity pattern. Pellerite and Brauman (1980, 1983) have proposed that the central energy barrier for an exothermic reaction (see Fig. 3) can be analysed in terms of a thermodynamic driving force, due to the exothermicity of the reaction, and an intrinsic energy barrier. The separation between these two components has been carried out by extending to S_N2 reactions the theory developed by Marcus for electron transfer reactions in solutions (Marcus, 1964). While the validity of the Marcus theory to atom and group transfer is open to criticism, the basic assumption of the proposed model is that the intrinsic barrier of reaction (38)

$$CH_3O^- + CH_3Cl \rightarrow Cl^- + CH_3OCH_3 \tag{38}$$

can be taken as the mean of the barriers for the individual component exchange reactions (39) and (40) as in (41). Since the theory of Marcus is

$$Cl^- + CH_3Cl \rightarrow ClCH_3 + Cl^- \tag{39}$$

$$CH_3O^- + CH_3OCH_3 \rightarrow CH_3OCH_3 + OCH_3^- \tag{40}$$

$$E_o'(38) = 1/2 [E_o'(39) + E_o'(40)] \tag{41}$$

applicable only to elementary processes, the energy barrier refers to the barrier for the step (42) involving the passage of one collision complex to

$$X^- . CH_3Y \rightarrow XCH_3 . Y^- \tag{42}$$

another. The Marcus equation then becomes (43), where E_o' is the barrier for an energy change E, and E_o^* is the intrinsic barrier for $E = 0$. The quantity E_o'

$$E_o' = [(E)^2/16\, E_o^*] + E_o^* + 1/2E \tag{43}$$

(the central barrier height) can be estimated for a given reaction by the RRKM method used to explain the efficiency of the reaction, and E is approximated as being equal to the exothermicity of the overall reaction. This procedure allows for the calculation of the intrinsic energy barrier for each reaction (E_o^*). The value of E_o^* thus obtained enables the calculation of the individual $E_o'(X^- + CH_3X)$ by the use of (41) if an estimate for the $E_o'(Cl^- + CH_3Cl)$ is derived directly from the efficiency of the reaction.

This approach yields a series of intrinsic energy barriers for the S_N2 self-exchange reactions which vary widely depending on the vibrational frequencies assumed for the transition states. Nevertheless, the numbers derived for a given model can be used to establish some interesting conclusions. For example, chloride- and bromide-exchange reactions display an E_o'

in the order of 10 to 11 kcal mol^{-1}, while fluoride- and alkoxide-exchange yield intrinsic energy barriers ranging from 26 to 30 kcal mol^{-1}. This means that, given this framework of reference, chloride and bromide are much better nucleophiles and leaving groups than alkoxides or fluorides. Because of the symmetry of the Marcus treatment, a good nucleophile becomes also a good leaving group and *vice versa*. The fact that reactions involving F$^-$ and CH$_3$O$^-$ can be very efficient in the gas phase is explained in this model on the basis of the large thermodynamic component which lowers the effective E_0'' of the reaction.

An attempt to establish a rate–equilibrium relationship shows that a plot of $E_0^*(X^- + CH_3X)$ as a function of the methyl cation affinity (MCA) of X$^-$ yields the linear correlation (44), where $IP(CH_3)$ represents the ionization

$$MCA(X^-) = D^\circ(CH_3\text{-}X) - EA(X) + IP(CH_3) \tag{44}$$

potential of the methyl radical. The resulting intrinsic trend of nucleophilicity and leaving group ability becomes

$$Cl^- \simeq Br^- > CH_3CO_2^- > CH_3S^- > F^- > CH_3O^- \simeq t\text{-BuO}^- > C_2H^- > H^-$$

It is early yet to decide whether this approach is the most appropriate to account for gas-phase S$_N$2 reactions. However it does provide a very useful approach to understanding and *predicting* the outcome of these reactions.

A very clear example of the application of this model can be seen in the case of the S$_N$2 reaction (45) which has not been detected either by icr or flow-

$$OH^- + (CH_3)_2O \rightarrow CH_3O^- + CH_3OH \quad \Delta H^\circ_{gas} = -6 \text{ kcal mol}^{-1} \tag{45}$$

ing afterglow. This fact can be rationalized by the large intrinsic barrier to CH$_3$O$^-$ displacement which is not offset by the low exothermicity of the process.

RECENT THEORETICAL DEVELOPMENTS

As mentioned in the previous section, theoretical calculations are consistent with the diagram (Fig. 3) proposed by Olmstead and Brauman (1977).

Further insight into gas-phase S$_N$2 reactions can be gained from recent calculations by Wolfe *et al.* (1981a,b). For the series of reactions (46), optimization of the transition-state geometry using a split Gaussian representation for the atomic orbitals reveals that the carbon—fluorine bond lengths vary

$$X^- + CH_3F \rightarrow XCH_3 + F^- \tag{46}$$

almost linearly with the exothermicity of the reaction. A similar relationship is found for the FCH angle in the transition state. The same behaviour is encountered in calculations involving CH_3OH and CH_4 as the substrate. These trends parallel the basic contention of the Bell–Evans–Polanyi principle (Bell, 1936; Evans and Polanyi, 1938) and the Hammond (1955) postulate which establish that the transition state structure progressively resembles the reagents more closely as the reaction becomes more exothermic.

On the other hand, the calculated energy barriers for interconversion (42) of the collision complexes, when plotted against the exothermicity, result in a meaningless graph. This is consistent with the experimental findings. By applying a Marcus-type treatment, analogous to that proposed by Pellerite and Brauman (1980, 1983), and computation of the intrinsic energy barrier for the thermoneutral reaction of a number of groups, it was shown (Wolfe et al., 1981b) that the calculations of the Marcus energy barriers, from the intrinsic self-exchange barriers and the exothermicity of the reaction, yield excellent agreement with the *ab initio* barriers calculated directly for each case. This result has been taken as a strong indication that the Marcus-type approach is indeed applicable to S_N2 reactions.

The above work has also led to the prediction that the so-called α-effect for nucleophiles containing a heteroatom adjacent to the reaction centre should display normal behaviour in the gas phase (Wolfe et al., 1981a,b, 1982). This prediction has been corroborated by recent experimental determinations that show little variation of reactivity between HOO^- and HO^- (DePuy et al., 1983), contrary to what is predicted by the application of frontier molecular orbital theory (Fleming, 1976).

An interesting approach to S_N2 reactions has been recently advanced by Shaik which leads to verifiable predictions (Shaik, 1981; Pross and Shaik, 1981; Shaik, 1982; Shaik and Pross, 1982). The basic contention of this theory is that the origin for the barrier of S_N2 reactions arises from an avoided curve-crossing between two curves containing the reactant and product Heitler–London VB forms. In this treatment, the reacting pair is analysed in terms of a nucleophile being the electron donor (D) and the substrate the electron acceptor (A). For the simple reactions of type (47), the

$$N^- + CH_3X \rightarrow NCH_3 + X^- \qquad (47)$$

two valence bond forms representing reagents and products correspond to $N:^- CH_3X$ and $NCH_3 : X^-$, where these moieties must be associated with the ion-dipole collision complexes. At the outset of the reaction co-ordinate the two valence bond forms differ in energy by the amount required to transfer an electron from the donor (D) to the acceptor (A) as given by (48), where

$$E(D_R^+ A_R^-) - E(D_R A_R) = I_{N^-} - A_{CH_3X} \qquad (48)$$

A_{CH_3X} represents the electron affinity of the substrate and the subscript refers to the reagent configuration. At the other end of the reaction co-ordinate, the difference in energy between the two valence bond forms of the product (P) will correspond to (49).

$$E(D_P^+ A_P^-) - E(D_P A_P) = I_{X^-} - A_{CH_3N} \tag{49}$$

A suitable graph of the variation of energy of these forms as a function of reaction co-ordinate shows that the energy barrier of the S_N2 reaction will be a fraction of the initial energy difference, $I_{N^-} - A_{CH_3X}$. The fraction of this quantity will depend on the slope of the energy curves associated with each VB form. This slope, or energy variation with the reaction co-ordinate, is determined by the relative stability of the three-electron bond in CH_3X^-. Groups which can effectively delocalize the negative charge result in slow variation of the energy of this species as a function of the reaction co-ordinate.

A detailed analysis of the resulting curves yields that, for the degenerate exchange reaction (N = X), the "intrinsic" barrier can be written as (50), where $0 < r < 1$ is the slope parameter, and β the avoided crossing constant of the two curves.

$$E_o^* = r(I_{X^-} - A_{CH_3X}) - \beta \tag{50}$$

The results obtained by this procedure reveal some very interesting predictions which parallel some of the conclusions of Pellerite and Brauman (1983). For example Cl^- and Br^- are much more reactive than F^- because of a considerably lower value of $(I_{X^-} - A_{CH_3X})$, whereas $CF_3CO_2^-$ is also predicted to be much more reactive than F^- on account of a low $(I_{X^-} - A_{CH_3X})$ and a much lower r-factor, associated with the fact that the negative charge is very poorly delocalized in $[CF_3COOCH_3]^-$.

The basic ideas advanced in this treatment are subject to experimental determination and can, in principle, be extended to exothermic reactions. The unique feature of this model is that the relevant parameters at stake in determining the reactivity are related to meaningful molecular parameters. Thus, reactions can be classified as *(a)* electron transfer controlled when the main contribution to the energy barrier is $(I_{X^-} - A_{CH_3X})$, or *(b)* three-electron bond controlled when the energy barrier is associated with strong delocalization of the negative charge in CH_3X^-.

5 Some examples of gas-phase S_N2 reactions involving positive ions

Gas-phase nucleophilic displacement reactions in aliphatic systems involving a neutral nucleophile and a positively charged substrate represent a much more heterogeneous set of reactions than that of negative ions. As for most

GAS-PHASE DISPLACEMENTS

systems of positive ions, the ion chemistry is in general very rich, and secondary or higher reactions must be accommodated in the reaction scheme especially when high pressures or long residence times are used experimentally. The overall effect is that in positive ions, competition between several mechanisms is the rule rather than the exception. Consequently, it is difficult to use an isolated approach as we did in Section 4. Because of the variety of these reactions, and the absence of a unique approach to the matter, this review will cover some general examples and establish empirical rules set forth for these systems.

The observation in the gas phase of positive ion-molecule reactions which can be interpreted as a nucleophilic displacement was first reported by Holtz et al. (1970). A typical example is the tertiary ion-molecule reaction in a mixture of HCl and CH_3F. Protonation of CH_3F as a result of a primary ion-molecule reaction is followed by reaction (51).

$$HCl + CH_3FH^+ \rightarrow CH_3ClH^+ + HF \qquad k_{icr} = 1.8 \times 10^{11} M^{-1} s^{-1} \qquad (51)$$

A particularly useful and interesting example of this type of reaction is with H_2O as the nucleophile. Reaction (52) is not detected within the dynamic range of icr, while reaction (53) occurs readily. On the other hand, nucleophilic displacement is observed in the case of C_2H_5Cl as shown in (54).

$$H_2O + CH_3ClH^+ \not\rightarrow CH_3OH_2^+ + HCl \qquad (52)$$

$$H_2O + CH_3ClH^+ \rightarrow H_3O^+ + CH_3Cl \qquad (53)$$

$$H_2O + C_2H_5ClH^+ \rightarrow C_2H_5OH_2^+ + HCl \qquad (54)$$

This observation has been rationalized on the basis of the relative proton affinities (PA) of the substrates and that of the nucleophile. In fact, the relative order of proton affinities reveals that $PA(C_2H_5Cl) > PA(H_2O) > PA(CH_3Cl)$. Thus, for the first case, rapid proton transfer dominates over nucleophilic displacement, since proton transfer involving species which do not require any electronic or bond reorganization will in general be much faster than a displacement.

On the basis of the above observations, it was proposed that nucleophilic displacement will take place in general subject to two conditions:

(a) proton transfer from the substrate to the nucleophile is endothermic;
(b) the displacement reaction is exothermic.

The first condition can be easily verified with the aid of the extensive data available on gas-phase proton affinities (Wolf et al., 1977). Likewise, the second condition can be verified with present knowledge of heats of formation of ions obtained by several mass spectrometric techniques. The chemistry of alcohols is one case where reactions (55; R = CH_3, C_2H_5, C_3H_7 and

$$ROH + ROH_2^+ \rightarrow R_2OH^+ + H_2O \tag{55}$$

C_4H_9) of the nucleophilic type become common (Beauchamp and Caserio, 1972). This case represents again a secondary ion-molecule reaction which occurs in a system with extremely rich chemistry.

Contrary to the case of anionic reactions, the formation of a strong proton-bound dimer for alcohols suggests that nucleophilic displacement may actually involve a frontside attack. Recent experiments carried out at atmospheric pressure by Speranza and Angelini (1980) using radiolytic techniques with isolation and glc analysis of neutral products reveal some interesting stereochemistry. For example, the reaction of protonated epoxy-*trans*-but-2-ene with H_2O results in 98% inversion of configuration, while a similar reaction with *cis*-1-chloro-4-methylcyclohexane results in approximately 80% of *trans*-4-methylcyclohexanol. With the high pressures utilized and with the possible participation of cluster ions a likelihood in this case, the data are consistent with a Walden inversion for these cases.

Further support for the idea that cationic nucleophilic displacement occurs with inversion of configuration has been advanced by Hall *et al.* (1981). The study of reaction (55) in an electron-bombardment flow reactor at reagent pressures below 10^{-3} torr, followed by neutral product analysis (Marinelli and Morton, 1978), reveals that these reactions also occur via backside attack. This is in disagreement with the original suggestion of Beauchamp *et al.* (1974) who proposed a frontside displacement in the case of t-butyl alcohol.

From these examples, one can conclude that anionic and cationic S_N2 reactions in the gas phase display similar stereochemistry.

6 Nucleophilic displacement reactions by negative ions in carbonyl systems

A discussion of nucleophilic reactions in carbonyl systems, and in particular in derivatives of carboxylic acids, is included in the present text because of the importance of these reactions and their analogy with well-known processes in solution. While the actual mechanism of these reactions is more adequately described in many cases as an addition–elimination, we shall restrict our comments to systems which formally behave as displacement reactions.

Further interest in these systems in the gas phase is generated by the fact that several nucleophilic reactions may become competitive allowing a comparison of relative reactivity of carbon centres.

GENERAL FEATURES

Reaction of negative ions with simple organic esters displays a rich chemistry in the gas phase as revealed in several icr studies. The most common case is reaction (56) which bears a formal resemblance to a hydrolysis reaction

$$OH^- + RCOOCH_3 \rightarrow RCOO^- + CH_3OH \tag{56}$$

(Faigle et al., 1976; Takashima and Riveros, 1978). Kinetic studies reveal this to be a fast reaction ($k_{icr} = 5 \times 10^{11} M^{-1} s^{-1}$ at room temperature for R = H). Mechanistic studies, based on the use of $^{18}OH^-$, are consistent with a reaction that proceeds substantially both by attack at the carbonyl centre ($B_{AC}2$), and at the methyl group (S_N2). The reactions of hydroxide ions and esters in the gas phase are also known to give rise to further reactions:

(i) Esters with β-hydrogens in the alkyl group react readily to yield $RCOO^-$ by an elimination mechanism, in the fashion shown in (57).

$$OH^- + RCOOC_2H_5 \rightarrow RCOO^- + C_2H_4 + H_2O \tag{57}$$

(ii) The presence of acidic hydrogens will result generally in proton abstraction rather than displacement.

(iii) For R = H, a very fast reaction (58) occurs which corresponds to a decarbonylation (Faigle et al., 1976)

Reaction (58) is an efficient way of generating a solvated ion in the gas phase at low pressures. The general requirements for this type of reaction, usually referred to as the Riveros reaction, have been advanced for a number of nucleophiles (Isolani and Riveros, 1975). A further example is shown in (64c).

$$OH^- + HCOOCH_3 \rightarrow CH_3O^-(H_2O) + CO \rightarrow CH_3O^- + H_2O \tag{58}$$

A comparison of the nucleophilic reactivity of methyl esters towards OH^- can be established in experiments using $^{18}OH^-$. This is particularly interesting because it affords a comparison of the relative reactivity of the carbonyl centre and an sp^3 carbon centre. Results of such experiments are shown in Table 7.

An interpretation of these trends is not straightforward since the modelling of reaction $B_{AC}2$ by a somewhat similar procedure used for the S_N2 reaction will result in an elaborate comparison of transition states and barriers for both channels. The results listed in Table 7 represent, nevertheless, a very abrupt departure from the situation in solution chemistry where the S_N2 process is several orders of magnitude slower than the $B_{AC}2$ mechanism. Another significant factor for reaction (56) is its high exothermicity, ranging from 45 kcal mol^{-1} for R = H to 67 kcal mol^{-1} for R = CF_3. Thus, the thermodynamic driving force for these reactions is expected to be large and can presumably overcome large "intrinsic" barriers.

TABLE 7

Percentage of the gas-phase hydrolysis reaction occurring by the $B_{AC}2$ and S_N2 mechanism[a]

	$B_{AC}2$	S_N2
$HCOOCH_3$	73%	27%
$(CH_3)_3CCOOCH_3$	90%	10%
$C_6H_5COOCH_3$	92%	8%
$(CH_3O)_2CO$	68%	32%
CF_3COOCH_3	24%	76%

[a]From Takashima and Riveros, 1978

The reactivity of other nucleophiles is also of unusual interest. For example, F^- reacts efficiently with methyl esters of carboxylic acids without β-hydrogens along an S_N2 pathway (59) (Takashima and Riveros, 1978) to

$$F^- + CF_3COOCH_3 \rightarrow CH_3F + CF_3COO^- \quad \Delta H° = -46 \text{ kcal mol}^{-1} \quad (59)$$

yield $RCOO^-$ [R = CF_3, $(CH_3)_3C$, CH_3O]. This same reaction becomes an order of magnitude slower for the case of R = H, CH_3, C_2H_5 (Faigle et al., 1976, José and Riveros, 1977) as exemplified in the results in (60). In the system $F^-/HCOOCH_3$, the preferred reaction is analogous to process (58) which yields $F^-(CH_3OH)$.

$$F^- + HCOOCH_3 \rightarrow CH_3F + HCOO^- \quad \Delta H° = -24 \text{ kcal mol}^{-1}$$
$$k_{icr} = 4.8 \times 10^{10} \text{ M}^{-1}\text{s}^{-1} \quad (60)$$

There is one very unusual example in which F^- reacts with an ester by what can be considered as a nucleophilic attack at the carbonyl system (61)

$$F^- + CH_3COOCH_3 \rightarrow CH_2CFO^- + CH_3OH \quad \Delta H° = -1 \text{ kcal mol}^{-1}$$
$$k_{icr} = 9 \times 10^{10} \text{ M}^{-1}\text{s}^{-1} \quad (61)$$

(José and Riveros, 1977). This reaction becomes overshadowed in higher alkyl acetates because attack at the β-hydrogen to induce elimination as in

$$F^- + CH_3COOC_2H_5 \rightarrow CH_3COO^- + C_2H_4 + HF \quad \Delta H° = -9 \text{ kcal mol}^{-1}$$
$$k_{icr} = 9.6 \times 10^{10} \text{ M}^{-1}\text{s}^{-1} \quad (62)$$

(62) becomes the most rapid reaction. Reaction (59) has been rationalized on the basis of the mechanism in (63). There is considerable controversy, which

$$F^- + CH_3COOCH_3 \rightleftharpoons [F^- \cdot CH_3COOCH_3] \rightleftharpoons [CH_3CFO \cdot CH_3O^-] \rightarrow$$
$$\text{Collision complex} \quad \text{Collision complex} \quad (63)$$
$$CH_2CFO^- + CH_3OH$$

will be discussed further below, as to whether the tetrahedral species $CH_3C(F)(OCH_3)O^-$ represents a stable intermediate or a local transition state along the reaction path. Nevertheless, an energy diagram similar to that in Fig. 3 is probably able to account for reaction (61) if proper consideration is given to the fact that, while the reaction is almost thermoneutral, not much is known about the possible depth of the wells for the collision complexes.

Other reactions of F^- with esters as a result of attack at the carbonyl centre are not expected because of the absence of exothermic reaction channels.

An even wider range of reactivity is observed for the reaction of alkoxide ions. For example, Blair *et al.* (1973) have reported the observations summarized in (64) in the gas phase. While reaction (64b) was reported to occur

$$CD_3O^- + HCOOCH_3 \rightarrow HCOO^- + CH_3OCH_3 \quad \Delta H° = -36 \text{ kcal mol}^{-1} \quad (64a)$$

$$\rightarrow HCOOCD_3 + CH_3O^- \quad \Delta H° \sim 0 \quad (64b)$$

$$\rightarrow CD_3O^-(CH_3OH) + CO$$
$$\rightarrow CD_3OH + CH_3O^- + CO \quad \Delta H° = 9 \text{ kcal mol}^{-1} \quad (64c)$$

as shown, later experiments, and in particular double resonance icr, showed that the appearance of CH_3O^- in reactions (64) was more consistent with pathway (64c), even though it is endothermic. As no analysis of neutral products has been performed on this system, there is some doubt as to the extent, if any, of nucleophilic displacement (64b).

Comisarow (1977) has shown that, in the gas phase, methoxide ions react readily with methyl trifluoroacetate and methyl benzoate by an S_N2 mechanism, while no reaction is observed as a result of nucleophilic displacement at the carbonyl centre. As for the case above, the S_N2 reaction is highly exothermic, while the same is not true for the equivalent of reaction (64b). There is at present no satisfactory explanation of why (64a) apparently proceeds very slowly in the gas phase.

A reversal of this trend has been observed in the reaction of RO^- with alkyl pivalates (Takashima and Riveros, 1978) as shown in (65).

$$CD_3O^- + (CH_3)_3CCOOR \begin{array}{c} \nearrow RO^- + (CH_3)_3CCOOCD_3 \quad (65a) \\ \searrow CD_3OR + (CH_3)_3CCOO^- \quad (65b) \end{array}$$

	$R = CH_3$	$R = C_2H_5$
k_{65a}/k_{65b}	1.5	2.5

This enormous variety of reactivity for the alkoxide ions has not been adequately explained at this stage, and remains as a challenge to researchers in the field. The fact that reaction (65a), an essentially thermoneutral nucleophilic displacement at the carbonyl group, can dominate over a highly exothermic reaction, finds its analogy in reactions of alkyl pivalates with

$$NH_2^- + (CH_3)_3CCOOCH_3 \longrightarrow \begin{array}{l} CH_3O^- + (CH_3)_3CCONH_2 \quad 24\% \\ (CH_3)_3CCONH^- + CH_3OH \quad 60\% \\ (CH_3)_3COO^- + CH_3NH_2 \quad 16\% \end{array} \quad (66)$$

other nucleophiles, for example, (66). In this case, strong preference is observed for nucleophilic displacement originating from attack at the carbonyl.

The behaviour of alkyl pivalates with OH^-, RO^- and NH_2^- in the gas phase is indeed remarkable since in solution these compounds are known to be extremely unreactive towards nucleophilic attack at the carbonyl centre because of steric requirements. For the gas-phase reaction, it is apparent that the steric requirements are of less importance than the driving force for the reaction at the carbonyl centre.

A very important contribution to the problem of nucleophilic displacement in carbonyl systems has been reported by Asubiojo and Brauman (1979). A detailed study by icr of reactions of type (67) results again in a wide

$$X^- + RCOY \rightarrow Y^- + RCOX \qquad (67)$$

X = F, Cl, CH_3O, CN, SH
R = CH_3, C_2H_5, $(CH_3)_2CH$, $(CH_3)_3C$, C_6H_5
Y = Cl, Br

range of efficiencies which are shown in Table 8. The reactivity of the different nucleophiles correlates closely with the exothermicity of the reaction, unlike the case of the S_N2 reaction, leading to a nucleophilicity scale of $CH_3O^- > F^- > CN^- \sim SH^- > Cl^-$.

The trends in the gas-phase reaction (67) follow closely the correlations encountered for the similar reaction in protic solvents (Jencks and Carriuolo, 1960). This observation, plus the fact that in the case of esters S_N2 reactions become competitive with attack at the carbonyl, has been rationalized by several authors (Asubiojo and Brauman, 1979; Comisarow, 1977; Takashima and Riveros, 1978) on the basis of the poor solvation expected for the S_N2 transition state due to charge delocalization. Thus, S_N2 reactions are expected to display much larger differences in the gas phase than in solution, when compared with nucleophilic displacement at carbonyl centres. This is reflected in a larger sensitivity of the activation parameters.

TABLE 8

Rate constants, reaction efficiencies and exothermicities of nucleophilic displacements at carbonyl centres[a]

	$k/10^{11} M^{-1} s^{-1}$	Efficiency	ΔH^0/kcal mol^{-1}
$^{37}Cl^- + CH_3CO^{35}Cl \rightarrow {}^{35}Cl^- + CH_3CO^{37}Cl$	0.7	0.05	0
$CH_3O^- + CH_3COCl \rightarrow Cl^- + CH_3COOCH_3$	3.9	0.26	-55
$F^- + CH_3COBr \rightarrow Br^- + CH_3COF$	5.7	0.33	-50
$CN^- + CH_3COBr \rightarrow Br^- + CH_3COCN$	1.6	0.11	-4
$F^- + (CH_3)_3CCOCl \rightarrow Cl^- + (CH_3)_3CCOF$	2.9	0.15	-39
$CH_3O^- + CF_3COCl \rightarrow Cl^- + CF_3COOCH_3$	1.9	0.21	-55

[a]From Asubiojo and Brauman, 1979

MODEL FOR NUCLEOPHILIC DISPLACEMENT REACTIONS AT CARBONYL CENTRES

A model similar to that used for S_N2 reactions has been advanced by Asubiojo and Brauman (1979) to account for the different aspects of reaction (67), and this model (68) is probably applicable to all systems discussed at the

$$X^- + RCOY \underset{k_{-1}}{\overset{k_1}{\rightleftharpoons}} X^- \ldots RCOY \underset{k_{-2}}{\overset{k_2}{\rightleftharpoons}} RCOX \ldots Y^- \underset{k_{-3}}{\overset{k_3}{\rightleftharpoons}} RCOX + Y^- \qquad (68)$$

beginning of this section. This scheme can be represented by a diagram similar to that shown in Fig. 3 and leads to the following limiting cases:

(a) For $X = Y$
$$k_{exp} = k_1 k_2 / (k_{-1} + 2k_2) \qquad (69)$$

(b) For exothermic reactions, $(k_3 \gg k_{-2})$
$$k_{exp} = k_1 k_2 / (k_{-1} + k_2) \qquad (70)$$

The procedure adopted here is to make use again of RRKM theory to calculate k_2/k_{-1} as a function of the relative barrier height. In this case, the transition state for the k_{-1} reaction is taken as the loose ion-molecule complex at the Langevin capture distance. The transition state for the reaction k_2 is taken as the tetrahedral intermediate $RCOYX^-$. By a suitable choice of the vibrational frequencies and moments of inertia, this type of calculation shows that $E'_0 - E_0$ for $Cl^- + CH_3COCl$ should be around -7 kcal mol^{-1} in order to reproduce the experimental efficiency. This amounts to an E'_0 of 4 kcal mol^{-1}.

The validity of this model is probably comparable to that used for S_N2 reactions except that it has not been extensively tested. Perhaps, the most fundamental question raised in these experiments is with respect to the location of the tetrahedral intermediate along the potential energy surface, and the nature of the tetrahedral species.

A number of theoretical calculations have been carried out on the stability of tetrahedral species. For example, Bürgi et al. (1974) show that addition of H^- to H_2CO leads to the formation of a stable CH_3O^-. A similar type calculation performed by Alagona et al. (1975) on $OH^- + HCONH_2$, predicted a very large stability for $HC(OH)(NH_2)O^-$. Analogous results were obtained for a series of related systems by Scheiner et al. (1976). Unlike the situation with S_N2 reactions, the available calculations do not support the idea of a double potential well for addition of a nucleophile to a carbonyl centre. On the other hand, thermochemical estimates for the solution reaction $OH^- + HCOOCH_3$ place the tetrahedral intermediate at a minimum but (4.5 kcal mol^{-1}) *above* the energy of the reagents (Guthrie, 1973).

The experimental evidence for the possible stability of the tetrahedral species in the gas phase comes from three independent studies. Bowie and Williams (1974) detected the formation of an adduct $CF_3CO_2^- \cdot (CF_3CO)_2O$ in the reaction of $CF_3CO_2^-$ with trifluoroacetic anhydride. It was concluded that this represented a tetrahedral intermediate. While this may be the case, in this kind of experiment the icr technique cannot distinguish whether the species is a loose adduct or a tetrahedral intermediate. Asubiojo et al. (1975) observed reaction (71). By careful selection of chlorine isotopes and multiple

$$COCl^- + CH_3COCl \rightarrow CH_3COCl_2^- + CO \quad (71)$$

resonance icr techniques, it was concluded that *both chlorines are equivalent in* $CH_3COCl_2^-$. This finding strongly argues for a stable tetrahedral intermediate. This view was subject to careful analysis by the same authors in 1979 and they concluded that thermochemical estimates were inconsistent with a tetrahedral $CH_3COCl_2^-$ of such stability. The question as to the exact nature of this species thus remains unresolved.

More recently, Takashima et al. (1983) have addressed the question of whether the tetrahedral species resulting from addition of OH^- to $HCOOCH_3$ is an intermediate or a local transition state but still below the energy of the reagents. The fact that no ^{18}O-exchange is observed between OH^- and $HC(^{18}O)(OCH_3)$ in an icr experiment has been used as an argument to propose that the tetrahedral species $HC(OH)(OCH_3)O^-$ is likely to be a local transition state and not a stable intermediate. However, because the reaction is highly exothermic, the interpretation of this experiment should be accepted with caution.

From the foregoing considerations, it can be concluded that there are still a number of points to be clarified regarding the energy surface for nucleophilic displacement at carbonyl centres. In particular, the role of the tetrahedral species needs further work. It is probable, in fact, that unlike the S_N2 reaction, the relative stability and location of the tetrahedral species may vary considerably from case to case.

7 Gas-phase nucleophilic reactions of carbonyl systems involving positive ions

Several of the systems reported in the previous section occur in solution under conditions of base or acid catalysis. Thus, it is not surprising that the positive ion chemistry of such systems will resemble the acid-catalysed process.

GENERAL FEATURES

Tiedemann and Riveros (1974) first discussed the gas-phase reactions which are equivalent to an esterification reaction. An icr study of alcohols and acetic acid revealed that the formation of protonated acetic acid by ion-molecule reactions of fragment ions with acetic acid is followed by the rapid reaction (72).

$$CH_3C(OH)_2^+ + ROH \rightarrow CH_3C(OR)(OH)^+ + H_2O \qquad (72)$$

In practice, (72) may also originate from the reaction of neutral acetic acid and protonated alcohol. As in the previous discussion of positive ion chemistry, this reaction takes place as a secondary ion-molecule reaction along with several other reactions typical of the ion chemistry of the isolated alcohol or acid. This reaction, when studied with $CH_3^{18}OH$ as the nucleophile, results in the formation of $CH_3C(OH)(^{18}OCH_3)^+$ in agreement with a mechanism in which cleavage occurs at the O-acyl position corresponding to the traditional $A_{AC}2$ mechanism (Pau et al., 1978). Both of these studies show that the nucleophilic reactivity trend in this reaction is $CH_3OH \sim C_2H_5OH > (CH_3)_2CHOH > (CH_3)_3COH$.

A small change in mechanism is suggested by an experiment with ^{18}O-enriched 2-propanol which results only in 80 to 88% protonated ^{18}O-ester, indicating that for branched alcohols the $A_{AC}2$ mechanism may not be the only one in operation.

The gas-phase esterification reaction has not been observed with formic acid. The reason for this behaviour has been explained on the basis of the proton affinities of the acid and alcohols as indicated in (73) and (74). Thus,

$$HC(OH)_2^+ + CH_3OH \rightarrow HCOOH + CH_3OH_2^+ \qquad \Delta H° = -2 \text{ kcal mol}^{-1} \qquad (73)$$

$$HC(OH)_2^+ + CH_3OH \rightarrow HC(OH)(OCH_3)^+ + H_2O \quad \Delta H° = -11 \text{ kcal mol}^{-1} \qquad (74)$$

rapid proton transfer to CH_3OH precludes the nucleophilic displacement reaction (74). It can therefore be concluded that a necessary condition for the esterification reaction to occur in the gas phase is that proton transfer from the substrate (acid) to the nucleophile (alcohol) is endothermic, a rule previously established in Section 5.

There is some doubt as to whether the formic acid-2-propanol system reacts according to reaction (72). Tiedemann and Riveros (1974) observed reaction (75) in their icr study claiming that this process occurs by alkyl-oxygen cleavage in the alcohol. This is a slow reaction, and due to the low abundance of the product ion, the ^{18}O experiment (Pau et al., 1978) was inconclusive.

$$HCOOH + (CH_3)_2CHOH_2^+ \rightarrow HC(OH)OCH(CH_3)_2^+ + H_2O$$
$$\Delta H° = -2.4 \text{ kcal mol}^{-1} \quad (75)$$

The corresponding reaction in esters (76), an alcoholysis, is surprisingly slow in the gas phase and has not been observed by icr techniques even though both conditions for a nucleophilic displacement are satisfied as shown in (77) and (78). There is at present no satisfactory explanation of why this type of reaction is slow in the gas phase.

$$RC(OH)OR'^+ + R''OH \nrightarrow RC(OH)OR'' + R'OH \quad (76)$$
$$CH_3C(OH)OCH_3^+ + C_2H_5OH \rightarrow CH_3C(OH)OC_2H_5 + CH_3OH$$
$$\Delta H° = -3 \text{ kcal mol}^{-1} \quad (77)$$
$$CH_3C(OH)OCH_3^+ + C_2H_5OH \rightarrow CH_3COOCH_3 + C_2H_5OH_2^+$$
$$\Delta H° = 9 \text{ kcal mol}^{-1} \quad (78)$$

While the alkyl group switching reaction is not observed in the gas phase, secondary and tertiary alcohols give rise to a totally different reaction with esters. Typical examples are shown in (79) and (80). These reactions display a reverse reactivity compared to the acids, viz., $t-C_4H_9OH > i-C_3H_7OH \gg$ primary alcohols (Table 9). In fact, these reactions have not been detected for primary alcohols.

TABLE 9

Kinetics of reaction (80) for different alcohols[a]

ROH	$k/10^{11} M^{-1} s^{-1}$
$(CH_3)_2CHOH$	0.54
$CH_3(C_2H_5)CHOH$	1.6
$(CH_3)_3COH$	4.1

[a] From Pau et al., 1978

$$HCOOCH_3 + (CH_3)_2CHOH_2^+ \rightarrow HC(OCH_3)OCH(CH_3)_2^+ + H_2O \quad (79)$$

$$CH_3C(OH)OCH_3^+ + (CH_3)_3COH \rightarrow CH_3C(OCH_3)OC(CH_3)_3^+ + H_2O \quad (80)$$

The corresponding reactions with [^{18}O] 2-propanol and [^{18}O]t-butyl alcohol show that the isotopic label is not incorporated in the product ion. This result clearly establishes that the alkyl—oxygen bond in the alcohol is cleaved in process (80). A similar situation is observed even for the case when the ester has a lower proton affinity than the alcohol as in the case of reaction (81).

$$CF_3CO_2CH_3 + (CH_3)_2CH^{18}OH \rightarrow CF_3C(OCH_3)OCH(CH_3)_2^+ + H_2^{18}O \quad (81)$$

The structure of the product ion has been assumed in all cases to correspond to a dialkoxy carbenium ion, a species similar to that observed in the mass spectra of orthoesters. The alternative structure, an oxonium ion, cannot be discarded but it is expected to be a higher energy form.

Reactions corresponding to acidic solvolysis of acid halides have also been studied in the gas phase (McMahon, 1978). The results which bear close resemblance to solution processes are illustrated in (82) and (83).

$$CH_3OH_2^+ + CH_3COCl \rightarrow CH_3C(OH)(OCH_3)^+ + HCl$$
$$\Delta H^\circ = -28 \text{ kcal mol}^{-1} \, k = 6.3 \times 10^{11} \text{M}^{-1}\text{s}^{-1} \quad (82)$$

$$C_2H_5OH_2^+ + CH_3COF \rightarrow CH_3C(OH)(OC_2H_5)^+ + HF$$
$$\Delta H^\circ = -23 \text{ kcal mol}^{-1} \, k = 9 \times 10^{11} \text{M}^{-1}\text{s}^{-1} \quad (83)$$

MECHANISM

Reactions (72), (82) and (83) are similar to well-known processes in solution. These reactions are assumed to proceed in solution by an addition–elimination mechanism through formation of a tetrahedral intermediate. For the typical case, it was initially assumed that such a mechanism is also responsible for the gas-phase reaction. This view has been challenged by Kim and

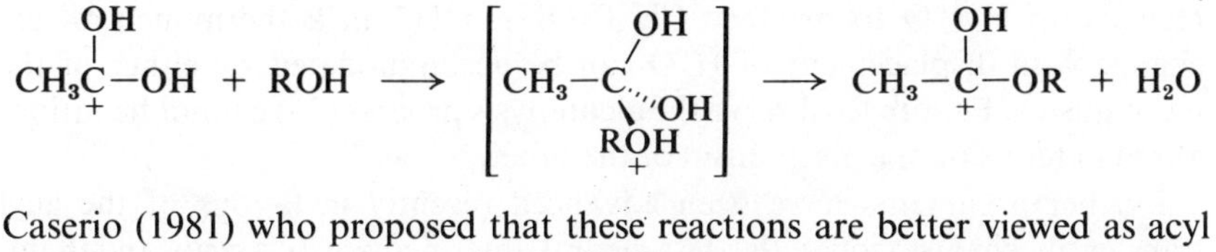

$$CH_3\overset{OH}{\underset{+}{C}}-OH + ROH \rightarrow \left[CH_3-C\overset{OH}{\underset{R\overset{+}{O}H}{\diagdown_{OH}}} \right] \rightarrow CH_3-\overset{OH}{\underset{+}{C}}-OR + H_2O$$

Caserio (1981) who proposed that these reactions are better viewed as acyl transfer reactions acting as direct displacement reactions. The key experiments leading to this conclusion involve reactions such as (84). In this case, it

$$(CH_3COSH)H^+ + CH_3OH \rightarrow (CH_3COOCH_3)H^+ + H_2S \quad (84)$$

was proposed that a long lived tetrahedral intermediate of the type shown in (85) would result in elimination of either H_2O or H_2S, with preference for

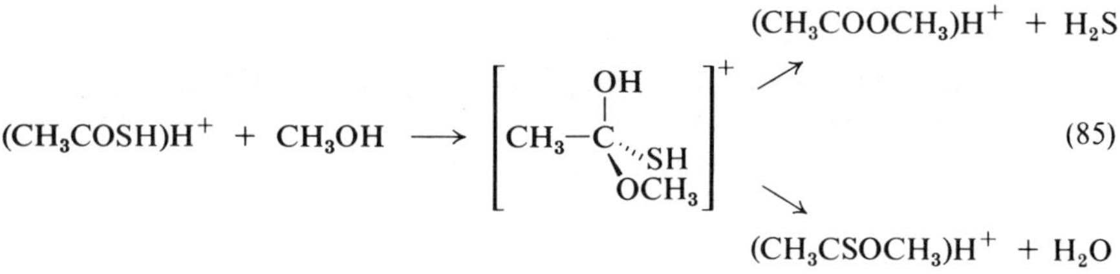

$$(CH_3COSH)H^+ + CH_3OH \longrightarrow \left[CH_3-\overset{\overset{\displaystyle OH}{|}}{\underset{\underset{\displaystyle OCH_3}{|}}{C}}{\cdots}SH \right]^+ \begin{matrix} \nearrow (CH_3COOCH_3)H^+ + H_2S \\ \\ \searrow (CH_3CSOCH_3)H^+ + H_2O \end{matrix} \quad (85)$$

elimination of water since this would be the more exothermic channel. Observation of reaction (84) exclusively is more consistent with a scheme in which the reaction proceeds by direct displacement (86).

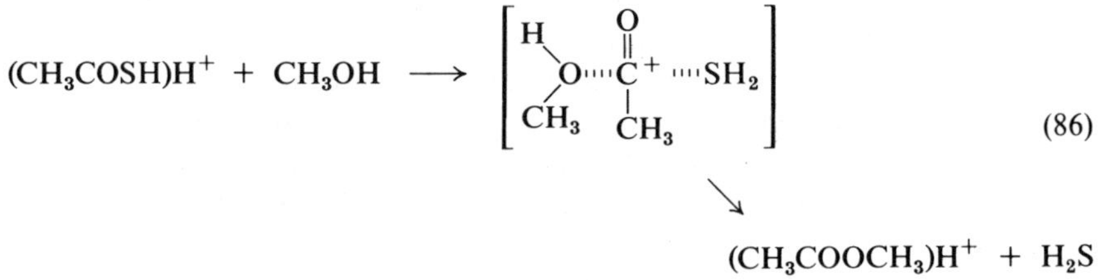

$$(CH_3COSH)H^+ + CH_3OH \longrightarrow \left[\begin{matrix} H & O \\ \diagdown & \| \\ O{\cdots}C^+{\cdots}SH_2 \\ \diagup & | \\ CH_3 & CH_3 \end{matrix} \right] \quad (86)$$

$$\searrow (CH_3COOCH_3)H^+ + H_2S$$

The same reasoning can be applied to the observation of reaction (87) and

$$CH_3COSH_2^+ + CH_3SH \rightarrow CH_3C(OH)SCH_3^+ + H_2S$$
$$\not\rightarrow CH_3C(SH)SCH_3^+ + H_2O \quad (87)$$

to the reaction with ethers and thioethers (88).

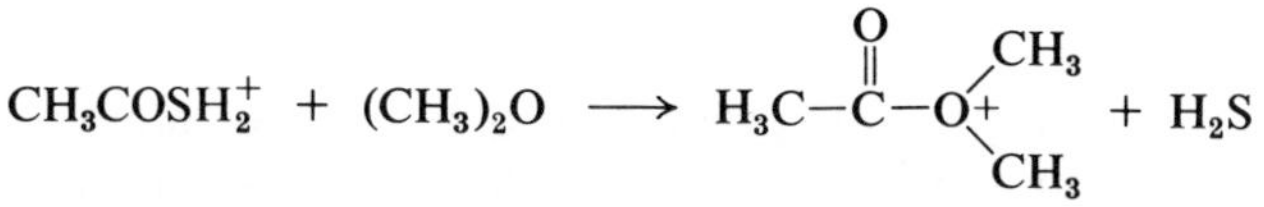

$$CH_3COSH_2^+ + (CH_3)_2O \longrightarrow H_3C-\overset{\overset{\displaystyle O}{\|}}{C}-\underset{\underset{\displaystyle CH_3}{}}{\overset{\overset{\displaystyle CH_3}{\diagup}}{O^+}} + H_2S \quad (88)$$

For the esterification reaction, there is of course no way of distinguishing between the two proposed mechanisms. The fact that protonated acetic acid reacts with $H_2^{18}O$ to produce $CH_3C(OH)^{18}OH^+$ in a thermoneutral exchange with displacement of H_2O can be accommodated by either of the mechanisms. Failure to observe the alcoholysis process (76) cannot be rationalized in terms of one mechanism or the other.

Further arguments have been advanced recently in favour of the acyl transfer mechanism over the tetrahedral intermediate (Caserio and Kim, 1983). The lack of ^{18}O-exchange in reaction (89) has been interpreted as additional evidence for non-participation of a tetrahedral intermediate. These conclusions are nevertheless ambiguous because both reactions are

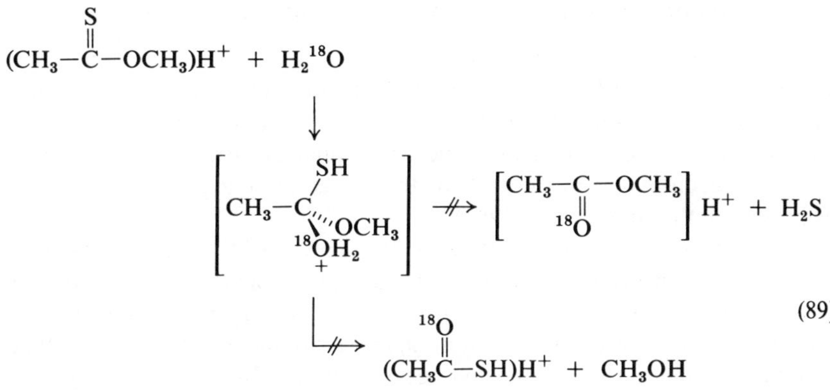

$$(CH_3-\overset{S}{\underset{\|}{C}}-OCH_3)H^+ + H_2{}^{18}O$$

$$\downarrow$$

$$\left[CH_3-\overset{SH}{\underset{{}^{18}OH_2}{\overset{|}{\underset{+}{C}}}}\!\!-\!OCH_3 \right] \xrightarrow{\not{}} \left[CH_3-\overset{C-OCH_3}{\underset{{}^{18}O}{\|}} \right] H^+ + H_2S$$

$$\downarrow\!\!\!\not{}\quad (CH_3\overset{{}^{18}O}{\underset{\|}{C}}-SH)H^+ + CH_3OH \qquad (89)$$

substantially endothermic so that it is unlikely that they would proceed at reasonable rates under icr conditions.

Reactions involving esters and secondary or tertiary alcohols proceed by a different mechanism as demonstrated by the experiments using ^{18}O-enriched alcohols. This mechanism entails essentially that the ester act as the nucleophile on a nascent carbenium ion produced from the protonated alcohol (Scheme 1). The initial association is described as a proton-bound cluster.

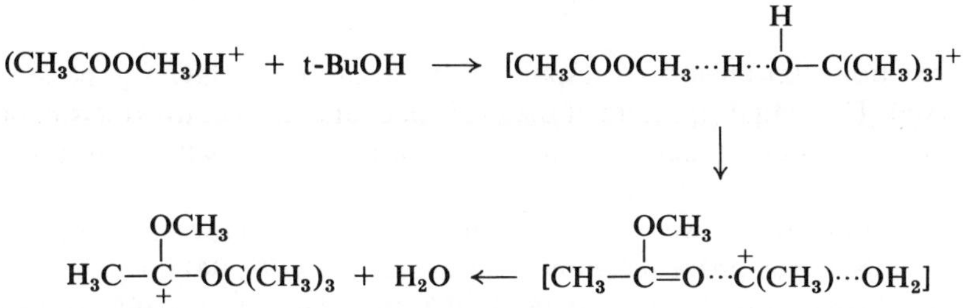

$$(CH_3COOCH_3)H^+ + \text{t-BuOH} \longrightarrow [CH_3COOCH_3\cdots H\cdots \overset{H}{\underset{|}{O}}-C(CH_3)_3]^+$$

$$\downarrow$$

$$H_3C-\overset{OCH_3}{\underset{+}{\overset{|}{C}}}-OC(CH_3)_3 + H_2O \longleftarrow [CH_3-\overset{OCH_3}{\underset{|}{C}}=O\cdots \overset{+}{C}(CH_3)\cdots OH_2]$$

Scheme 1

These species, for oxygen-containing compounds, have a large stability (in excess of 20 kcal mol^{-1}) and require a very loose association. This stability allows for partial proton transfer to the alcohol, even when the ester has a larger proton affinity than the alcohol. The scheme is also consistent with the picture that reactions will take place only at centres prone to form a carbenium ion, namely secondary or tertiary carbon centres.

An analogous mechanism is capable of explaining similar reactions, for example (90), which occur in systems involving protonated esters and t-butyl halides (Riveros *et al.*, 1979).

$$(HCOOCH_3)H^+ + (CH_3)_3CCl \rightarrow HC(OCH_3)OC(CH_3)_3{}^+ + HCl \qquad (90)$$

8 Nucleophilic displacement reactions in aromatic systems

Gas-phase nucleophilic displacement reactions in aromatic systems have until recently received considerably less attention than those in aliphatic systems. This situation is not unlike reactions in solution where these reactions are less common than displacement at aliphatic centres. It is well known that these reactions occur only when strongly activating groups (e.g. nitro) are present in the ring (Miller, 1968).

An early report (Briscese and Riveros, 1975) revealed that in the gas phase, alkoxide ions can displace fluoride from fluorobenzene (91). Hydroxide ion fails to react because C_6H_5F is more acidic than H_2O and thus proton transfer becomes the most important channel. Similar reactions with other monohalobenzenes are complicated because these substrates usually generate halide ions directly by dissociative electron attachment.

$$CH_3O^- + C_6H_5F \rightarrow C_6H_5OCH_3 + F^- \quad k_{icr} = 2.4 \times 10^{10} M^{-1} s^{-1} \quad (91)$$

The presumption that these reactions proceed by way of a Meisenheimer-type complex is supported by the data of Bowie and Stapleton (1977).

$$Cl^- + \underset{NO_2}{\underset{|}{C_6H_4}}{-NO_2} \longrightarrow NO_2^- + \underset{Cl}{\underset{|}{C_6H_4}}{-NO_2} \quad (92)$$

For reaction (92) an ion corresponding to a $Cl^-[C_6H_4(NO_2)_2]$ species is observed. The actual structure of the ion is of course subject to criticism, but neutral analysis experiments confirm chloronitrobenzene as the neutral product.

Polyfluorobenzenes were the first compounds to be shown to undergo a rather unusual gas-phase reaction. Typical examples (93, 94) involve alkoxide ions (Briscese and Riveros, 1975). In (93) the relative yields of $C_6H_4FO^-$ for different isomers of difluorobenzene are influenced by the alkyl group (R) since the different isomers display acidities comparable to the aliphatic alcohols. Reaction (93) is particularly important for *p*-difluorobenzene, the least

$$RO^- + C_6H_4F_2 \rightarrow RF + C_6H_4FO^- \quad (93)$$

$$RO^- + C_6F_6 \rightarrow RF + C_6F_5O^- \quad k_{icr} \sim 9 \times 10^{11} M^{-1} s^{-1} \quad (94)$$

acidic of these compounds. In the case of (94), the process is extremely fast and it was initially suggested that internal return by a nascent F^- results in the formation of the highly stable pentafluorophenoxide ion (Briscese and Riveros, 1975). This mechanism is consistent with similar observations in fluorinated ethylenes which give rise to products which can be accounted by such a scheme (Riveros and Takashima, 1976; Sullivan and Beauchamp, 1977).

Further examples of reactions which occur under nucleophilic attack in aromatic systems have also received consideration, with particular emphasis on phenyl ethers (Kleingeld and Nibbering, 1980). For example, reactions (95a) and (95b) give rise to a competition between an S_N2 and an S_NAr

$$^{18}OH^- + C_6H_5OCH_3 \nearrow C_6H_5O^- + CH_3{}^{18}OH \quad 85\% \quad (95a)$$
$$\searrow C_6H_5{}^{18}O^- + CH_3OH \quad 15\% \quad (95b)$$

mechanism. A similar ratio is observed for reaction (96). As in the previous example of polyfluorinated benzene, the reaction of OH^- with $C_6F_5OCH_3$ also gives rise to a product ion formed by a pathway similar to that of reactions (93) and (94).

$$OH^- + C_6F_5OCH_3 \rightarrow C_6F_5O^- + CH_3OH \quad (96a)$$
$$OH^- + C_6F_5OCH_3 \rightarrow CH_3OC_6F_4O^- + HF \quad (96b)$$

To account for these reactions as well as more recent observations (Ingemann et al., 1982), reaction (97) has been proposed for reactions of a number

$$HY^- + C_6F_5OCH_3 \nearrow C_6F_5O^- + CH_3YH \quad S_N2$$
$$\rightarrow C_6F_5Y^- + CH_3OH \quad IPSO \quad (97)$$
$$\searrow CH_3OC_6F_4Y^- + HF \quad S_NAr$$

TABLE 10

Relative importance of reaction channels in $C_6F_5OCH_3{}^a$

Nucleophile	$S_N2\%$	IPSO(%)	S_NAr
OH^-	22	5	73
CH_3O^-	16		84
$C_2H_5O^-$	17		83
SH^-	78		22
CH_3S^-	31		69
NH_2^-	23	6	71

^{a}From Ingemann et al., 1982.

of nucleophiles (HY$^-$) with $C_6F_5OCH_3$. The results are presented in Table 10. The reaction labelled IPSO substitution is only applicable to species like OH^- and NH_2^- and corresponds to a special case of the S_NAr mechanism.

An extensive study was also carried out for the reaction of anions with 2-, 3- and 4-fluoroanisole (Ingemann and Nibbering, 1983). The use of NH_2^-, OH^- and RO^- results in most cases in proton transfer, except for the case of 2-fluoroanisole. Nevertheless, competition is observed between an S_N2 reaction resulting from attack at the methyl group and a nucleophilic reaction initiated by attack at the fluorine bearing carbon. Fluoride displacement is only observed in reaction (98). These results have led to the proposal that

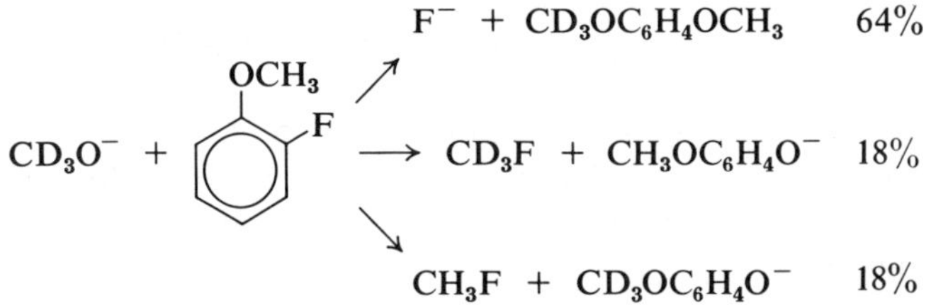

$$CD_3O^- + \text{(2-fluoroanisole)} \longrightarrow \begin{cases} F^- + CD_3OC_6H_4OCH_3 & 64\% \\ CD_3F + CH_3OC_6H_4O^- & 18\% \\ CH_3F + CD_3OC_6H_4O^- & 18\% \end{cases} \quad (98)$$

such reactions proceed through long-lived ion–molecule collision complexes which can undergo secondary reactions within the complex. This idea, advanced by DePuy and coworkers (Stewart *et al.*, 1977; DePuy *et al.*, 1978, 1980) to account for extensive hydrogen–deuterium exchange and the appearance of products which cannot be explained by simple mechanistic considerations, can probably be extended to many different gas-phase ionic reactions.

For the case of the 2-fluoroanisole, the proposed mechanism is shown in Scheme 2. This scheme predicts the formation of products originating from

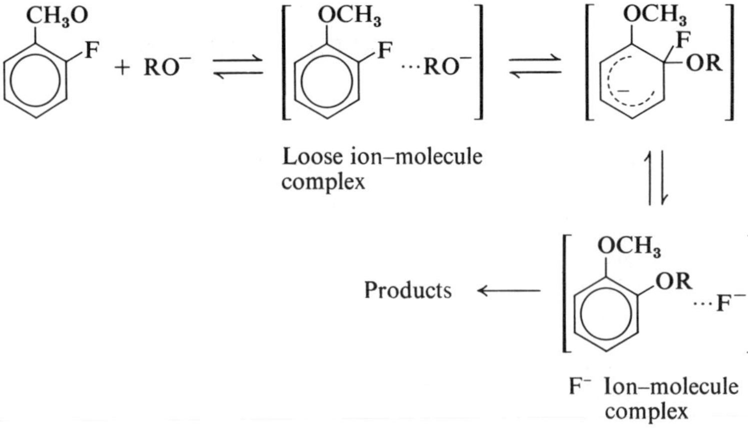

[Scheme 2: 2-fluoroanisole + RO$^-$ ⇌ Loose ion–molecule complex ⇌ [intermediate] ⇌ F$^-$ Ion–molecule complex → Products]

Scheme 2

attack of F^- on the neutral in the complex by proton transfer, S_N2 or elimination mechanisms.

While the explanation is of a qualitative nature, it is consistent with the patterns that are observed by icr. This approach constitutes an initial point of view which needs further exploration. There remain some very pertinent unanswered questions relating to the fact that these reactions seem to involve primarily fluorinated benzenes. Furthermore, fluoride ion itself is seldom observed as a product of aromatic nucleophilic displacement.

The number of cases which have been investigated in the last three years allows us to foresee that an adequate model can be set up if further data for some of the other halogens are obtained.

9 Conclusions

The views presented here show that gas-phase nucleophilic displacement reactions constitute a mature field of physical organic chemistry. It is a field that provides some unusual insights into the world of solvent-free chemistry, although many experiments remain to be rationalized on a more quantitative basis. The extent to which the understanding of gas-phase reactivity provides an insight into solution behaviour by appropriate extrapolation of solvation effects has been under attack recently, and it is a subject of lively discussion. Yet, the most important lesson that can be derived from these examples is that significantly different chemistry can be generated in ionic reactions in the gas phase. This aspect is one which most theoretical and experimental chemists find intellectually stimulating and a challenge to be pursued.

References

Alagona, G., Scrocco, E. and Tomasi, J. (1975). *J. Am. Chem. Soc.* **97**, 6976
Arshadi, M., Yamdagni, R. and Kebarle, P. (1970). *J. Phys. Chem.* **74**, 1475
Asubiojo, O. I., Blair, L. K. and Brauman, J. I. (1975). *J. Am. Chem. Soc.* **97**, 6685
Asubiojo, O. I. and Brauman, J. I. (1979). *J. Am. Chem. Soc.* **101**, 3715
Baldeschwieler, J. D. (1968). *Science* **159**, 263
Bartmess, J. E. and McIver, R. T. (1979). *In* "Gas Phase Ion Chemistry" (Bowers, M. T. ed.) Vol. 2, Ch. 11. Academic Press, New York
Bartmess, J. E., Scott, J. A. and McIver, R. T. (1979). *J. Am. Chem. Soc.* **101**, 6046
Beauchamp, J. L. and Caserio, M. C. (1972). *J. Am. Chem. Soc.* **94**, 2638
Beauchamp, J. L., Caserio, M. C. and McMahon, T. B. (1974). *J. Am. Chem. Soc.* **96**, 6243
Bell, R. P. (1963). *Proc. R. Soc., Ser. A.* **154**, 414
Blair, L. K., Isolani, P. C. and Riveros, J. M. (1973). *J. Am. Chem. Soc.* **95**, 1057
Bohme, D. K. and Mackay, G. I. (1981). *J. Am. Chem. Soc.* **103**, 979
Bohme, D. K. and Young, L. B. (1970). *J. Am. Chem. Soc.* **92**, 7354
Bohme, D. K., Hemsworth, R. S., Rundle, H. W. and Schiff, H. I. (1973). *J. Chem. Phys.* **58**, 3504

Bohme, D. K., Mackay, G. I. and Payzant, J. D, (1974). *J. Am. Chem. Soc.* **96**, 4027
Bowie, J. H. (1980). *Acc. Chem. Res.* **13**, 76
Bowie, J. H. and Williams, B. D. (1974). *Aust. J. Chem.* **27**, 1923
Bowie, J. H. and Stapleton, B. J. (1977). *Aust. J. Chem.* **30**, 795
Brauman, J. I. and Blair, L. K. (1968). *J. Am. Chem. Soc.* **90**, 6561
Brauman, J. I. and Blair, L. K. (1970). *J. Am. Chem. Soc.* **92**, 5986
Brauman, J. I., Riveros, J. M. and Blair, L. K. (1971). *J. Am. Chem. Soc.* **93**, 3914
Brauman, J. I., Olmstead, W. N. and Lieder, C. A. (1974). *J. Am. Chem. Soc.* **96**, 4030
Briscese, S. M. J. and Riveros, J. M. (1975). *J. Am. Chem. Soc.* **97**, 230
Bürgi, H. B., Lehn, J. M. and Wipff, G. (1974). *J. Am. Chem. Soc.* **96**, 1956
Caserio, M. C. and Kim, J. K. (1983). *J. Am. Chem. Soc.* **105**, 6896
Comisarow, M. (1977). *Can. J. Chem.* **55**, 171
Comisarow, M. (1978). *In* "Ion Cyclotron Resonance Spectrometry" (Hartman, H. and Wanczek, K.-P., eds). Springer-Verlag, Berlin
Cumming, J. B. and Kebarle, P. (1978). *Can. J. Chem.* **56**, 1
Cunningham, A. J., Payzant, J. D. and Kebarle, P. (1972). *J. Am. Chem. Soc.* **94**, 7627
Dedieu, A. and Veillard, A. (1972). *J. Am. Chem. Soc.* **94**, 6730
DePuy, C. H. and Bierbaum, V. M. (1981). *Acc. Chem. Res.* **14**, 146
DePuy, C. H., Bierbaum, V. M., King, G. K. and Shapiro, R. H. (1978). *J. Am. Chem. Soc.* **100**, 2921
DePuy, C. H., Bierbaum, V. M., Flippin, L. A., Grabowski, J. J., King, G. K., Schimitt, R. J. and Sullivan, S. A. (1980). *J. Am. Chem. Soc.* **102**, 5012
DePuy, C. H., Della, E. W., Filley, J., Grabowski, J. J. and Bierbaum, V. M. (1983). *J. Am. Chem. Soc.* **105**, 2481
Dillard, J. G. (1973). *Chem. Rev.* **73**, 589
Dougherty, R. C. (1974). *Org. Mass Spectrom.* **8**, 85
Dougherty, R. C. and Roberts, J. D. (1974). *Org. Mass Spectrom.* **8**, 81
Dougherty, R. C., Dalton, J. and Roberts, J. D. (1974). *Org. Mass Spectrom.* **8**, 77
Duke, A. J. and Bader, R. F. W. (1971). *Chem. Phys. Lett.* **10**, 631
Emerson, M. T., Grunwald, E. and Kromhout, R. A. (1960). *J. Chem. Phys.* **33**, 547
Evans, M. G. and Polanyi, M. (1938). *Trans. Faraday Soc.* **34**, 11
Faigle, J. F. G., Isolani, P. C. and Riveros, J. M. (1976). *J. Am. Chem. Soc.* **98**, 2049
Ferguson, E. E., Fehsenfeld, F. C. and Schmeltekopf, A. L. (1969). *Adv. At. Mol. Phys.* **5**, 1
Fleming, I. (1976). "Frontier Orbitals and Organic Chemical Reactions". John Wiley and Sons, New York.
Franklin, J. L. (1972). *In* "Ion-Molecule Reactions (Franklin, J. L., ed.). Plenum Press, New York
Giomousis, G. and Stevenson, D. P. (1958). *J. Chem. Phys.* **29**, 294
Guthrie, J. P. (1973). *J. Am. Chem. Soc.* **95**, 6999
Hall, D. G., Gupta, C. and Morton, T. H. (1981). *J. Am. Chem. Soc.* **103**, 2416
Hammond, G. S. (1955). *J. Am. Chem. Soc.* **77**, 334
Harrison, A. G., Myher, J. J. and Thynne, J. C. J. (1966). *In* "Ion-Molecule Reactions in the Gas Phase", Advances in Chemistry Series No. 58 (Gould, R. F., ed.), American Chemical Society, Washington
Hemsworth, R. S., Payzant, J. D., Schiff, H. I. and Bohme, D. K. (1974). *Chem. Phys. Lett.* **26**, 417
Henchmann, M., Paulson, J. F. and Hierl, P. M. (1983). *J. Am. Chem. Soc.* **105**, 5509
Holtz, D., Beauchamp, J. L. and Woodgate, S. D. (1970). *J. Am. Chem. Soc.* **92**, 7484
Ingemann, S. and Nibbering, N. M. M. (1983). *J. Org. Chem.* **48**, 183

Ingemann, S., Nibbering, N. M. M., Sullivan, S. A. and DePuy, C. H. (1982). *J. Am. Chem. Soc.* **104**, 6520
Ingold, C. K. (1957). *Quart. Revs.* **11**, 1
Isolani, P. C. and Riveros, J. M. (1975). *Chem. Phys. Lett.* **33**, 362
Janousek, B. K. and Brauman, J. I. (1979). *In* "Gas Phase Ion Chemistry" (Bowers, M. T., ed.) Vol. 2, Ch. 10. Academic Press, New York
Jencks, W. P. and Carriuolo, J. (1960). *J. Am. Chem. Soc.* **82**, 1778
José, S. M. (1976). Doctoral thesis, Institute of Chemistry, University of São Paulo, Brazil
José, S. M. and Riveros, J. M. (1977). *Nouv. J. Chim.* **1**, 113
Kim, J. K. and Caserio, M. C. (1981). *J. Am. Chem. Soc.* **103**, 2124
Kleingeld, J. C. and Nibbering, N. M. M. (1980). *Tetrahedron Lett.* **21**, 1687
Langevin, P. (1905). *Ann. Chim. Phys.* **5**, 245
Lehman, T. A. and Bursey, M. M. (1976). "Ion Cyclotron Resonance Spectrometry". John Wiley and Sons, New York
Lieder, C. A. and Brauman, J. I. (1974). *J. Am. Chem. Soc.* **96**, 4028
Lieder, C. A. and Brauman, J. I. (1975). *Int. J. Mass Spectrom. Ion Phys.* **16**, 307
Mackay, G. I., Betowski, L. D., Payzant, J. D., Schiff, H. I. and Bohme, D. K. (1976). *J. Phys. Chem.* **80**, 2919
Marcus, R. A. (1964). *Ann. Rev. Phys. Chem.* **15**, 155
Marinelli, W. J. and Morton, T. H. (1978). *J. Am. Chem. Soc.* **100**, 3536
McDaniel, E. W., Cermak, V., Dalgarno, A., Ferguson, E. E. and Friedman, L. (1970). "Ion-Molecule Reactions". Wiley-Interscience, New York
McIver, R. T. and Dunbar, R. C. (1971). *Int. J. Mass Spectrom. Ion Phys.* **7**, 471
McIver, R. T. (1978). *Rev. Sci. Instrum.* **49**, 111
McMahon, T. B. (1978). *Can. J. Chem.* **56**, 670
McMahon, T. B. and Beauchamp, J. L. (1972). *Rev. Sci. Instrum.* **43**, 509
McMahon, T. B. and Kebarle, P. (1977). *J. Am. Chem. Soc.* **99**, 2222
Meot-Ner, M. and Field, F. H. (1975). *J. Am. Chem. Soc.* **97**, 5339
Meot-Ner, M. (1979a). *J. Am. Chem. Soc.* **101**, 2389
Meot-Ner, M. (1979b). *In* "Gas Phase Ion Chemistry" (Bowers, M. T., ed.) Vol 1, Ch. 6. Academic Press, New York
Miller, J. (1968). "Aromatic Nucleophilic Substitution". Elsevier, Amsterdam
Morokuma, K. (1982). *J. Am. Chem. Soc.* **104**, 3732
Olmstead, W. N. and Brauman, J. I. (1977). *J. Am. Chem. Soc.* **99**, 4219
Pau, J. K., Kim, J. K. and Caserio, M. C. (1978). *J. Am. Chem. Soc.* **100**, 3831
Pearson, R. G., Sobel, H. and Songstad, J. (1968). *J. Am. Chem. Soc.* **90**, 319
Pellerite, M. J. and Brauman, J. I. (1980). *J. Am. Chem. Soc.* **102**, 5993
Pellerite, M. J. and Brauman, J. I. (1983). *J. Am. Chem. Soc.* **105**, 2672
Pross, A. and Shaik, S. S. (1981). *J. Am. Chem. Soc.* **103**, 3702
Riveros, J. M. and Takashima, K. (1976). *Can. J. Chem.* **54**, 1840
Riveros, J. M., Breda, A. C. and Blair, L. K. (1973). *J. Am. Chem. Soc.* **95**, 4066
Riveros, J. M., Tiedemann, P. W., de Melo, B. C. and Faigle, J. F. G. (1979). *J. Phys. Chem.* **83**, 1488
Scheiner, S., Lipscomb, W. N. and Kleier, D. A. (1976). *J. Am Chem. Soc.* **98**, 4770
Shaik, S. S. (1981). *J. Am. Chem. Soc.* **103**, 3692
Shaik, S. S. (1982). *Nouv. J. Chim.* **6**. 159
Shaik, S. S. and Pross, A. (1982). *J. Am. Chem. Soc.* **104**, 2708
Smith, D. and Adams, N. G. (1979). *In* "Gas Phase Ion Chemistry" (Bowers, M. T., ed.) Vol. 1, Ch. 1. Academic Press, New York

Smith, M. A., Barkley, R. M. and Ellison, G. B. (1980). *J. Am. Chem. Soc.* **102**, 6851
Solomon, J. J., Meot-Ner, M. and Field, F. H. (1974). *J. Am. Chem. Soc.* **96**, 3727
Speranza, M. and Angelini, G. (1980). *J. Am. Chem. Soc.* **102**, 3115
Stewart, J. H., Shapiro, R. H., DePuy, C. H. and Bierbaum, V. M. (1972). *J. Am. Chem. Soc.* **99**, 7650
Su, T. and Bowers, M. T. (1973a). *J. Chem. Phys.* **58**, 3027
Su, T. and Bowers, M. T. (1973b). *Int. J. Mass Spectrom. Ion Phys.* **12**, 347
Su, T. and Bowers, M. T. (1979). *In* "Gas Phase Ion Chemistry" (Bowers, M. T., ed.) Vol. 1, Ch. 3. Academic Press, New York
Su, T., Su, E. C. F. and Bowers, M. T. (1978). *J. Chem. Phys.* **69**, 2243
Sullivan, S. A. and Beauchamp, J. L. (1977). *J. Am. Chem. Soc.* **99**, 5017
Takashima, K. and Riveros, J. M. (1978). *J. Am. Chem. Soc.* **100**, 6128
Takashima, K., José, S. M., do Amaral, A. T. and Riveros, J. M. (1983). *J. Chem. Soc. Chem. Comm.* 1255
Tanaka, K., Mackay, G. I. Payzant, J. D. and Bohme, D. K.(1976). *Can. J. Chem.* **54**, 1643
Tiedemann, P. W. and Riveros, J. M. (1974). *J. Am. Chem. Soc.* **96**, 185
Winstein, S., Savedoff, L. G., Smith, S., Stevens, D. R. änd Gall, J. S. (1960). *Tetrahedron Lett.* 24
Wolf, J. F., Staley, R. H., Koppel, I, Taagepera, M., McIver, R. T., Beauchamp, J. L. and Taft, R. W. (1977). *J. Am. Chem. Soc.* **99**, 5417
Wolfe, S., Mitchell, D. J. and Schlegel, H. B. (1981a). *J. Am. Chem. Soc.* **103**, 7692
Wolfe, S., Mitchell, D. J. and Schlegel, H. B. (1981b). *J. Am. Chem. Soc.* **103**, 7694
Wolfe, S., Mitchell, D. J., Schlegel, H. B., Minot, C. and Eisenstein, O. (1982). *Tetrahedron Lett.* **23**, 615
Young, L. B., Lee-Ruff, E. and Bohme, D. K. (1973). *J. Chem. Soc. Chem. Comm.* 35

Author Index

Numbers in italics refer to the pages on which references are listed at the end of each article

Abeles, R. H., 13, 25, *30*, *35*
Abraham, M. H., 189, *191*
Abrahamson, E. W., 102, 139, 140, *194*
Adams, N. G., 204, *239*
Addink, R., 42, *97*
Adolfsson, L., 45, *97*
Adrian, E. D., 13, *30*
Agadzhanyan, Z. E., 41, *97*
Agmon, N., 148, 151, 168, 169, 172, 185, *191*
Ahmad, M., 53, 54, 55, 60, 62, 65, 68, 78, 83, 84, 85, 87, *94*, *97*
Alber, T., 30, *31*
Alberty, R. A., 24, *31*
Albery, W. J., 182, 184, *191*
Aldrich, F. L., 17, *31*
Aldridge, W. N., 14, *31*
Aldwin, L., 71, 72, 73, 78, 83, *95*
Algona, G., 228, *237*
Alibhai, M., 71, 72, 87, *97*
Allewell, N. M., 22, *35*
Amatore, C., 133, 177, *193*
Anderson, R., 45, *97*
Ando, T., 155, *196*
Anfinsen, C. B., 21, 22, 23, *31*, *34*
Angelini, G., 222, *239*
Antonov, V. K., 41, *94*, *97*
Appel, B., 128, *196*
Arato, H., 10, *34*
Arnett, E. M., 146, 149, *191*
Arshadi, M., 204, *237*
Asubiojo, O. I., 226, 227, 228, *237*
Augustinsson, K. B., 14, *31*
Auwers, K., 27, *31*
Aviram, K., 145, 154, *193*, *195*

Bach, R. D., 145, 154, *195*
Bader, R. F. W., 216, *238*
Baker, R., 128, *196*
Baldeschwieler, J. D., 201, *237*

Ballistreri, F. P., 155, *191*
Balls, A. K., 14, 17, *31*, *33*
Balzani, V., 182, *195*
Bank, S., 147, *191*
Barker, H. A., 12, *32*
Barkley, R. M., 204, 209, *239*
Barlet, R., 38, *95*
Barnard, E. A., 21, *31*, *35*
Barnett, R., 39, *94*
Bartmess, J. E., 206, *237*
Bates, H. A., 45, *95*
Baughan, E. C., 102, *191*
Beauchamp, J. L., 202, 221, 222, 235, *237*, *238*, *239*, *240*
Bell, R. P., 124, 148, 150, 168, 169, 177, 179, *191*, 219, *237*
Bello, J., 23, *33*
Beltrame, P., 166, *191*
Bender, M. L., 18, 27, *31*, *34*, 38, 40, 66, *94*
Benkovic, S. J., 27, *31*
Benner, D. W., 30, *31*
Bentley, T. W., 153, *192*
Berger, A., 27, *31*, 42, *97*
Bergman, F., 14, *31*, *35*
Bergstrom, R. G., 53, 54, 55, 60, 62, 65, 68, 78, 83, *94*
Bernard, S. A., 27, *31*
Bernasconi, C. F., 169, *192*
Bernstein, R. B., 116, *194*
Betowski, L. D., 206, *239*
Biale, G., 166, *191*
Bierbaum, V. M., 203, 219, 236, *238*, *240*
Blackburn, G. M., 39, *94*
Bladon, P., 40, *95*
Blair, L. K., 198, 207, 214, 225, 228, *237*, *238*, *239*
Bloomer, A. C., 30, *31*
Blow, D. M., 18, *31*, *33*

Bobrański, B., 42, *95*
Bodansky, O., 14, *33*
Bohme, D. K., 199, 203, 206, 207, 208, 211, 212, 215, *237*, *238*, *239*, *240*
Bohonek, J., 53, 60, 62, 63, 78, 83, 84, 85, 87, *97*
Bonsignore, A., 20, *34*
Bordner, J., 45, *95*
Bordwell, F. G., 149, 168, 169, 172, 178, *192*
Bose, S., 45, *95*
Bowen, C. T., 153, *192*
Bowers, M. T., 205, 206, *240*
Bowie, J. H., 199, 228, 234, *238*
Boyd, R., 9, *34*
Boyle, W. J., Jr., 67, *98*, 168, 169, 172, 178, *192*
Brader, W. H., 146, *193*
Brant, S. R., 151, 181, *193*
Brauman, J. I., 178, 184, *194*, *195*, 198, 206–211, 215–220, 226–228, *237*, *238*, *239*
Braunstein, A. E., 4, *31*, *33*
Breda, A. C., 214, *239*
Bredt, J., 18, *31*
Brecher, A. S., 17, *31*
Breslow, R., 9, 10, 30, *31*
Briggs, G. E., 24, *31*
Briscese, S. M. J., 234, *238*
Brønsted, J. N., 177, *192*
Brown, A. J., 24, *31*
Brown, J. F., 25, *35*
Brown, R. F., 27, *31*
Bruice, T. C., 9, 25, 27, *31*, *34*, 38, 42, 89, *95*, *97*
Bryson, J. A., 178, *194*
Buchanan, J. M., 23, *31*, *32*
Bull, H. G., 68, *95*
Bunnett, J. F., 28, *31*, 67, *98*, 147, *192*
Bunton, C. A., 44, *95*
Burgi, H. B., 228, *238*
Burt, R. A., 55, 66, 71, *95*
Bursey, M. M., 201, *239*
Byers, L. D., 44, *96*

Capon, B., 45, 48–50, 52, 57, 58, 61, 64–66, 69, 70, 72, 75, 76, 78, 80, 83, 84, 92, 94, *95*
Campbell-Crawford, A. N., 184, *191*
Carrington, T., 129, *192*
Carriuolo, J., 226, *239*

Caserio, M. C., 222, 229, 230, 231, 232, *237*, *238*, *239*
Cashen, M. J., 53, 54, 55, 60, 62, 65, 68, 78, 83, *94*
Cermak, V., 198, *239*
Cerrini, S., 42, *95*
Challand, S. R., 45, *95*
Chahine, J. M., El H., 42, 91, *95*
Chatterjee, A., 45, *95*
Chaturvedi, R. K., 38, 39, *95*
Cheng, T., 20, *32*
Cheriyan, U. O., 38, *95*
Cheung, M. F., 38, *96*
Chiang, Y., 53, 54, 55, 59, 60, 62, 65, 66, 67, 68, 69, 70, 71, 78, 83, *94*, *95*
Cipiciani, A., 44, *95*
Clardy, J., 45, *97*
Cohen, A. O., 184, *192*
Cohen, J. A., 14, *34*
Cohen, L. A., 67, *97*
Collet, H., 47, *95*
Comisarow, M., 202, 225, 226, *238*
Comeryras, A., 47, *95*
Conn, E. E., 7, *32*, 35
Cooke, J. P., 22, *31*
Cordes, E. H., 68, *95*
Cox, B. G., 168, *192*
Cramer, F., 29, *31*, 35
Cramm, D., 63, 78, 83, *97*
Crestfield, A. M., 21, *32*
Critchlow, J. E., 161, *192*
Crosby J., 26, *32*
Cullimore, P. A., 69, 70, *96*
Cullis, A. F., 3, *34*
Cumming, J. B., 206, *238*
Cunningham, A. J., 204, *238*
Cunningham, B. A., 39, *97*
Curran, J. S., 184, *191*

Dahlqvist, K. I., 44, *95*
Dalgarno, A., 198, *239*
Dalton, J., 213, 214, *238*
Dauben, W. G., 103, 139, 140, 143, *192*
Davies, D. R., 3, *33*
Dawid, I. B., 23, *32*
Day, J. N. E., 38, *95*
Dedieu, A., 216, *238*
Della, E. W., 219, *238*
de Melo, B. C., 233, *239*
DePuy, C. H., 203, 219, 235, 236, *238*, *240*

AUTHOR INDEX

Deslongchamps, P., 38, 39, *95*
Desmond, K. M., 128, *195*
Devaquet, A., 143, *192*
Dewar, M. J. S., 124, 148, *192*
DeWolfe, R. H., 45, 66, *95*
Diaz, A., 128, *196*
Dickerson, R. E., 3, *33*
Dillard, J. G., 200, *238*
Dillon, R. L., 168, *195*
Dixon, G. H., 18, *32*
Dixon, M., 2, *32*
do Amaral, A. T., 228, *240*
Dobbert, N. N., 4, *33*
Dokowa, T., 10, *35*
Dolphin, D., 25, *30*
Dorsey, G. E., 38, *96*
Dosunmu, M. I., 58, 69, 75, 78, 84, *95*
Doudoroff, M., 12, *32*
Dougherty, R. C., 213, 214, *238*
Dounce, A. L., 2, *35*
Dreyer, W. J., 18, *32*
Dubois, J. E., 42, 91, *95*
Dubois, J. T., 131, *193*
Duke, A. J., 216, *238*
Dunbar, R. C., 202, *239*
Dunning, T. H., Jr., 114, 115, *193*

Eberson, L., 182, *192*
Edwards, L. J., 27, *32*
Eigen, M., 168, *192*
Eisenstein, O. O., 219, *240*
Ellison, G. B., 204, 209, *239*
El-Nasr, M. M., 161, *194*
Emerson, M. T., 199, *238*
Engelman, D. M., 30, *34*
Engler, E. M., 146, *192*
Epiotis, N. D., 103, 106, 123, 130, 174, 177, *192*
Evans, M. G., 102, 123, 124, 148, 177, *192*, 219, *238*
Eyring, H., 28, *32*

Faigle, J. F. G., 210, 223, 224, *238*, *239*
Falkner, I. J., 25, *33*
Fastrez, J., 42, 72, 88, *95*
Fedeli, W., 42, *95*
Fedor, L. R., 38, *95*
Fehsenfeld, F. C., 203, *238*
Feld, E. A., 14, *34*
Feldberg, W., 13, *30*
Feldt, R. J., 39, *98*

Fendrich, G., 151, 181, *193*
Ferguson, E. E., 198, 203, *238*, *239*
Field, F. H., 204, *239*
Fieser, L., 29, *32*
Fife, T. H., 24, *32*, 70, 71, *95*, *97*
Filley, J., 219, *238*
Findlay, D., 22, *32*
Fischer, E., 3, 28, *32*
Fischer, E. H., 19, *32*
Fisher, H. F., 7, 8, *32*, *33*, *35*
Fitting, C., 12, *32*
Fleming, I., 100, *192*, 219, *238*
Flippin, L. A., 236, *238*
Fodor, G., 45, *95*
Forrest, G. C., 40, *95*
Forsén, S., 44, *95*
Forster, W., 154, *192*
Fraenkel, G., 41, *95*
Franklin, J. L., 198, *238*
French, D., 29, *32*
French, T. C., 23, *32*
Frey, A. J., 41, *96*
Fridovich, I., 20, *32*
Friedman, L., 198, *239*
Fry, A., 155, *193*
Fry, J. L., 146, *192*
Fuchs, R., 169, *192*
Fukui, K., 103, 173, *192*
Fukuzumi, S., 133, 135, 137, 176, *192*, *193*
Funderburk, L. H., 71, 72, 73, 78, 83, *95*

Gaetjens, E., 27, *32*
Gall, J. H., 50, *95*
Gall, J. S., 212, *240*
Gandler, J. R., 151, 181, 189, *193*
Gandour, R. W., 173, *193*
Gedge, S., 53, 60, 62, 63, 78, 83, 84, 85, 87, *97*
Gensmantel, N. P., 39, *96*
Gerhartz, W., 120, *193*
Germain, A., 47, *95*
Ghazarossian, V. E., 45, *97*
Ghosh, A. K., 48, 57, 58, 78, 80, 83, 92, *95*
Ghosh, C., 45, *95*
Gibson, A., 168, *192*
Giese, B., 161, *193*
Gilchrist, M., 38, *96*
Giomousis, G., 205, *238*
Goddard, W. A., III, 114, 115, *193*

Gougoutas, J. Z., 40, *98*
Goto, T., 40, *96*
Grace, M. E., 42, *96*
Grabowski, J. J., 219, 236, *238*
Gravitz, N., 42, 80, 91, 92, *96*
Grazi, E., 20, *32*
Green, E., 67, *97*
Grellier, P. L., 189, *191*
Grieve, D. McL. A., 48, 50, 54, 59, 92, *95*, *96*
Grimsrud, E. P., 155, *193*
Griot, R. G., 41, *96*
Grodowski, M., 156, *196*
Grunwald, E., 126, 178, *194*, 199, *238*
Guida, A., 38, *95*
Gundlach, H. G., 21, *32*
Gupta, C., 222, *238*
Gutfreund, H., 15, *32*
Guthrie, J. P., 40, 45, 50, 69, 70, 87, 88, 93, 94, *96*, 228, *238*

Haber, E., 23, *31*
Haldane, J. B. S., 24, *31*
Hall, D. G., 222, *238*
Hallas, M. D., 39, *96*
Hamilton, G., 18, *32*
Hammett, L. P., 2, *32*, 38, *96*
Hammond, G. S., 148, *193*, 219, *238*
Handler, P., 10, *34*
Hanke, M. E., 4, *33*
Hardman, K. D., 23, *35*
Harker, D., 23, *33*
Harris, J. C., 161, *193*
Harrison, A. G., 198, *238*
Harron, J., 72, 88, *96*
Hartley, B. S., 14, *32*
Hartsuck, J. A., 30, *34*
Hassid, W. Z., 12, *32*
Hautala, J. A., 168, 169, 178, *192*
Hay, J. P., 114, 115, *193*
Heck, H. d'A., 38, *94*
Hehre, W. J., 173, *195*
Heitler, W., 103, *193*
Hemsworth, R. S., 199, 203, 206, *237*, *238*
Henchmann, M., 213, *238*
Henderson, R., 18, *33*
Henglein, F. M., 29, *31*
Henri, V., 24, *32*
Herbert, R. B., 45, *95*
Herries, D. G., 22, *32*

Hershfield, R., 39, 67, *96*
Herzberg, G., 116, 129, *193*
Hibbert, F., 38, *96*
Hierl, P. M., 213, *238*
Hilinski, E. F., 133, 177, *193*
Hill, J. W., 155, *193*
Hine, J., 41, 70, *96*, 146, 167, *193*
Hinshelwood, C. N., 67, *96*, *97*
Hirata, Y., 40, *96*
Hirs, C. H. W., 21, *32*
Hoffmann, R., 100, 102, 106, 140, 173, *194*
Hoit, R. G., 3, *33*
Holliman, F. G., 45, *95*
Holtz, D., 221, *238*
Horecker, B. L., 20, *32*, *34*
Horiuti, J., 124, *193*
Houk, K. N., 173, *193*
Hughes, D. L., 149, *192*
Huisgen, R., 28, *32*
Hunt, H., 40, *97*
Hunt, W. J., 114, 115, *193*
Hupe, D. J., 151, 181, 189, *193*

Ikawa, M., 6, *32*, *34*
Inagami, T., 22, *35*
Ingemann, S., 235, 236, *238*
Ingold, C. K., 27, *32*, 38, *95*, 210, *239*
Ingraham, L. L., 26, *35*
Inoue, S., 40, *97*
Ishihara, R., 45, *98*
Isolani, P. C., 210, 223, 224, 225, *237*, *238*, *239*
Iwasaki, T., 156, *193*

Jandorf, B. J., 16, *35*
Jang, R., 14, *33*
Janousek, B. K., 206, *239*
Jansen, E. F., 14, *33*
Jao, L. K., 70, *95*
Jayaraman, H., 38, *97*
Jencks, D. A., 161, *193*
Jencks, W. P., 2, 25, 27, *33*, *34*, 38, 39, 42, 44, 71, 72, 73, 78, 80, 83, 91, 92, *94*, *95*, *96*, 101, 151, 153, 161, 189, *193*, 226, *239*
Jenkins, W. T., 6, *33*
Jenson, J. L., 66, *95*
Johnson, C. D., 161, *193*
Johnson, L. N., 22, *35*
Johnson, S. L., 38, *96*

AUTHOR INDEX

Jones, D. S., 42, *96*
Jones, W. A., 18, *35*
José, S. M., 210, 224, 228, *239*
Jutting, G., 11, *33*

Kaloustian, M. K., 45, *96*
Kamada, H., 156, *193*
Kanavarioti, A., 169, *192*
Kanchuger, M. S., 44, *96*
Kankaanperä, A., 58, *96*
Kartha, G., 23, *33*
Katchalski, E., 27, *31*, 42, *97*
Kebarle, P., 204, 206, *237*, *238*, *239*
Keeffe, J. R., 168, 169, *193*
Kellmann, A., 131, *193*
Kendrew, J. C., 3, *33*
Kennedy, E. P., 17, *33*
Kenner, G. W., 42, *96*
Kent, A. B., 19, *32*
Kershner, L., 38, *97*
Kevill, D. N., 146, *193*
Khouri, F., 45, *96*
Kilby, B. A., 13, 14, *30*, *32*
Kim, J. K., 229, 230, 231, 232, *238*, *239*
Kimura, C., 154, *194*
King, G. K., 236, *238*
Kirby, A. J., 24, *33*, 38, *96*
Kirsch, J. F., 38, *97*
Kiselev, V. D., 177, *193*
Kishi, Y., 40, *96*
Kitai, R., 3, *34*
Kleier, D. A., 228, *239*
Kleingeld, J. C., 235, *239*
Klix, R. C., 145, 154, *195*
Knappe, J., 11, *33*
Knier, B. L., 151, *193*
Knox, J. R., 23, *35*
Kochi, J. K., 133, 135, 137, 176, 177, *192*, *193*
Koehler, K., 68, *95*
Konasewich, D. E., 184, *193*
Konikova, A. S., 4, *33*
Koppel, I., 221, *240*
Koppelman, R., 4, *33*
Kornblum, N., 147, *193*
Koshland, D. E., Jr., 13, 17, 30, *33*
Kost, D., 145, 154, *193*, *195*
Krampitz, L. O., 11, *33*
Krebs, E. G., 19, *32*
Kreevoy, M. M., 151, 182, 184, *191*, *193*

Kresge, A. J., 45, 53, 54, 55, 59, 60, 62, 65, 66, 67, 68, 69, 70, 71, 78, 83, *94*, *95*, *96*, *97*, 168, 169, *193*, *194*
Kritzmann, M. G., 4, *31*
Kromhout, R. A., 199, *238*
Krueger, G., 13, *33*
Kuhn, R., 45, *96*
Kukes, S., 149, *194*
Kunitz, M., 2, 21, *33*, *34*
Kunst, P., 14, *34*
Kurz, J. L., 161, 185, *193*, *194*
Kwart, H., 68, *96*

Labbé, C., 45, *95*
Lachance, J.-P., 11, *33*
Lahti, M. O., 54, 65, 69, 70, *95*
Laidler, K. J., 102, 123, 139, *194*
Laird, R. M., 154, *192*
Lane, C. A., 38, *96*
Lange, W., 13, *33*
Langevin, P., 205, *239*
Lapworth, A., 10, *33*
Latowski, T., 156, *196*
Leach, S. J., 27, *33*
Lechtken, P., 119, *195*
Lee, B., 23, *35*
Lee, H. A., Jr., 25, *30*
Lee, I.-S. H., 151, *193*
Lee, J. C., 168, 169, *193*
Lee-Ruff, E., 211, *240*
Leeson, P., 42, *96*
Leffler, J. E., 126, 177, 178, *194*
Leforestier, C., 103, 106, 116, 129, 130, 132, 133, *195*
Legard, A. R., 67, *96*
Lehn, J. M., 228, *238*
Lehman, T. A., 201, *239*
Letourneau, F., 45, *95*
Levesque, G., 45, *96*
Levine, R. D., 116, *194*
Lewis, E. S., 148, 149, 159, 169, *192*, *194*
Libit, L., 100, 106, *194*
Lieder, C. A., 208, 209, 210, 211, *238*, *239*
Lienhard, G. E., 26, *32*, 44, *96*
Linda, P., 44, *95*
Lindley, H., 27, *33*
Lipscomb, W. N., 30, *34*, 228, *239*
Loewus, F. A., 7, 8, *33*
London, F., 103, *193*

Long, F. A., 26, *34*
Longenecker, J. B., 6, *32, 33*
Longuet-Higgins, H. C., 102, 129, 139, 140, *193, 194*
Loosemore, M. J., 42, *96*
Lorch, E., 11, *33*
Lowry, T. M., 25, *33*, 38, *96*
Loyd, D. J., 166, *191*
Lucente, G., 42, *96*
Ludwig, M. L., 30, *34*
Lumry, R., 28, *32, 33*
Lynen, F., 11, *33*

Maccarone, E., 155, *191*
Maeda, K., 45, *97*
MacInnes, I., 9, *33*
Mackay, G. I., 206, 208, 211, 212, 215, *237, 238, 239, 240*
MacMahon, A. E., 38, *95*
Mager, H. I. X., 42, *97*
Magnoli, D. E., 184, *194*
Mahjoub, A., 45, *96*
Mamo, A., 155, *191*
Mandava, N., 45, *95*
Mandeles, S., 4, *33*
Marinelli, W. J., 210, 222, *239*
March, J., 128, *194*
Marcus, R. A., 101, 150, 151, 168, 182, 184, 185, *192, 194*, 217, *239*
Martin, J. C., 128, *195*
Martin, R. B., 39, *96*
Masnovici, J. M., 133, 177, *193*
Mathias, A. P., 22, *32*
Matloubi, H., 44, *97*
Matsudu, T., 45, *98*
Matthews, B. W., 18, *33*
May, S. C., 14, *34*
Mazur, A., 14, *33*
Mazza, F., 42, *95*
McCapra, F., 42, *96*
McClelland, R. A., 38, 45, 53, 54, 55, 56, 60, 62, 63, 65, 66, 68, 69, 71, 72, 73, 78, 81, 83, 84, 85, 87, 88, 89, 91, 92, *94, 96, 97, 98*
McDaniel, E. W., 198, *239*
McDonald, C. E., 17, *31*
McDonald, M. R., 21, *33*
McIver, R. T., 202, 206, 221, *237, 239, 240*
McKay, D. B., 30, *34*
McKee, R. L., 42, *98*

McKinney, M. A., 55, 66, 71, *95*
McLennan, D. J., 101, 145, 153, 154, 181, 184, *194*
McMahon, T. B., 203, 204, 222, 231, *237, 239*
McMillen, D. F., 178, *194*
McVey, J., 119, *195*
Melander, L., 2, *33*
Menten, M. L., 24, *34*
Meot-Ner, M., 204, 205, *239*
Metiu, H., 126, *194*
Metzler, D. E., 6, *33, 34*
Meyer, V., 27, *31*
Michaelis, L., 24, *34*
Michl, J., 103, 120, 139, *193, 194*
Miller, J., 234, *239*
Miller, J. G., 177, *193*
Milstein, S., 67, *97*
Minot, C., 219, *240*
Mislow, K., 8, *34*
Mitchell, D. J., 184, *196*, 218, 219, *240*
Mizuhara, S., 10, *34*
Mochida, K., 133, *193*
Moffat, A., 40, *97*
Moiseyev, Y. V., 47, *98*
Montgomery, J. A., 42, *98*
Moore, S., 21, *32, 35*
Morawetz, H., 27, *32*
More O'Ferrall, R. A., 101, 161, 162, 164, 168, 169, *194*
Morey, J., 168, 169, *193*
Morokuma, K., 216, *239*
Morrison, R., 9, *34*
Morten, D. H., 153, *192*
Morton, T. H., 210, 222, *238, 239*
Muirhead, H., 3, *34*
Mulliken, R. S., 102, 107, 110, 123, *194*
Murai, K., 154, *194*
Murdoch, J. R., 151, 178, 181, 184, 185, *194*
Myher, J. J., 198, *238*

Nachmansohn, D., 14, *31, 34, 35*
Nagakura, S., 102, 123, *194*
Nakamura, C., 151, 181, *193*
Nakamura, K., 18, *34*
Nakamura, S., 45, *97*
Negelein, E., 2, *34*
Neurath, H., 18, *32*
Neveu, M. C., 27, *31*
Newman, M. S., 29, *32*

AUTHOR INDEX

Nibbering, N. M. M., 235, 236, *238*, *239*
Nicolaides, N., 2, *35*
Nimmo, K., 49, 70, *95*
Nonhebel, D. C., 9, *33*
Norrish, R. G. W., 131, *194*
North, A. C. T., 3, *34*
Northrop, J. H., 2, *33*, *34*
Noyd, D. A., 147, *191*
Nutting, M.-D. F., 14, *33*

Oesterlin, R., 45, *98*
Offner, P., 7, 8, *33*
Ogg, R. A. Jr., 102, *194*
Ogston, A. G., 8, *34*
Oh, S., 184, *193*
Okada, K., 40, *97*
Okamoto, T., 28, *31*
Okuyama, M., 156, *193*
Olivard, J., 6, *34*
Olmstead, W. N., 208, 210, 211, 215, 216, 218, *238*, *239*
Olsson, K., 45, *97*
Oosterbaan, R. A., 14, *34*
Oosterhoff, L. J., 102, 103, 139, 140, 142, *195*
Ortiz, J. J., 68, *95*
Orszulik, S. T., 9, *33*
O'Sullivan, C., 24, *34*
Ott, H., 28, *32*

Pachter, I. J., 45, *97*
Page, M. I., 27, *34*, 39, *96*
Palmer, C. A., 168, 169, *193*
Palmer, D. A., 166, *195*
Pandit, U. K., 27, *31*
Paquette, L. A., 173, *193*
Parker, A. J., 155, 166, 189, *191*, *194*, *195*
Patchornik, A., 42, *97*
Patel, G., 56, 66, 68, 69, 71, 73, 81, 84, 87, *97*
Pau, J. K., 229, 230, *239*
Pauling, L., 25, *34*, 113, 114, *195*
Paulson, J. F., 213, *238*
Payzant, J. D., 199, 204, 206, 208, 211, 212, 215, *238*, *239*, *240*
Pearson, R. E., 128, *195*
Pearson, R. G., 129, 168, *195*, 212, *239*
Pedersen, K. J., 18, *34*, 177, *192*
Pellerite, M. J., 184, *195*, 208, 217, 219, 220, *239*

Person, W. B., 102, 107, 110, *194*
Perutz, M. F., 3, *34*
Perz, R., 41, 70, *96*
Petsko, G. A., 30, *31*
Pezzanite, J. O., 45, *97*
Phillips, D. C., 3, 30, *31*, *33*
Pickover, C. A., 30, *34*
Pletcher, T. C., 68, *95*
Pocker, Y., 67, *97*
Polanyi, M., 102, 123, 124, 148, 177, *191*, *192*, *193*, *194*, 219, *238*
Pontremoli, S., 20, *32*, *34*
Poshusta, R. D., 120, *193*
Powell, M. F., 9, *34*, 53, 54, 55, 60, 62, 65, 68, 78, 83, *94*
Prandini, B. D., 20, *34*
Pratt, R. F., 42, *96*
Price, M. B., 68, *96*
Pross, A., 101, 110, 124, 145, 147, 151, 152, 153, 154, 155, 157, 161, 162, 166, 167, 170, 178, 179, 184, *194*, *195*, 219, *239*
Przystas, T. J., 71, *97*

Quiocho, F. A., 30, *34*

Raban, M., 8, *34*
Rabin, B. R., 22, *32*
Radom, L., 124, 147, *195*
Ramamurthy, V., 119, *195*
Ramunni, G., 173, *195*
Rapoport, H., 45, *95*, *98*
Readio, P. D., 128, *195*
Reed, P. B., 39, *96*
Reeke, G. N., 30, *34*
Reich, R., 146, 149, *191*
Reid, G. P., 49, *95*
Reinbolt, H., 29, *34*
Rentzepis, P. M., 133, 177, *193*
Ricard, D., 41, 70, *96*
Ricci, C., 20, *32*
Richards, F. M., 21, 23, *34*, *35*
Rieger, A. L., 128, *195*
Ringelmann, E., 11, *33*
Riveros, J. M., 198, 210, 214, 223, 224, 225, 226, 228, 229, 230, 233, 234, 235, 237, *238*, *239*, *240*
Rivers, P. S., 30, *31*
Roberts, J. D., 213, 214, *238*
Robinson, D. R., 47, 48, 91, *97*
Robinson, R. A., 78, *97*

Rogers, G. A., 42, 89, *97*
Rolfe, A. C., 67, *97*
Romeo, A., 42, *96*
Rondan, N. G., 173, *193*
Rosenberg, H. R., 4, *34*
Ross, C. A., 22, *32*
Ross, J., 126, *194*
Rothe, M., 41, *97*
Rossman, M. G., 3, *34*
Rothenberg, M. H., 14, *34*
Rowland, C., 120, *195*
Ruane, M., 166, *191*, *195*
Rundle, H. W., 203, *237*
Russell, G. A., 128, *195*
Rycroft, D. S., 45, 52, 94, *95*
Ryle, A. P., 3, 21, *34*

Sakuma, H., 40, *97*
Salem, L., 102, 103, 106, 116, 120, 129, 130, 131, 132, 133, 139, 141, 143, 145, 173, *192*, *195*
Salomaa, P., 59, 65, 66, 67, *95*
Samuel, D., 38, *97*
Sanchez, M. de N. de M., 61, 64, 65, 66, 72, 75, 76, 78, 83, *95*
Sanger, F., 3, *34*
Santry, L. J., 53, 55, 84, 85, 87, *97*
Satchell, D. P. N., 38, *96*
Sauer, J., 176, *195*
Savedoff, L. G., 212, *240*
Savelli, G., 44, *95*
Sawada, T., 156, *193*
Scandola, F., 182, *195*
Schaffer, N. K., 14, *34*
Schaleger, L. L., 26, *34*
Schantz, E. J., 45, *97*
Scheiner, S., 228, *239*
Scherowsky, G., 44, *97*
Scheuer, P. J., 45, *97*
Schiff, H. I., 199, 203, 206, *237*, *238*, *239*
Schimitt, R. J., 236, *238*
Schlegel, H. B., 184, *196*, 218, 219, *240*
Schleyer, P. v. R., 146, 153, *192*
Schmeltekopf, A. L., 203, *238*
Schmir, G. L., 25, 27, *31*, *34*, 38, 39, 67, *95*, *96*, *97*
Schnoes, H. K., 45, *97*
Schoellman, G., 16, *34*
Schonbaum, G. R., 18, *31*, *34*
Schowen, R. L., 38, *97*
Schuster, G. B., 182, *195*

Scott, J. A., 206, *237*
Scrocco, E., 228, *237*
Seaman, N. E., 63, 78, 83, 89, 91, 92, *97*, *98*
Segal, G., 103, 106, 116, 129, 130, 132, 133, 173, *195*
Segal, H. J., 24, *35*
Seibles, T. S., 21, *35*
Sela, M., 22, 27, *31*
Semmel, M. L., 42, *96*
Senyavina, L. B., 41, *97*
Shaik, S. S., 101, 110, 120, 123, 145, 147, 157, 162, 166, 167, 170, 174, 177, 184, *192*, *195*, 219, *239*
Shain, S. A., 38, *97*
Shapiro, R. H., 236, *240*
Shaw, E., 16, *34*
Shchelokov, V. I., 41, *97*
Sheinker, Y. N., 41, *97*
Shemyakin, M. M., 4, *31*, 41, *94*, *97*
Sheppard, R. C., 42, *96*, *97*
Shkrob, A. M., 41, *94*, *97*
Shore, V. C., 3, *33*
Shuler, K. E., 102, 123, 139, *194*
Sigler, P. B., 18, *33*
Silver, B. L., 38, *97*
Sizer, I. W., 6, *33*
Skell, P. S., 128, *195*
Sladowska, M., 42, *95*
Slater, C. D., 149, *194*
Smith, D., 204, *239*
Smith, L. F., 3, *34*
Smith, M. A., 204, 209, *239*
Smith, S., 212, *240*
Smith, S. G., 39, *98*
Snell, E. E., 4, 6, *32*, *33*, *34*, *35*
Snyder, E. R., 19, *32*
Sobel, H., 212, *239*
Solomon, J. J., 204, *239*
Somani, R., 45, 56, *97*
Songstad, J., 212, *239*
Sorenson, P. E., 179, *191*
Sorge, H., 29, *35*
Spackman, D. H., 21, *35*
Speranza, M., 222, *239*
Spikes, J. D., 28, *32*
Springer, J. P., 45, *97*
Staley, R. H., 221, *240*
Stapleton, B. J., 234, *238*
Stauffer, C. E., 45, 47, *98*
Stein, W. D., 21, *31*, *35*

AUTHOR INDEX

Stein, W. H., 21, *32*, *35*
Steinberger, R., 21, *35*, 41, *97*
Steitz, T. A., 30, *34*
Stevens, D. R., 212, *240*
Stevenson, D. P., 205, *238*
Stewart, J. H., 236, *238*, 240
Stokes, R. H., 78, *97*
Strandberg, B. E., 3, *33*
Straub, T. S., 59, *96*, *98*
Streitwieser, A. Jr., 146, *195*
Strong, F. M., 45, *97*
Strozier, R. W., 173, *193*
Sturtevant, J. M., 15, 27, *31*, *32*
Su, E. C. F., 206, *240*
Su, T., 205, 206, *240*
Suckling, C. J., 9, *33*
Suld, G., 45, *97*
Sullivan, S. A., 235, 236, *238*, *240*
Summerson, W. H., 14, *34*
Sumner, J. B., 2, *35*
Swain, C. G., 25, *35*

Taagepera, M., 221, *240*
Taillefer, R. J., 38, *95*
Takagi, M., 45, *98*
Takahashi, S., 40, *96*
Takashima, K., 223, 224, 225, 226, 228, 235, *240*
Takeuchi, S., 154, *194*
Tamura, R., 10, *34*
Tanaka, K., 208, 211, 212, 215, *240*
Tanaka, R., 10, *35*
Tanner, D. D., 128, *195*
Taylor, J. W., 155, *193*
Taylor, P. J., 39, *98*
Tee, O. S., 89, 91, 92, *98*
Teller, E., 129, *195*
Temple, C., 42, *98*
Thankachan, C., 72, 88, *97*
Thiessen, W. E., 45, *95*
Thompson, E. O. P., 3, *34*
Thornton, E. R., 70, *98*, 101, 155, 161, *195*
Thuillier, A., 45, *96*
Thynne, J. C. J., 198, *238*
Tidwell, T. T., 72, 88, *96*
Tiedemann, P. W., 229, 230, 233, *239*, *240*
Tomasi, J., 228, *237*
Tomoto, N., 67, *98*
Tompson, F. W., 24, *34*

Townshend, R. E., 173, *195*
Trani, M., 89, 91, 92, *98*
Tsernoglou, A. W., 23, *35*
Tuleen, D. L., 128, *195*
Tuominen, H., 58, *96*
Tuppy, H., 3, *34*
Turro, N. J., 103, 119, 139, 143, *192*, *195*

Ukai, T., 10, *35*
Umezawa, H., 45, *97*

van der Lugt, W. Th. A. M., 102, 103, 139, 140, 142, *195*
van Gulick, N., 27, *31*
Veillard, A., 216, *238*
Vennesland, B., 7, 8, *32*, *33*, *35*
Vinnik, M. I., 47, *98*
Vithayathil, P. J., 21, *34*
Voet, J., 13, *35*
von Dietrich, H., 29, *35*

Waley, S. G., 24, *35*
Walling, C., 128, *195*
Wang, J. T., 115, *195*
Warren, S., 20, *35*
Warshel, A., 103, 123, 127, *196*
Warhurst, E., 102, 123, *192*
Waszczylo, Z., 155, *196*
Watson, D., 41, *95*
Watson, T. W., 52, 94, *95*
Webb, E. C., 2, *32*
Weeks, D. P., 54, 65, 67, 69, 70, *95*, *98*
Weil, L., 21, *35*
Weiser, D., 45, *96*
Weiss, K., 156, *196*
Weiss, R. M., 103, 123, 127, *196*
Westaway, K. C., 155, *196*
Westheimer, F. H., 2, 7, 8, 18, 19, 20, 21, 26, 27, *32*, *33*, *35*
Wetmore, R., 103, 106, 116, 129, 130, 132, 133, *195*
Whitaker, J. R., 16, *35*
White, F. H., Jr., 23, *35*
Whitesides, G. M., 126, *194*
Wieland, H. J., 29, *35*
William, F., 115, *195*
Williams, B. D., 228, *238*
Will, G., 3, *34*
Wilson, E. B., Jr., 113, 114, *195*
Wilson, I. A., 30, *31*
Wilson, I. B., 14, *31*, *34*, *35*

Winstein, S., 128, 166, *191*, *195*, *196*, 212, *240*
Wipff, G., 228, *238*
Wolf, J. F., 221, *240*
Wolfe, S., 184, *196*, 218, 219, *240*
Wolfenden, R., 25, *35*
Wong, C. L., 133, 135, *193*
Wong, J. L., 45, *98*
Wong, R. J., 181, *194*
Wood, H. N., 14, 17, *31*
Woodgate, S. D., 221, *238*
Woodward, R. B., 40, *98*, 102, 140, 173, *196*
Wu, D., 151, 181, *193*
Wulff, H.-J., 2, *34*
Wurtz, A., 24, *35*

Wyckoff, H. W., 21, 22, 23, *34*, *35*
Wyrzykowska, K., 156, *196*

Yamataka, H., 155, *196*
Yamdagni, R., 204, *237*
Yeager, M. J., 39, *96*
Yee, K. C., 168, 169, 178, *192*
Young, C. I., 59, 65, 66, 67, *95*
Young, L. B., 206, 207, 208, 211, *237*, *240*

Zander, W., 123, 174, *192*
Zerner, B., 18, 20, *31*, *35*
Zimmerman, H. E., 139, *196*
Zucco, C., 52, *95*

Cumulative Index of Authors

Ahlberg, P., **19**, 223
Albery, W. J., **16**, 87
Allinger, N. L., **13**, 1
Anbar, M., **7**, 115
Arnett, E. M., **13**, 83
Bard, A. J., **13**, 155
Bell, R. P., **4**, 1
Bennett, J. E., **8**, 1
Bentley, T. W., **8**, 151; **14**, 1
Berger, S., **16**, 239
Bethell, D., **7**, 153: **10**, 53
Blandamer, M.J., **14**, 203
Brand, J. C. D., **1**, 365
Brändström, A., **15**, 267
Brinkman, M. R., **10**, 53
Brown, H. C., **1**, 35
Buncel, E., **14**, 133
Cabell-Whiting, P. W., **10**, 129
Cacace, F., **8**, 79
Capon, B., **21**, 37
Carter, R. E., **10**, 1
Collins, C. J., **2**, 1
Cornelisse, J., **11**, 225
Crampton, M. R., **7**, 211
Davidson, R. S., **19**, 1; **20**, 191
Desvergne, J. P., **15**, 63
de Gunst, G. P., **11**, 225
de Jong, F., **17**, 279
Dosunmu, M. I., **21**, 37
Eberson, L., **12**, 1: **18**, 79
Engdahl, C., **19**, 223
Farnum, D. G., **11**, 123
Fendler, E. J., **8**, 271
Fendler, J. H., **8**, 271; **13**, 279
Ferguson, G., **1**, 203
Fields, E. K., **6**, 1
Fife, T. H., **11**, 1
Fleischmann, M., **10**, 155
Frey, H. M., **4**, 147
Gilbert, B. C., **5**, 53
Gillespie, R. J., **9**, 1
Gold, V., **7**, 259
Goodin, J. W., **20**, 191
Gould, I. R., **20**, 1
Greenwood, H. H., **4**, 73
Hammerich, O., **20**, 55
Havinga, E., **11**, 225
Hine, J., **15**, 1
Hogen-Esch, T. E., **15**, 153
Hogeveen, H., **10**, 29, 129
Ireland, J. F., **12**, 131
Johnson, S. L., **5**, 237
Johnstone, R. A. W., **8**, 151
Jonsäll, G., **19**, 223
José, S. M., **21**, 197
Kemp, G., **20**, 191
Kice, J. L., **17**, 65
Kirby, A. J., **17**, 183
Kohnstam, G., **5**, 121
Kramer, G. M., **11**, 177
Kreevoy, M. M., **6**, 63; **16**, 87
Kunitake, T., **17**, 435
Ledwith, A., **13**, 155
Liler, M., **11**, 267
Long, F. A., **1**, 1
Maccoll, A., **3**, 91
McWeeny, R., **4**, 73
Melander, L., **10**, 1
Mile, B., **8**, 1
Miller, S. I., **6**, 185
Modena, G., **9**, 185
More O'Ferrall, R. A., **5**, 331
Morsi, S. E., **15**, 63
Neta, P., **12**, 223
Norman, R. O. C., **5**, 33
Nyberg, K., **12**, 1
Olah, G. A., **4**, 305
Parker, A. J., **5**, 173
Parker, V. D., **19**, 131; **20**, 55
Peel, T. E., **9**, 1
Perkampus, H. H., **4**, 195
Perkins, M. J., **17**, 1
Pittman, C. U. Jr., **4**, 305
Pletcher, D., **10**, 155
Pross, A., **14**, 69; **21**, 99
Ramirez, F., **9**, 25
Rappoport, Z., **7**, 1
Reeves, L. W., **3**, 187
Reinhoudt, D. N., **17**, 279
Ridd, J. H., **16**, 1
Riveros, J. M., **21**, 197
Robertson, J. M., **1**, 203
Rosenthal, S. N., **13**, 279
Samuel, D., **3**, 123
Sanchez, M. de N. de M., **21**, 37
Schaleger, L. L., **1**, 1
Scheraga, H. A., **6**, 103
Schleyer, P. von R., **14**, 1
Schmidt, S. P., **18**, 187
Schuster, G. B., **18**, 187
Scorrano, G., **13**, 83
Shatenshtein, A. I., **1**, 156
Shine, H. J., **13**, 155
Shinkai, S., **17**, 435
Silver, B. L., **3**, 123
Simonyi, M., **9**, 127
Stock, L. M., **1**, 35
Symons, M. C. R., **1**, 284
Takashima, K., **21**, 197
Tedder, J. M., **16**, 51
Thomas, A., **8**, 1
Thomas, J. M., **15**, 63
Tonellato, U., **9**, 185
Toullec, J., **18**, 1
Tüdös, F., **9**, 127
Turner, D. W., **4**, 31
Turro, N. J., **20**, 1
Ugi, I., **9**, 25
Walton, J. C., **16**, 51

Ward, B., **8,** 1
Westheimer, F. H., **21,** 1
Whalley, E., **2,** 93
Williams, D. L. H., **19,** 381
Williams, J. M. Jr., **6,** 63
Williams, J. O., **16,** 159
Williamson, D. G., **1,** 365
Wilson, H., **14,** 133
Wolf, A. P., **2,** 201
Wyatt, P. A. H., **12,** 131
Zimmt, M. B., **20,** 1
Zollinger, H., **2,** 163
Zuman, P., **5,** 1

Cumulative Index of Titles

Abstraction, hydrogen atom, from O–H bonds, **9,** 127
Acid solutions, strong, spectroscopic observation of alkylcarbonium ions in, **4,** 305
Acid-base properties of electronically excited states of organic molecules, **12,** 131
Acids, reactions of aliphatic diazo compounds with, **5,** 331
Acids, strong aqueous, protonation and solvation in, **13,** 83
Activation, entropies of, and mechanisms of reactions in solution, **1,** 1
Activation, heat capacities of, and their uses in mechanistic studies, **5,** 121
Activation, volumes of, use for determining reaction mechanisms, **2,** 93
Addition reactions, gas-phase radical, directive effects in, **16,** 51
Aliphatic diazo compounds, reactions with acids, **5,** 331
Alkylcarbonium ions, spectroscopic observation in strong acid solutions, **4,** 305
Ambident conjugated systems, alternative protonation sites in, **11,** 267
Ammonia, liquid, isotope exchange reactions of organic compounds in **1,** 156
Aqueous mixtures, kinetics of organic reactions in water and, **14,** 203
Aromatic photosubstitution, nucleophilic, **11,** 225
Aromatic substitution, a quantitative treatment of directive effects in, **1,** 35
Aromatic substitution reactions, hydrogen isotope effects in, **2,** 163
Aromatic systems, planar and non-planar, **1,** 203
Aryl halides and related compounds, photochemistry of, **20,** 191
Arynes, mechanisms of formation and reactions at high temperatures, **6,** 1
A-S_E2 reactions, developments in the study of, **6,** 63

Base catalysis, general, of ester hydrolysis and related reactions, **5,** 237
Basicity of unsaturated compounds, **4,** 195
Bimolecular substitution reactions in protic and dipolar aprotic solvents, **5,** 173

^{13}C N.M.R. spectroscopy in macromolecular systems of biochemical interest, **13,** 279
Carbene chemistry, structure and mechanism in, **7,** 163
Carbanion reactions, ion-pairing effects in, **15,** 153
Carbocation rearrangements, degenerate, **19,** 223
Carbon atoms, energetic, reactions with organic compounds, **3,** 201
Carbon monoxide, reactivity of carbonium ions towards, **10,** 29
Carbonium ions (alkyl), spectroscopic observation in strong acid solutions, **4,** 305
Carbonium ions, gaseous, from the decay of tritiated molecules, **8,** 79
Carbonium ions, photochemistry of, **10,** 129
Carbonium ions, reactivity towards carbon monoxide, **10,** 29
Carbonyl compounds, reversible hydration of, **4,** 1
Carbonyl compounds, simple, enolisation and related reactions of, **18,** 1
Carboxylic acids, tetrahedral intermediates derived from, spectroscopic detection and investigation of their properties, **21,** 37
Catalysis by micelles, membranes and other aqueous aggregates as models of enzyme action, **17,** 435
Catalysis, enzymatic, physical organic model systems and the problem of, **11,** 1

Catalysis, general base and nucleophilic, of ester hydrolysis and related reactions, **5**, 237
Catalysis, micellar, in organic reactions; kinetic and mechanistic implications, **8**, 271
Catalysis, phase-transfer by quaternary ammonium salts, **15**, 267
Cation radicals in solution, formation, properties and reactions of, **13**, 155
Cation radicals, organic, in solution, kinetics and mechanisms of reaction of, **20**, 55
Cations, vinyl, **9**, 135
Charge density—N.M.R. chemical shift correlations in organic ions, **11**, 125
Chemically induced dynamic nuclear spin polarization and its applications, **10**, 53
Chemiluminescence of organic compounds, **18**, 187
CIDNP and its applications, **10**, 53
Conduction, electrical, in organic solids, **16**, 159
Configuration mixing model: a general approach to organic reactivity, **21**, 99
Conformations of polypeptides, calculations of, **6**, 103
Conjugated, molecules, reactivity indices, in, **4**, 73
Crown-ether complexes, stability and reactivity of, **17**, 279

D_2O–H_2O mixtures, protolytic processes in, **7**, 259
Degenerate carbocation rearrangements, **19**, 223
Diazo compounds, aliphatic, reactions with acids, **5**, 331
Diffusion control and pre-association in nitrosation, nitration, and halogenation, **16**, 1
Dimethyl sulphoxide, physical organic chemistry of reactions, in, **14**, 133
Dipolar aprotic and protic solvents, rates of bimolecular substitution reactions in, **5**, 173
Directive effects in aromatic substitution, a quantitative treatment of, **1**, 35
Directive effects in gas-phase radical addition reactions, **16**, 51
Discovery of the mechanisms of enzyme action, 1947–1963, **21**, 1
Displacement reactions, gas-phase nucleophilic, **21**, 197

Effective molarities of intramolecular reactions, **17**, 183
Electrical conduction in organic solids, **16**, 159
Electrochemical methods, study of reactive intermediates by, **19**, 131
Electrochemistry, organic, structure and mechanism in, **12**, 1
Electrode processes, physical parameters for the control of, **10**, 155
Electron spin resonance, identification of organic free radicals by, **1**, 284
Electron spin resonance studies of short-lived organic radicals, **5**, 23
Electron-transfer reactions in organic chemistry, **18**, 79
Electronically excited molecules, structure of, **1**, 365
Electronically excited states of organic molecules, acid-base properties of, **12**, 131
Energetic tritium and carbon atoms, reactions of, with organic compounds, **2**, 201
Enolisation of simple carbonyl compounds and related reactions, **18**, 1
Entropies of activation and mechanisms of reactions in solution, **1**, 1
Enzymatic catalysis, physical organic model systems and the problem of, **11**, 1
Enzyme action, catalysis by micelles, membranes and other aqueous aggregates as models of, **17**, 435
Enzyme action, discovery of the mechanisms of, 1947–1963, **21**, 1
Equilibrium constants, N.M.R. measurements of, as a function of temperature, **3**, 187

Ester hydrolysis, general base and nucleophilic catalysis, **5,** 237
Exchange reactions, hydrogen isotope, of organic compounds in liquid ammonia, **1,** 156
Exchange reactions, oxygen isotope, of organic compounds, **2,** 123
Excited complexes, chemistry of, **19,** 1
Excited molecules, structure of electronically, **1,** 365

Force-field methods, calculation of molecular structure and energy by, **13,** 1
Free radicals, identification by electron spin resonance, **1,** 284
Free radicals and their reactions at low temperature using a rotating cryostat, study of, **8,** 1

Gaseous carbonium ions from the decay of tritiated molecules, **8,** 79
Gas-phase heterolysis, **3,** 91
Gas-phase nucleophilic displacement reactions, **21,** 197
Gas-phase pyrolysis of small-ring hydrocarbons, **4,** 147
General base and nucleophilic catalysis of ester hydrolysis and related reactions, **5,** 237

H_2O–D_2O mixtures, protolytic processes in, **7,** 259
Halogenation, nitrosation, and nitration, diffusion control and pre-association in, **16,** 1
Halides, aryl, and related compounds, photochemistry of, **20,** 191
Heat capacities of activation and their uses in mechanistic studies, **5,** 121
Heterolysis, gas-phase, **3,** 91
Hydrated electrons, reactions of, with organic compounds, **7,** 115
Hydration, reversible, of carbonyl compounds, **4,** 1
Hydrocarbons, small-ring, gas-phase pyrolysis of, **4,** 147
Hydrogen atom abstraction from O—H bonds, **9,** 127
Hydrogen isotope effects in aromatic substitution reactions, **2,** 163
Hydrogen isotope exchange reactions of organic compounds in liquid ammonia, **1,** 156
Hydrolysis, ester, and related reactions, general base and nucleophilic catalysis of, **5,** 237

Intermediates, reactive, study of, by electrochemical methods, **19,** 131
Intermediates, tetrahedral, derived from carboxylic acids, spectroscopic detection and investigation of their properties, **21,** 37
Intramolecular reactions, effective molarities for, **17,** 183
Ionization potentials, **4,** 31
Ion-pairing effects in carbanion reactions, **15,** 153
Ions, organic, charge density–N.M.R. chemical shift correlations, **11,** 125
Isomerization, permutational, of pentavalent phosphorus compounds, **9,** 25
Isotope effects, hydrogen, in aromatic substitution reactions, **2,** 163
Isotope effects, magnetic, magnetic field effects and, on the products of organic reactions, **20,** 1
Isotope effects, steric, experiments on the nature of, **10,** 1
Isotope exchange reactions, hydrogen, of organic compounds in liquid ammonia, **1,** 150
Isotope exchange reactions, oxygen, of organic compounds, **3,** 123

Isotopes and organic reaction mechanisms, **2,** 1

Kinetics and mechanisms of reactions of organic cation radicals in solution, **20,** 55
Kinetics, reaction, polarography and, **5,** 1
Kinetics of organic reactions in water and aqueous mixtures, **14,** 203

Least nuclear motion, principle of, **15,** 1

Macromolecular systems of biochemical interest, ^{13}C N.M.R. spectroscopy in, **13,** 279
Magnetic field and magnetic isotope effects on the products of organic reactions, **20,** 1
Mass spectrometry, mechanisms and structure in: a comparison with other chemical processes, **8,** 152
Mechanism and structure in carbene chemistry, **7,** 153
Mechanism and structure in mass spectrometry: a comparison with other chemical processes, **8,** 152
Mechanism and structure in organic electrochemistry, **12,** 1
Mechanisms and reactivity in reactions of organic oxyacids of sulphur and their anhydrides, **17,** 65
Mechanisms, nitrosation, **19,** 381
Mechanisms, organic reaction, isotopes and, **2,** 1
Mechanisms of reaction in solution, entropies of activation and, **1,** 1
Mechanisms of solvolytic reactions, medium effects on the rates and, **14,** 10
Mechanistic applications of the reactivity–selectivity principle, **14,** 69
Mechanistic studies, heat capacities of activation and their use, **5,** 121
Medium effects on the rates and mechanisms of solvolytic reactions, **14,** 1
Meisenheimer complexes, **7,** 211
Methyl transfer reactions, **16,** 87
Micellar catalysis in organic reactions: kinetic and mechanistic implications, **8,** 271
Micelles, membranes and other aqueous aggregates, catalysis by, as models of enzyme action, **17,** 435
Molecular structure and energy, calculation of, by force-field methods, **13,** 1

Nitration, nitrosation, and halogenation, diffusion control and pre-association in, **16,** 1
Nitrosation mechanisms, **19,** 381
Nitrosation, nitration, and halogenation, diffusion control and pre-association in, **16,** 1
N.M.R. chemical shift–charge density correlations, **11,** 125
N.M.R. measurements of reaction velocities and equilibrium constants as a function of temperature, **3,** 187
N.M.R. spectroscopy, ^{13}C, in macromolecular systems of biochemical interest, **13,** 279
Non-planar and planar aromatic systems, **1,** 203
Norbornyl cation: reappraisal of structure, **11,** 179
Nuclear magnetic relaxation, recent problems and progress, **16,** 239
Nuclear magnetic resonance, *see* N.M.R.
Nuclear motion, principle of least, **15,** 1

Nucleophilic aromatic photosubstitution, **11**, 225
Nucleophilic catalysis of ester hydrolysis and related reactions, **5**, 237
Nucleophilic displacement reactions, gas-phase, **21**, 197
Nucleophilic vinylic substitution, **7**, 1

OH—bonds, hydrogen atom abstraction from, **9**, 127
Oxyacids of sulphur and their anhydrides, mechanisms and reactivity in reactions of organic, **17**, 65
Oxygen isotope exchange reactions of organic compounds, **3**, 123

Permutational isomerization of pentavalent phosphorus compounds, **9**, 25
Phase-transfer catalysis by quaternary ammonium salts, **15**, 267
Phosphorus compounds, pentavalent, turnstile rearrangement and pseudorotation in permutational isomerization, **9**, 25
Photochemistry of aryl halides and related compounds, **20**, 191
Photochemistry of carbonium ions, **9**, 129
Photosubstitution, nucleophilic aromatic, **11**, 225
Planar and non-planar aromatic systems, **1**, 203
Polarizability, molecular refractivity and, **3**, 1
Polarography and reaction kinetics, **5**, 1
Polypeptides, calculations of conformations of, **6**, 103
Pre-association, diffusion control and, in nitrosation, nitration, and halogenation, **16**, 1
Products of organic reactions, magnetic field and magnetic isotope effects on, **30**, 1
Protic and dipolar aprotic solvents, rates of bimolecular substitution reactions in, **5**, 173
Protolytic processes in H_2O–D_2O mixtures, **7**, 259
Protonation and solvation in strong aqueous acids, **13**, 83
Protonation sites in ambident conjugated systems, **11**, 267
Pseudorotation in isomerization of pentavalent phosphorus compounds, **9**, 25
Pyrolysis, gas-phase, of small-ring hydrocarbons, **4**, 147

Radiation techniques, application to the study of organic radicals, **12**, 223
Radical addition reactions, gas-phase, directive effects in, **16**, 51
Radicals, cation in solution, formation, properties and reactions of, **13**, 155
Radicals, organic application of radiation techniques, **12**, 223
Radicals, organic cation, in solution, kinetics and mechanisms of reaction of, **20**, 55
Radicals, organic free, identification by electron spin resonance, **1**, 284
Radicals, short-lived organic, electron spin resonance studies of, **5**, 53
Rates and mechanisms of solvolytic reactions, medium effects on, **14**, 1
Reaction kinetics, polarography and, **5**, 1
Reaction mechanisms, use of volumes of activation for determining, **2**, 93
Reaction mechanisms in solution, entropies of activation and, **1**, 1
Reaction velocities and equilibrium constants, N.M.R. measurements of, as a function of temperature, **3**, 187
Reactions of hydrated electrons with organic compounds, **7**, 115
Reactions in dimethyl sulphoxide, physical organic chemistry of, **14**, 133
Reactive intermediates, study of, by electrochemical methods, **19**, 131
Reactivity indices in conjugated molecules, **4**, 73
Reactivity, organic, a general approach to: the configuration mixing model, **21**, 99

Reactivity-selectivity principle and its mechanistic applications, **14,** 69
Rearrangements, degenerate carbocation, **19,** 223
Refractivity, molecular, and polarizability, **3,** 1
Relaxation, nuclear magnetic, recent problems and progress, **16,** 239

Short-lived organic radicals, electron spin resonance studies of, **5,** 53
Small-ring hydrocarbons, gas-phase pyrolysis of, **4,** 147
Solid-state chemistry, topochemical phenomena in, **15,** 63
Solids, organic, electrical conduction in, **16,** 159
Solutions, reactions in, entropies of activation and mechanisms, **1,** 1
Solvation and protonation in strong aqueous acids, **13,** 83
Solvents, protic and dipolar aprotic, rates of bimolecular substitution-reactions in, **5,** 173
Solvolytic reactions, medium effects on the rates and mechanisms of, **14,** 1
Spectroscopic detection of tetrahedral intermediates derived from carboxylic acids and the investigation of their properties, **21,** 37
Spectroscopic observations of alkylcarbonium ions in strong acid solutions, **4,** 305
Spectroscopy, ^{13}C N.M.R., in macromolecular systems of biochemical interest, **13,** 279
Spin trapping, **17,** 1
Stability and reactivity of crown-ether complexes, **17,** 279
Stereoselection in elementary steps of organic reactions, **6,** 185
Steric isotope effects, experiments on the nature of, **10,** 1
Structure and mechanisms in carbene chemistry, **7,** 153
Structure and mechanism in organic electrochemistry, **12,** 1
Structure of electronically excited molecules, **1,** 365
Substitution, aromatic, a quantitative treatment of directive effects in, **1,** 35
Substitution, nucleophilic vinylic, **7,** 1
Substitution reactions, aromatic, hydrogen isotope effects in, **2,** 163
Substitution reactions, bimolecular, in protic and dipolar aprotic solvents, **5,** 173
Sulphur, organic oxyacids of, and their anhydrides, mechanisms and reactivity in reactions of, **17,** 65
Superacid systems, **9,** 1

Temperature, N.M.R. measurements of reaction velocities and equilibrium constants as a function of, **3,** 187
Tetrahedral intermediates derived from carboxylic acids, spectroscopic detection and the investigation of their properties, **21,** 37
Topochemical phenomena in solid-state chemistry, **15,** 63
Tritiated molecules, gaseous carbonium ions from the decay of, **8,** 79
Tritium atoms, energetic, reactions with organic compounds, **2,** 201
Turnstile rearrangements in isomerization of pentavalent phosphorus compounds, **9,** 25

Unsaturated compounds, basicity of, **4,** 195

Vinyl cations, **9,** 185
Vinylic substitution, nucleophilic, **7,** 1
Volumes of activation, use of, for determining reaction mechanisms, **2,** 93

Water and aqueous mixtures, kinetics of organic reactions in, **14,** 203